Lecture Notes in Geoinformation and Cartography

The Lecture Notes in Geoinformation and Cartography series provides a contemporary view of current research and development in Geoinformation and Cartography, including GIS and Geographic Information Science. Publications with associated electronic media examine areas of development and current technology. Editors from multiple continents, in association with national and international organizations and societies bring together the most comprehensive forum for Geoinformation and Cartography.

The scope of Lecture Notes in Geoinformation and Cartography spans the range of interdisciplinary topics in a variety of research and application fields. The type of material published traditionally includes:

- proceedings that are peer-reviewed and published in association with a conference;
- post-proceedings consisting of thoroughly revised final papers; and
- research monographs that may be based on individual research projects.

The Lecture Notes in Geoinformation and Cartography series also includes various other publications, including:

- tutorials or collections of lectures for advanced courses;
- contemporary surveys that offer an objective summary of a current topic of interest; and
- emerging areas of research directed at a broad community of practitioners.

Alias Abdul-Rahman · Ivin A. Musliman ·
Imzan Hassan · Ainn Zamzuri
Editors

Advances in Geoinformation Sciences

Springer

Editors
Alias Abdul-Rahman
Department of Geoinfirmation
Universiti Teknologi Malaysia
Johor Bahru, Malaysia

Ivin A. Musliman
Department of Geoinformation
Universiti Teknologi Malaysia
Johor Bahru, Johor, Malaysia

Imzan Hassan
Department of Geoinformation
Universiti Teknologi Malaysia
Johor Bahru, Johor, Malaysia

Ainn Zamzuri
3D GIS Research Lab
Universiti Teknologi Malaysia
Johor Bahru, Johor, Malaysia

ISSN 1863-2246 ISSN 1863-2351 (electronic)
Lecture Notes in Geoinformation and Cartography
ISBN 978-3-031-86653-1 ISBN 978-3-031-86654-8 (eBook)
https://doi.org/10.1007/978-3-031-86654-8

This Springer imprint is published by the registered company Springer Nature Switzerland AG
The registered company address is: Gewerbestrasse 11, 6330 Cham, Switzerland

Preface

First, we would like to thank all contributors to this book. The book contains chapters deal with the most recent research and development in geoinformation science and accepted for presentation at the Geoinformation Week 2024 International Conference as part of the Joint International Conference event (in conjunction with FIG Commission 5, FIG Commission 7, FIG LADM & 3D LA, and UN-Habitat STDM Conferences). In brief, themes of 34 chapters are 3D Buildings Modelling, UAV based Mapping, Geo-based Artificial Intelligence / Deep Learning, Marine Information System, Satellite based Image Processing & Analysis, and GIS & Land Administration. All accepted chapters were reviewed by the experts of the disciplines. We would like to thank all reviewers for their insightful comments and suggestions where their inputs elevate the quality of the chapters. We also hope these contributions able to provide new knowledge in geospatial science and trigger more interesting research in the near future.

September 2024

Alias Abdul Rahman
Muhammad Imzan Hassan
Ivin Amri Musliman
Ainn Zamzuri

Local Organizing Committee

Abd Halim Hamzah
Ainn Zamzuri
Alias Abdul Rahman
Ivin Amri Musliman
Kelvin Tang Kang Wee
Muhammad Imzan Hassan
Othman Zainon
Ramzi Abdillah
Stephen Ling Jin Huat
Suzanna Noor Azmy
Tajul Ariffin Musa
Wan Anom Wan Aris

Reviewers

Alias Abdul-Rahman (Malaysia)
Abd Halim Hamzah (Malaysia)
Abdelhamid Fadil (Morocco)
Ainn Zamzuri (Malaysia)
Alexis Richard Claridades (Philippines)
Ariel Blanco (Philippines)
Ayin Tamondong (Philippines)
Ben Gorte (Australia)
Chien Zheng Yong (New Zealand)
Gurcan Buyuksalih (Turkiye)
Hassan Rhinane (Morocco)
Ihab Hijazi (Germany)
Ivin Amri Musliman (Malaysia)
Jeark Principe (Philippines)
Jesper Paasch (Sweden)
Kamal Agharroud (Morocco)
Kean Huat Soon (Singapore)
Kelvin Kang Wee Tang (Malaysia)
Khalid Al Kalbani (Oman)
Mehdi Maanan (Morocco)
Mehmet Alkan (Turkiye)
Mohsen Kalantari (Australia)
Muhammad Imzan Hassan (Malaysia)
Muhammad Zulkarnain Abd Rahman (Malaysia)

Omar Alami (Morocco)
Othman Zainon (Malaysia)
Pawel Boguslawski (Poland)
Peter van Oosterom (Netherlands)
Rosalie Reyes (Philippines)
Serdar Bayburt (Turkiye)
Sisi Zlatanova (Australia)
Suhaibah Azri (Malaysia)
Trias Aditya (Indonesia)
Uznir Ujang (Malaysia)
Volker Coors (Germany)

Contents

List of Contributors

Amir Sharifuddin Ab Latip College of Built Environment, Universiti Teknologi MARA, Shah Alam, Malaysia

Ramzi Abdillah Lands and Surveys Department Sarawak, Kuching, Malaysia

Alias Abdul Rahman 3D GIS Research Lab, Faculty of Built Environment & Surveying, Universiti Teknologi Malaysia, Johor Bahru, Malaysia;
Department of Geoinformation, Faculty of Built Environment and Surveying, Universiti Teknologi Malaysia, Johor Bahru, Malaysia

Rizqi Abdulharis Bandung Institute of Technology, Bandung, Indonesia

Muhammad Luqman Ahmad Affandi Geospatial Imaging and Information Research Group (GI2RG), Faculty of Built Environment and Surveying, Universiti Teknologi Malaysia, Johor Bahru, Johor, Malaysia

Mahiuddin Alamgir Geoscience and Digital Earth Centre (INSTeG), Research Institute of Sustainable Environment (RISE), Universiti Teknologi Malaysia (UTM), 81310 Johor Bahru, Malaysia

Nur Surayatul Atikah Alihan Geospatial Imaging and Information Research Group (Gi2RG), Faculty of Built Environment and Surveying, Universiti Teknologi Malaysia, Johor Bahru, Johor, Malaysia

Awan Amal Lands and Surveys Department Sarawak, Kuching, Malaysia

Madeline Anak Ambit Lands and Surveys Department Sarawak, Kuching, Malaysia

Michael Moses Apeh Department of Geoinformation, Faculty of Built Environment and Surveying, Universiti Teknologi Malaysia, 81310 UTM Skudai, Johor Bahru, Malaysia;
Department of Surveying and Geoinformatics, Federal Polytechnic, Idah, Nigeria

Abg Ahmad Azlan Abg Ariffin Lands and Surveys Department Sarawak, Kuching, Malaysia

Wan Anom Wan Aris Geomatic Innovation Research Group, Faculty of Built Environment and Surveying, Universiti Teknologi Malaysia, Iskandar Puteri, Malaysia

Suzanna Noor Azmy Geoinformation, Faculty of Built Environment and Surveying, Universiti Teknologi Malaysia, 81310 Johor Bahru, Johor, Malaysia;
Geoscience and Digital Earth Centre, Faculty of Built Environment and Surveying, Universiti Teknologi Malaysia, 81310 Johor Bahru, Johor, Malaysia

Mohamed Benayad Geoscience Laboratory, Department of Geology, Faculty of Science Aïn Chock, University Hassan II, Casablanca, Morocco

Erica Benya Billy School of Geomatics Science and Natural Resources, College of Built Environment, Seattle, US;
Hurricane Geo Inspection Survey Sdn Bhd (HGIS), Kuala Belait, Darussalam, Brunei

Sulaiman Budin Lands and Surveys Department Sarawak, Sarawak, Malaysia

Gurcan Buyuksalih Institute of Marine Sciences and Management, Istanbul University, Istanbul, Türkiye

Marouane Ouazani Chahidi Department of Geology, Faculty of Sciences Aïn Chock, Hassan II University, Casablanca, Morocco

Sim Choon Cheak SD Guthrie Berhad, Telok Panglima Garang, Selangor, Malaysia

Musa Muhammad Chindo Faculty of Built Environment & Surveying, Universiti Teknologi Malaysia, 81310 Johor Bahru, Johor, Malaysia;
Department of Survey & Geoinformatics, Federal Polytechnic Nasarawa, Nasarawa, 962101 Nigeria

Pei-Cheng Chen National Yang Ming Chiao Tung University, Hsinchu, Taiwan

Malumbo Chipofya Faculty ITC, University of Twente, Enschede, The Netherlands

Walter Dachaga Technical University of Munich, Munich, Germany

Walter Timo de Vries Technical University of Munich, Munich, Germany

Ami Hassan Md Din Geospatial Imaging and Information Research Group (GI2RG), Faculty of Built Environment and Surveying, Universiti Teknologi Malaysia, Johor Bahru, Johor, Malaysia

Saad Farah Geoscience Laboratory, Department of Geology, Faculty of Science Aïn Chock, University Hassan II, Casablanca, Morocco

Zineb El Faraj Department of Geology, Faculty of Sciences Aïn Chock, Hassan II University, Casablanca, Morocco

Nursyafiqah Jessica Georgeson Department of Water and Environmental Engineering, Faculty of Civil Engineering, Universiti Teknologi Malaysia, UTM Skudai, Johor, Malaysia

Rosmadi Ghazali School of Geomatics Science and Natural Resources, College of Built Environment, Shah Alam, Malaysia

Cassendra Gilbert Faculty of Sport Science and Recreation, Shah Alam, Malaysia

Serene Anak Joseph Goyang School of Geomatics Science and Natural Resources, College of Built Environment, Shah Alam, Malaysia

Mohammad Hanif Hamden Geospatial Imaging and Information Research Group (GI2RG), Faculty of Built Environment and Surveying, Universiti Teknologi Malaysia, Johor Bahru, Johor, Malaysia

Laaziza Hammoumi Geoscience Laboratory, Department of Geology, Faculty of Science Aïn Chock, University Hassan II, Casablanca, Morocco

Dzuren Hamzah Lands and Surveys Department Sarawak, Sarawak, Malaysia

Alfita Puspa Handayani Bandung Institute of Technology, Bandung, Indonesia

Mohd Faizury Abol Hassan Lands and Surveys Department Sarawak, Kuching, Malaysia

Noordyana Hassan Geoinformation, Faculty of Built Environment and Surveying, Universiti Teknologi Malaysia, 81310 Johor Bahru, Johor, Malaysia;
Geoscience and Digital Earth Centre, Faculty of Built Environment and Surveying, Universiti Teknologi Malaysia, 81310 Johor Bahru, Johor, Malaysia

Mazlan Hashim Geoscience and Digital Earth Centre (INSTeG), Research Institute of Sustainable Environment (RISE), Universiti Teknologi Malaysia (UTM), 81310 Johor Bahru, Malaysia;
Faculty of Built Environment and Surveying, Universiti Teknologi Malaysia (UTM), 81310 Johor Bahru, Malaysia

Nouriddine Houran Green Energy Park (GEP), Benguerir, Morocco

Stephen Ling Jin Huat Lands and Surveys Department Sarawak, Kuching, Malaysia

Nurul Hazrina Idris Geoinformation, Faculty of Built Environment and Surveying, Universiti Teknologi Malaysia, 81310 Johor Bahru, Johor, Malaysia;
Geoscience and Digital Earth Centre, Faculty of Built Environment and Surveying, Universiti Teknologi Malaysia, 81310 Johor Bahru, Johor, Malaysia

Zamri Ismail Geospatial Imaging and Information Research Group (Gi2RG), Faculty of Built Environment and Surveying, Universiti Teknologi Malaysia, Johor Bahru, Johor, Malaysia

Syarif Abdul Asa'ad Jairin Geospatial Imaging and Information Research Group (Gi2RG), Faculty of Built Environment and Surveying, Universiti Teknologi Malaysia, Johor Bahru, Johor, Malaysia

Ismail Rakip Karas Department of Computer Engineering, Karabuk University, Karabuk, Türkiye

Chaymae Khaloua Geoscience Laboratory, Department of Geology, Faculty of Science Aïn Chock, University Hassan II, Casablanca, Morocco

Mayuree Nasa Khan Faculty of Tourism and Hotel Management, Mahasarakham University, Kantharawichai, Thailand

Muhammad Ghaly Kurniawan Faculty ITC, University of Twente, Hallenweg 8, 7522 NH Enschede, The Netherlands;
Ministry of Agrarian Affairs and Spatial Planning/National Land Agency, Jakarta, Indonesia

Rafeah Legino College of Creative Arts, Universiti Teknologi MARA, Shah Alam Selangor, Malaysia

Claudia Lindner Kadaster International, Amsterdam, The Netherlands

Safia Loulad Department of Geology, Faculty of Sciences Aïn Chock, Hassan II University, Casablanca, Morocco

Mehdi Maanan Geosciences Laboratory, Faculty of Sciences-Ain Chock, Hassan II University, Casablanca, Morocco

Mohd Razali Mahmud GeoCoastal Research Unit (Gi2RG), Universiti Teknologi Malaysia, Johor Bahru, Malaysia

Mohd. Rizaludin Mahmud Geoinformation, Faculty of Built Environment and Surveying, Universiti Teknologi Malaysia, 81310 Johor Bahru, Johor, Malaysia;
Geoscience and Digital Earth Centre, Faculty of Built Environment and Surveying, Universiti Teknologi Malaysia, 81310 Johor Bahru, Johor, Malaysia

Nursuhaili Najwa Masrol SD Guthrie Berhad, Telok Panglima Garang, Selangor, Malaysia

Myriame Virginie Matius University Teknologi Malaysia, Skudai, Johor, Malaysia

Wong Miliw Lands and Surveys Department Sarawak, Kuching, Malaysia

Ahmad Hamidi Mohamed College of Built Environment, Universiti Teknologi MARA, Shah Alam, Malaysia

Nur Hafizah Mohammed SD Guthrie Berhad, Telok Panglima Garang, Selangor, Malaysia

Mohd Radhie Mohd Salleh Department of Water and Environmental Engineering, Faculty of Civil Engineering, Universiti Teknologi Malaysia, UTM Skudai, Johor, Malaysia

Mickcy Mandang Anak Morgan Lands and Surveys Department Sarawak, Sarawak, Malaysia

Muhammad Abdul Hakim Muhamad School of Geomatics Science and Natural Resources, College of Built Environment, Shah Alam, Malaysia

Ivin Amri Musliman University Teknologi Malaysia, Skudai, Johor, Malaysia;
Department of Geoinformation, Faculty of Built Environment and Surveying, Universiti Teknologi Malaysia, Johor Bahru, Malaysia;
GIS Research Lab, Faculty of Built Environment and Surveying, Universiti Teknologi Malaysia, Johor Bahru, Malaysia

Wahyu Marta Mutiarasari 3D GIS Research Lab, Department of Geoinformation, Faculty of Built Environment and Surveying, Universiti Teknologi Malaysia, Johor Bahru, Malaysia

Ilya Khairanis Othman Department of Water and Environmental Engineering, Faculty of Civil Engineering, Universiti Teknologi Malaysia, UTM Skudai, Johor, Malaysia

Hanis Rashidan 3D GIS Research Lab, Faculty of Built Environment & Surveying, Universiti Teknologi Malaysia, Johor Bahru, Malaysia;
GIS Research Lab, Faculty of Built Environment and Surveying, Universiti Teknologi Malaysia, Johor Bahru, Malaysia

Abd Wahid Rasib Geospatial Imaging and Information Research Group (GI2RG), Faculty of Built Environment and Surveying, Universiti Teknologi Malaysia, Johor Bahru, Johor, Malaysia

Mohamad Hezri Razali College of Built Environment, Universiti Teknologi MARA, Shah Alam, Malaysia

Hassan Rhinane Geosciences Laboratory, Department of Geology, Faculty of Sciences-Ain Chock, Hassan II University, Casablanca, Morocco;
Geoscience Laboratory, Faculty of Science Aïn Chock, University Hassan II, Casablanca, Morocco;
Department of Geology, Faculty of Sciences Aïn Chock, Hassan II University, Casablanca, Morocco

Fatin Nabihah Syahira Ridzuan Geospatial Imaging and Information Research Group (Gi2RG), Faculty of Built Environment and Surveying, Universiti Teknologi Malaysia, Johor Bahru, Johor, Malaysia

Abdelilah Rochd Green Energy Park (GEP), Benguerir, Morocco

Hajah Siti Romiza Romali Lands and Surveys Department Sarawak, Kuching, Malaysia

Nur Nadhirah Rusyda Rosnan SD Guthrie Berhad, Telok Panglima Garang, Selangor, Malaysia

Siti Hayatunissa Saminan Lands and Surveys Department Sarawak, Sarawak, Malaysia

Bakeri Samsudin Lands and Surveys Department Sarawak, Sarawak, Malaysia

Mohd Zunica Mohd Sapi'ee Lands and Surveys Department Sarawak, Kuching, Malaysia

Radzuan Sa'ari Department of Water and Environmental Engineering, Faculty of Civil Engineering, Universiti Teknologi Malaysia, UTM Skudai, Johor, Malaysia

Shamsuddin Shahid Department of Water and Environmental Engineering, Faculty of Civil Engineering, Universiti Teknologi Malaysia (UTM), Johor, Malaysia

Ching Yen Sim Faculty of Built Environment and Surveying, Universiti Teknologi Malaysia, Johor Bahru, Malaysia

Mohamed Rabii Simou Geosciences Laboratory, Faculty of Sciences-Ain Chock, Hassan II University, Casablanca, Morocco

Vijaya Subramaniam SD Guthrie Berhad, Telok Panglima Garang, Selangor, Malaysia

Saiful Aman Sulaiman College of Built Environment, Universiti Teknologi MARA, Shah Alam, Malaysia

Abdul Rauf Abdul Rasam School of Geomatics Science and Natural Resources, College of Built Environment, Seattle, US

Mohd. Nadzri Md. Reba Geoinformation, Faculty of Built Environment and Surveying, Universiti Teknologi Malaysia, 81310 Johor Bahru, Johor, Malaysia;
Geoscience and Digital Earth Centre, Faculty of Built Environment and Surveying, Universiti Teknologi Malaysia, 81310 Johor Bahru, Johor, Malaysia

Kelvin Kang Wee Tang Geomatic Innovation Research Group (GnG), Universiti Teknologi Malaysia, Johor Bahru, Malaysia

Tee-Ann Teo National Yang Ming Chiao Tung University, Hsinchu, Taiwan

Dimo Todorovski Faculty ITC, University of Twente, Enschede, The Netherlands

Mat Nizam Uti Kolej Alam Bina, Universiti Teknologi Mara Cawangan Perak, Seri Iskandar, Perak, Malaysia

Dudy Darmawan Wijaya Geodesy Research Group, Faculty of Earth Sciences and Technology, Institute ofTechnology Bandung, Bandung, Indonesia;
Geodetic Science, Engineering and Innovation Research Group, Faculty of Earth Science and Technology, Institute of Technology Bandung, Bandung, Indonesia

Nurul Nadiah Yahya Geoinformation, Faculty of Built Environment and Surveying, Universiti Teknologi Malaysia, 81310 Johor Bahru, Johor, Malaysia;
Geoscience and Digital Earth Centre, Faculty of Built Environment and Surveying, Universiti Teknologi Malaysia, 81310 Johor Bahru, Johor, Malaysia

Hao Yu National Yang Ming Chiao Tung University, Hsinchu, Taiwan

Mohd Iskandar Yusof Lands and Surveys Department Sarawak, Kuching, Malaysia

Ahmad Razali Yusoff Department of Geotechnics and Transportation, Faculty of Civil Engineering, Universiti Teknologi Malaysia, UTM Skudai, Johor, Malaysia

Hakima Zair Department of Geology, Faculty of Sciences Aïn Chock, Hassan II University, Casablanca, Morocco

Ainn Zamzuri 3D GIS Research Lab, Faculty of Built Environment & Surveying, Universiti Teknologi Malaysia, Johor Bahru, Malaysia

Samir Zouhri Department of Geology, Faculty of Sciences Aïn Chock, Hassan II University, Casablanca, Morocco

3D Buildings Modelling

The Third Dimension of GIS and LADM – Current Status and Perspectives

Alias Abdul Rahman(✉), Ainn Zamzuri, Hanis Rashidan, and Wahyu Marta Mutiarasari

3D GIS Research Lab, Faculty of Built Environment & Surveying, Universiti Teknologi Malaysia, Johor Bahru, Malaysia
alias@utm.my

Abstract. This paper describes the overall trends in 3D geoinformation and the Land Administration Domain Model (LADM). Several aspects of these domains, such as data collection, processing, modeling, databasing, and visualization, are under investigation by researchers in academia or industry. Their works have been reported in international meetings and conferences, which have discussed increasingly advanced methods and techniques. In general, 3D geoinformation has advanced considerably, as is seen in current discussions within the 3D GIS community and the LADM and 3D Land Administration (LA) group. Conferences, workshops, and specialized user group meetings worldwide provide the latest research and development on the subject. This paper highlights the most recent findings on these two major domains in 3D geoinformation and provides insights into possible future developments.

Keywords: 3D geoinformation · LADM · 3D data fusion · Semantic modeling

1 Introduction

Geoinformation science has existed since the 1960s as a simple system to manage geographic objects for environmental research in North America, specifically Canada [1]. Later, additional systems were researched and developed by the community. Initially, these systems were categorized as 2D due to the nature of data at that time. These days, systems can handle spatial data in more than two dimensions; they also have more possibilities in terms of processing and visualization, as reported by [2]. The research and development of 3D GIS domains began around the mid-1990s when several major spatial object primitives were explored and recognized as 3D spatial objects; thus, a proper data model could be established. Several data models have been reported since then, such as spatial theory by [3] and other models by [4]. These models have since been further developed for greater flexibility in object and relationship connections. These connections between objects provide a better understanding of the 3D situations we perceive in the real world. An increasing number of research groups in academia and industry (vendors) have pursued greater performance in the processing and databasing stages. Here, a database's data structure and object indexing play a vital role in making the related processes and analyses faster, as reported by [5].

A. Abdul-Rahman et al. (Eds.): GeoWeek 2024, LNGC, pp. 3–16, 2025.
https://doi.org/10.1007/978-3-031-86654-8_1

Another important aspect of 3D GIS is semantic labeling (assigning and validation) of 3D buildings, which provides an accurate and fast way of recognizing information about the objects. The trend indicates the utilization of machine learning and deep learning to assist in the labeling process, as mentioned by [6]. This process requires large datasets to develop a robust segmentation model to achieve the desired outcome. Technological advancements have opened doors for further research on the automation and refinement of semantic labeling processes, making it possible to handle complex and large-scale 3D spatial datasets more efficiently. Moreover, integrating artificial intelligence with 3D GIS has facilitated improvements in data acquisition, generation, and the overall quality of geospatial processes.

3D building generation has been investigated for quite some time, especially in the context of acquisition techniques. However, research on combining techniques in one computing environment is lacking, as reported by [7]. Recent research shows that the accuracy of the combined techniques (fusion) could be improved by using a Laplacian algorithm. The optimal configuration of this fusion method is a subject of significant discussion, as reported by [8]. Ensuring high accuracy and completeness in 3D models is both essential and challenging when utilizing different datasets. One of the key challenges lies in developing effective data integration methods to generate 3D models from these varied datasets, such as the integration of multi-source point clouds as proposed by [9]. In the future, these valuable 3D models can be used in various applications such as urban planning and land management.

The integration of land and 3D spatial objects, as offered by the Land Administration Domain Model (LADM), results in better data modeling for managing and disseminating information. Thus, this integrated model (with well-defined classes) is an effective and reliable land management system, as reported by several researchers such as [10–12]. The LADM has the potential to be further developed to address more advanced situations and other international standards and formats, including CityGML, CityJSON, BIM/IFC, and INTERLIS.

In general, four main aspects of the 3D GIS domain have been investigated: spatial data acquisition, spatial data processing, spatial databases, and visualization, similar to what has been researched thus far in the 2D domain. These investigations aim to generate more advanced spatial data modeling and visualization concepts and techniques. The community is also investigating interoperability and standards related to 3D spatial data—for example, to share and disseminate data and information. 3D geoinformation research has been investigated in academia and by major software vendors.

Section 2 describes the status and perspective of 3D GIS. Section 3 describes the status and perspective of LADM. Finally, Sect. 4 concludes the paper.

2 3D GIS

The third dimension of geoinformation has been explored and investigated since the mid-1990s [13–15]. GIS improves upon simple 2D data capability by being able to handle and manage point cloud data (from LiDAR and other similar data collection systems). The goal is better processing with fast algorithms; databases should also be able to perform searches quickly. Smart indexing for spatial records and objects is of prime importance for many research groups in academia and industry (software development houses). As reported by [2], digital twins and 3D city models are a significant research trend in the community. In Germany, TU Munich and HFT Stuttgart researchers have presented on CityGML 3.0, while the University of Melbourne group has investigated the same model for 3D underground land administration. These proceedings address point clouds and 3D building reconstruction, and BIM and GIS continue to be research topics in the community. One particularly interesting paper on the use of CityJSON for energy demand calculation examined JSON as an alternative (or a complement) for CityGML, and other works have described 3D indoor navigation for pedestrian interaction purposes (all of the preceding papers are discussed in [16]). We also acknowledge that several groups in the 3D geoinformation domain are investigating 3D data fusion and the semantic modeling of 3D buildings.

2.1 Data Fusion for 3D Models

3D models representing 3D objects can be generated from datasets like images and point clouds. Achieving accurate and complete 3D models is challenging due to the weaknesses of each spatial data acquisition method and the presence of occlusions. Thus, the fusion of several datasets has been utilized to overcome these weaknesses and subsequently generate complete 3D models, as demonstrated in [17, 18]. Point cloud datasets are commonly integrated with iterative closest point (ICP) registration. However, post-processing of this ICP registration reveals some limitations, including stratified redundancies [19] and gaps within the interconnected regions [9]. Consequently, enhancing 3D fused models to achieve more accurate representations is a key research focus, with methods such as the Laplacian fusion offering potential improvements. We believe that developing and improving data fusion methods could provide a better outcome. Specifically, the advancements should focus on handling large point clouds with varying accuracy and scale and address the complex geometrical holes in 3D fused models. 3D fused models, which are point cloud models, can be reconstructed into 3D models based on faces and could be developed for larger areas like 3D city models, as reported by [20], the potential applications of which include energy demand estimation, the classification of building types, and 3D cadastre. Examples of 3D models of major cities include Singapore [21], Helsinki [22], Rotterdam [23], and several Japanese cities [24].

Figure 1 illustrates the development of 3D building models. Spatial data can be obtained using several acquisition techniques, such as airborne laser scanning (ALS), terrestrial laser scanning (TLS), and drone photogrammetry. These datasets are integrated and enhanced through data fusion into 3D fused building models, and subsequently, 3D building models are reconstructed based on these fused datasets. Developing 3D representations for an urban environment results in 3D city models, including buildings, streets, and other infrastructure. These models are useful for urban applications, such as 3D cadastre and heat demand analysis for buildings. In the next five years, the development of data fusion for point cloud datasets will likely involve automated processes (to quickly and efficiently integrate big data).

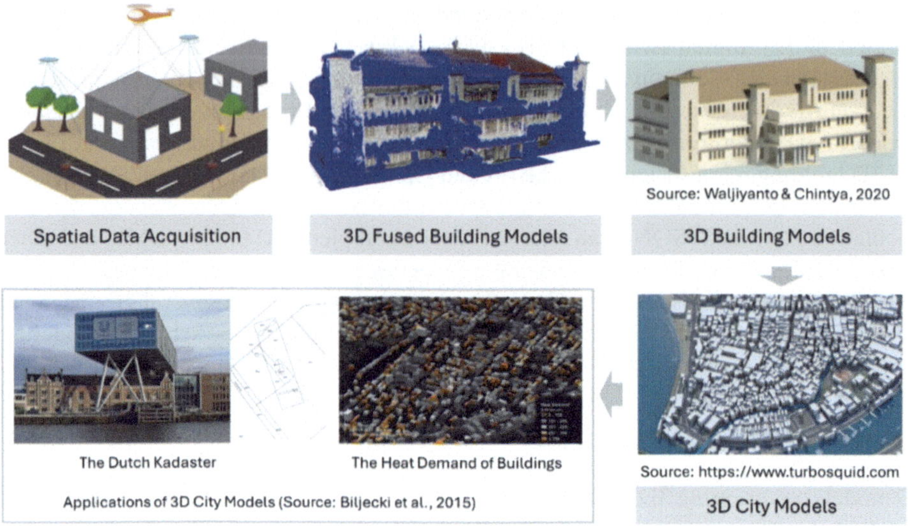

Fig. 1. Development of 3D models through data fusion.

2.2 Semantic Segmentation of 3D Models

As urban environments grow in complexity, detailed and accurate 3D city models become increasingly crucial for urban planning, GIS, and the development of smart cities. CityGML has emerged as a standard for modeling and exchanging 3D city models, encompassing various aspects such as buildings, transportation, vegetation, bodies of water, and more [25]. The semantic segmentation of building models using deep learning within the context of CityGML has opened possibilities for automating and improving certain processes. Semantic segmentation involves classifying each element of a building model into predefined categories. Efficiency and scalability are the motivation for integrating deep learning with CityGML for semantic segmentation. Traditional manual methods for segmenting building models are labor-intensive, making them challenging for large datasets. In contrast, deep learning techniques can process vast amounts of data with minimal human intervention, leading to significant time and cost savings.

Moreover, the semantic segmentation of building models is crucial for creating detailed databases of building components and structures. By automatically identifying and classifying elements such as walls, roofs, and windows, deep learning methods facilitate the creation of comprehensive semantic labels. These labels can then be converted into structured data for databases, cataloging every aspect of building assets and structural elements. This structured data is invaluable for various applications, including building management, smart city development, and geospatial analysis. Figure 2 illustrates the concept of building segmentation within CityGML to locate and generate databases of building components.

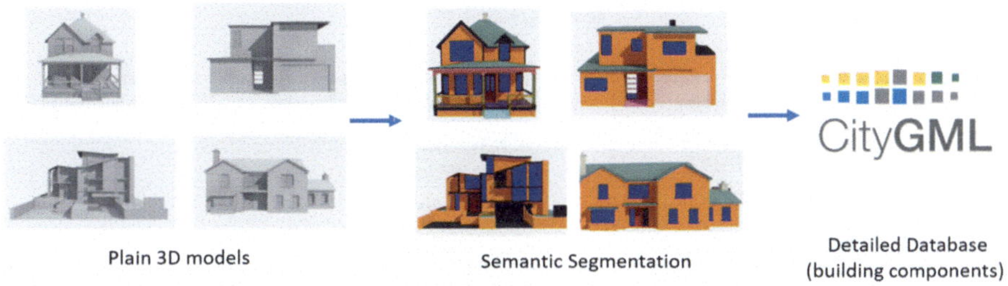

Fig. 2. The overall concept of building segmentation within CityGML.

3 The LADM

The Land Administration Domain Model (LADM) has been utilized in various countries, providing a useful structure or concept for managing land administration (LA). The flexibility of this standard allows it to be adapted for different systems worldwide and integrated with 3D spatial units. [26] stated that ISO 19152:2012 LADM serves as the official model for 3D parcel representation in the provincial city of Shenzhen in China, as well as in Finland, Malaysia, and Scotland. Additionally, approximately 35% of the countries involved in the adoption of LADM have indicated that their cadastral databases are either fully or partially based on this standard. The integration of 3D data into the LADM is a significant and ongoing development aimed at enhancing the capacity of land administration systems (LAS) to handle complex spatial information. Several countries, including the Netherlands [27], Sweden [28], Singapore [29], and Australia [30], have conducted pilot projects and implemented prototypes to test and refine these capabilities.

LADM has now been extended (i.e., LADM Edition II) to cover both land and marine environments, including information on land value and use [31]. This second edition has multiple parts with different purposes, as shown in Fig. 3. Part 1 represents LADM's generic conceptual model. Part 2 extends on LADM Edition I's land registration with a refined survey model, 3D spatial profiles, and semantically enriched code lists. Part 3 focuses on the rights, restrictions, and responsibilities (RRRs) of marine georegulation, aligning them with land concepts to provide seamless land and marine administration; this part is refined from the IHO S-121 Standards for Maritime Limits and Boundaries. Part 4 focuses on valuation information as it pertains to LA, detailing the characteristics and semantics of valuation registries that are commonly managed by government agencies. Part 5 represents the RRRs regarding spatial plan information, including land use, and Part 6 concerns the implementation of LADM (i.e., the methodology of developing country profiles and frameworks representing LA processes).

These domains are crucial for better administration, as reported by [32]. This standard is designed with flexibility, enabling the consolidation of a variety of data sources such as BIM/IFC, GIS information, property valuations, and cadastral records. As illustrated in Fig. 4, this seamless integration paves the way for versatile applications and services. In line with this trend, the integration of 3D building models with LADM appears particularly promising, especially as the number of complex buildings continues to grow. These complex scenarios require proper handling through comprehensive data models. The utilization of building information modeling (BIM) as a data source is increasingly

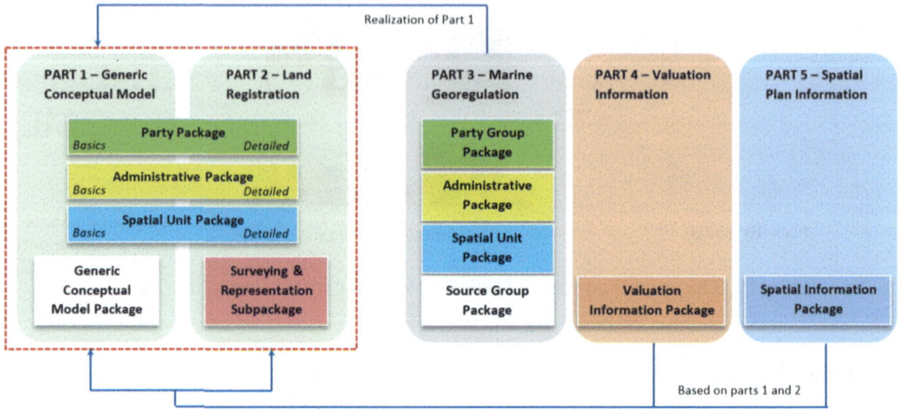

Fig. 3. The multiple parts of LADM Edition II (adopted from [32]).

attracting attention from researchers around the globe [33–36], who aim to merge it with LADM to enhance 3D land and property management. The authors of this paper are currently conducting an initiative to combine LADM Edition II with BIM for better 3D land management in Sarawak, a Malaysian state. Through such integration, authorities could enjoy a unified administration system that encompasses legal, spatial, and technical information, all of which support the development of smart city initiatives.

Table 1 (in Appendix) contains an overview of LADM implementations associated with various disciplines, including integration with BIM/IFC and CityGML, 2D and 3D underground mapping, 3D cadastre, building valuation information, RRR mapping for strata buildings, web-based applications, applied marine cadaster, and incorporating temporal elements (3D + time). It details the researchers responsible for these implementations, the year in which the research was conducted, the researchers' contributions, challenges, improvements, and recommendations from various countries that use LADM as part of their development.

Characterizing multifunctional structures is challenging, particularly where information about legal boundaries and rights, restrictions, and responsibilities (RRRs) must be identified. Apart from that, there are also challenges in integrating LADM with other standards, including CityGML, CityJSON, IndoorGML, and INTERLIS. These standards within LADM are under development and consideration by the relevant authorities. This initiative to incorporate various standards is in line with the UN's Sustainable Development Goal (SDG) 9 (industry, innovation, and infrastructure) and SDG 11 (sustainable cities and communities). By leveraging advanced models and data integration, land administration systems offer improved sustainable development and urban planning initiatives.

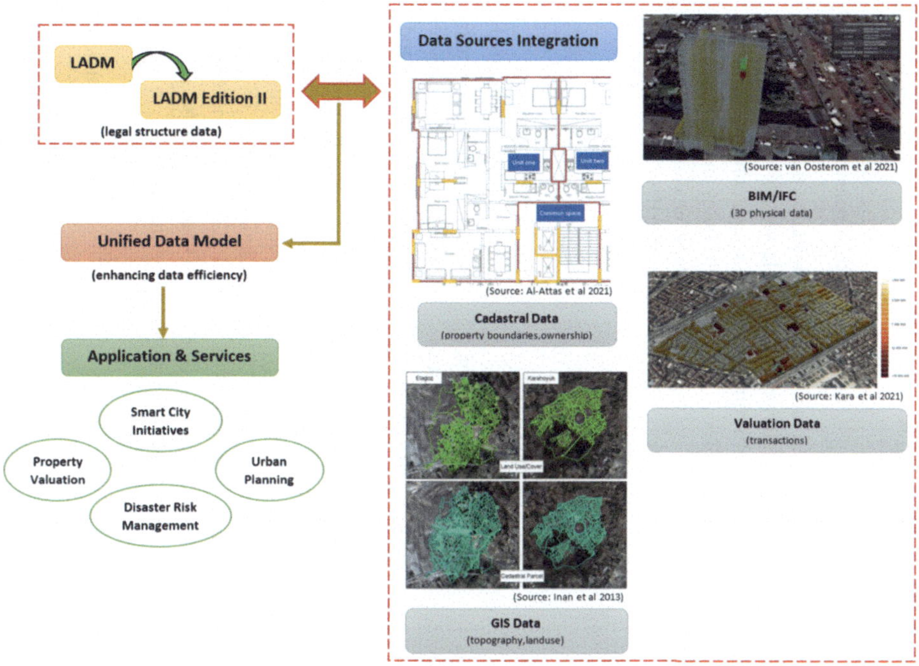

Fig. 4. Integration of data sources through the unified data model for applications and services.

4 Conclusion

This paper discussed the overall trends and development in 3D geoinformation, particularly in the disciplines of 3D GIS, 3D data fusion, semantic modeling, and LADM. Recent research has revealed that the domain is highly linked with computing, mathematics, databases, and visualization. Web and desktop computing platforms also play crucial roles in visualizing and disseminating spatial objects and information in 3D. Researchers in 3D geoinformation and the Land Administration Domain Model (LADM) have made significant advancements in techniques, processing, computing tools, algorithms, and standards. Currently, research communities are investigating better solutions and services, as seen in many major GIS vendors' advertisements and information. Above all, the 3D geoinformation domain will surely remain active, strong, and innovative.

Appendix

Table 1. Overview of LADM implementations in some countries.

Researcher(s)/Year	Contributions	Improvements	Challenges	Recommendation
Lemmen et al. (2013) [37]	The publication of land standard ISO 19152:2012 LADM provides a generic representation model which covers both spatial (main class: LA_SpatialUnit) and non-spatial aspects (main classes: LA_Party, LA_RRR, and LA_BAUnit)	A descriptive standard applicable worldwide for a variety of spatial units (e.g., land parcels, legal spaces around buildings, utilities, and networks)	Complexity in modeling and building an international consensus for refined country profiles	Selecting the relevant classes from LADM and adding country-specific attributes and associations between classes for country profile development
Trinidad & Tobago Griffth-Charles and Edwards (2014) [38]; Griffith-Charles et al. (2018) [39]	Developing a 3D LADM-based system accommodating existing legal components and extending to marine and fiscal cadastre details	The extended LADM country profile also incorporates STDM as informal value parcels	New systems can be hindered by a lack of technical development, maintenance skills, and effective user engagement	The planning database needs to be aligned with the property taxation system to ensure consistency between designated and actual land use
Poland Góźdź et al. (2014) [40]; Góźdź and van Oosterom (2016) [41]; Bydłosz (2020) [42]	Investigating the possibility of integrating CityGML for cadastral purposes by integrating the conceptual level of LADM and developing national information infrastructure	Incorporating the spatial elements of the LADM into the CityGML data model, highlighting the interrelation between these standards	Some classes (e.g., parties, administrative units, and rights) are difficult to introduce because of missing counterparts in CityGML. Lack of clear data structures defined in UML and databases	LOD2 is recommended as the most convenient level of detail. The application of solutions such as IFC, CityGML, or LandInfra in a 3D cadastre can be considered

(*continued*)

Table 1. *(continued)*

Researcher(s)/Year	Contributions	Improvements	Challenges	Recommendation
Malaysia Zulkifli et al. (2014) [43]; Rajabifard et al. (2018) [44]; Zulkifli et al. (2019) [45]; Hanafi and Hassan (2019) [46]; Zamzuri et al. (2021) [47]; Atulukwu et al. (2024) [48]; Zamzuri et al. (2024) [49]	Extending the LADM country profile with 3D representations (buildings and lots), full version management, historical information, the possibility to group multiple spatial units in one basic administrative unit with the same rights attached, and legal spaces around utilities (in 3D). Realizing LADM-based 3D cadastral registration and implementing 3D-NCDB. Further developing the country profile to encompass strata management, marine cadastre, and 4D (3D + time)	The LADM-based extensions provide a comprehensive structure for administration in Malaysia, encompassing multiple domains, including land, strata buildings, marine, and indigenous RRRs	Lacking available data causes difficulties in realizing real-world situations and ensuring data compatibility and seamless integration without data loss or inconsistencies	Need for coordination among agencies for effective implementation

(continued)

Table 1. *(continued)*

Researcher(s)/Year	Contributions	Improvements	Challenges	Recommendation
The Netherlands Kara et al. (2019) [50]; Hagemans et al. (2022) [51]	One of the earliest initiatives to implement LADM into land administration to extend the country profile with 3D cadastre and valuation information. Newly surveyed cadastral boundary geometry is documented by applying a sub-package (Survey and Representation) for a solid solution to the process	Developed and implemented advanced technologies to support LADM, such as integrating LADM with 3D cadastre systems and CityGML	Ensuring high-quality and consistent data across different regions and systems	Continue to invest in technological advancements that can enhance the capabilities of LADM, such as incorporating AI and machine learning for data analysis and decision-making. Promote the use of cloud-based solutions for data sharing among stakeholders
Cyprus Elia et al. (2011) [52]; Demetriades et al. (2023) [53]	Proposed a valuation model and demonstrated how BIM can enhance the valuation process by comprehensively representing land characteristics in a 3D environment. Showed the potential for BIM to streamline valuation processes and improve compliance with industry regulations	Future capability to represent legal spaces related to apartments or utility infrastructures as 3D spatial units	The non-one-to-one correspondence between the existing valuation model requires adjustments. The absence of certain attributes in ISO 19152-4 that are present in Cyprus's existing valuation model necessitates creative modeling solutions	The establishment of a database for the proposed valuation model in alignment with ISO 19152-4. The development of 3D visualization of the model to support land value queries and present results visually. The bidirectional exchange of data between the IFC model and the database to automate valuation processes

References

1. Burrough, P.A.: Principles of Geographical Information Systems for Land Resources Assessment (Monographs on Soil and Resources Survey No. 12). Oxford University Press, Oxford, UK (1986)
2. Rahman, A.A.: The 3D geoinfo journey from 2006 to 2022—a reflection. In: Kolbe, T.H., Donaubauer, A., Beil, C. (eds.) Recent Advances in 3D Geoinformation Science. 3DGeoInfo 2023. Lecture Notes in Geoinformation and Cartography, pp. 853–861. Springer, Cham (2024). https://doi.org/10.1007/978-3-031-43699-4_51
3. Molenaar, M.: An Introduction to the Theory of Spatial Object Modelling for GIS. Taylor & Francis, Milton Park (1998)
4. Zlatanova, S., Rahman, A.A., Shi, W.Z.: Topological models and frameworks for 3D spatial objects. Comput. Geosci. **30**(4), 419–428 (2004). https://doi.org/10.1016/j.cageo.2003.06.004
5. Azri, S., Ujang, U., Castro, F.A., Abdul-Rahman, A., Mioc, D.: Classified and clustered data constellation: An efficient approach of 3D urban data management. ISPRS J. Photogramm. Remote. Sens. **113**, 30–42 (2016)
6. Rashidan, H., Musliman, I.A., Sani, M.J., Abdul Rahman, A.: Semantic labeling of 3D buildings by using graph neural network (GNN). Int. Arch. Photogramm. Remote. Sens. Spat. Inf. Sci. **48**, 307–311 (2024). https://doi.org/10.5194/isprs-archives-XLVIII-4-W9-2024-307-2024
7. Mutiarasari, W.M., Rahman, M.Z.A., Rahman, A.A.: Investigating data fusion from three different point cloud datasets using iterative closest point (ICP) registration. In: Kolbe, T.H., Donaubauer, A., Beil, C. (eds.) Recent Advances in 3D Geoinformation Science. 3DGeoInfo 2023. Lecture Notes in Geoinformation and Cartography, pp. 299–308. Springer, Cham (2023). https://doi.org/10.1007/978-3-031-43699-4_18
8. Maset, E., Valente, R., Iamoni, M., Haider, M., Fusiello, A.: Integration of photogrammetry and portable mobile mapping technology for 3D modeling of cultural heritage sites: The case study of the Bziza Temple. Int. Arch. Photogramm. Remote Sens. Spat. Inf. Sci. **XLIII-B2**, 831–837 (2022). https://doi.org/10.5194/isprs-archives-XLIII-B2-2022-831-2022
9. Li, S., Ge, X., Hu, H., Zhu, Q.: Laplacian fusion approach of multi-source point clouds for detail enhancement. ISPRS J. Photogramm. Remote. Sens. **171**, 385–396 (2021). https://doi.org/10.1016/j.isprsjprs.2020.11.021
10. Kara, A., van Oosterom, P., Lemmen, C., Kalogianni, E.: Design of the new structure and capabilities of LADM edition II including 3D aspects. Land Use Pol., **137**, 107003 (2024). https://doi.org/10.1016/j.landusepol.2023.107003
11. Zamzuri, A., Abdul-Rahman, A., Hassan, M. I., van Oosterom, P.: 3D land administration system for Wilayah Sarawak based on LADM edition II—preliminary works. In: van Oosterom, P., Paasch, J. M. (eds.) 11th International FIG Land Administration Domain Model/3D Land Administration Workshop, pp. 149–166 (2023). https://doi.org/10.4233/uuid:b617293f-e902-4c08-ba93-b72aff2ee30d
12. Guler, D.: Implementation of 3D spatial planning through the integration of the standards. Trans. GIS **27**(8), 2252–2277 (2023). https://doi.org/10.1111/tgis.13122
13. Abdul-Rahman, A.: The design and implementation of a two and three-dimensional triangular irregular network based GIS. PhD Thesis. University of Glasgow, UK (2000). https://theses.gla.ac.uk/4069/
14. Zlatanova, S.: 3D GIS for urban development. PhD Thesis. ITC The Netherlands & Technical University of Graz, Austria (2000). https://zlatanova.xyz/PhDthesis/
15. Pilouk, M.: Integrated modelling for 3D GIS. PhD Thesis. Wageningen Agricultural University and ITC (1996). https://edepot.wur.nl/210498

16. Xie, R., Zlatanova, S., Lee, J.B.: 3D indoor environments in pedestrian evacuation simulations. Autom. Constr. **144**, 104593 (2022). https://doi.org/10.1016/j.autcon.2022.104593
17. Hua, W., Qiao, Y., Hou, M.: The Great Wall 3D documentation and application based on multi-source data fusion—a case study of no. 15 enemy tower of the New Guangwu Great Wall. Int. Arch. Photogram. Remote Sens. Spat. Inf. Sci. **XLIII-B2**, 1465–1470 (2020). https://doi.org/10.5194/isprs-archives-XLIII-B2-2020-1465-2020
18. Castilla, F.J., Ramón, A., Adán, A., Trenado, A., Fuentes, D.: 3D Sensor-fusion for the documentation of rural heritage buildings. Remote Sens. **13**(7), 1337 (2021). https://doi.org/10.3390/rs13071337
19. Li, W., et al.: A volumetric fusing method for TLS and SfM point clouds. IEEE J. Selected Topics Appl. Earth Obs. Remote Sens. **11**(9), 3349–3357 (2018)
20. Biljecki, F., Stoter, J., Ledoux, H., Zlatanova, S., Çöltekin, A.: Applications of 3D city models: state of the art review. ISPRS Int. J. Geo Inf. **4**(4), 2842–2889 (2015). https://doi.org/10.3390/ijgi4042842
21. Soon, K.H., Khoo, V.H.S.: CityGML modelling for Singapore 3D national mapping. Int. Arch. Photogramm. Remote Sens. Spat. Inf. Sci. **XLII-4-W7**, 37–42 (2017). https://doi.org/10.5194/isprs-archives-XLII-4-W7-37-2017
22. Ruohomaki, T., Airaksinen, E., Huuska, P., Kesaniemi, O., Martikka, M., Suomisto, J.: Smart city platform enabling digital twin. In: 9th International Conference on Intelligent Systems 2018: Theory, Research and Innovation in Applications, IS 2018, pp. 155–161. IEEE, Piscataway, New Jersey (2018)
23. Coumans, F.: The Rotterdam 3D city model. https://www.gim-international.com/content/article/digital-city-rotterdam, last modified 2019/11/21
24. Seto, T., Furuhashi, T., Uchiyama, Y.: Role of 3D city model data as open digital commons: A case study of openness in Japan's digital twin "project plateau." Int. Arch. Photogramm. Remote Sens. Spat. Inf. Sci. **XLVIII-4/W7**, 201–208 (2023). https://doi.org/10.5194/isprs-archives-XLVIII-4-W7-2023-201-2023
25. Kutzner, T., Chaturvedi, K., Kolbe, T.H.: CityGML 3.0: New functions open up new applications. PFG–J. Photogramm. Remote Sens. Geoinform. Sci. **88**(1), 43–61 (2020)
26. Kalogianni, E., Lemmen, C., van Oosterom, P., Ploeger, H.: 3D land administration: Current status (2022) and expectation for the near future (2026)—initial analysis. Paper presented at FIG Working Week 2023 (2023) https://pure.tudelft.nl/ws/portalfiles/portal/180900130/TS02I_kalogianni_van_oosterom_et_al_12246.pdf
27. Stoter, J., van Oostero, P., Ploeger, H.: The phased 3D cadastre implementation in the Netherlands. In: 3rd International Workshop on 3D Cadastres: Developments and Practices, pp. 203–217 (2012). https://www.fig.net/resources/proceedings/2012/2012_3dcadastre/3Dcad_2012_42.pdf
28. Mekawy, M. S. A., Paulsson, J., Paasch, J.: Integration of 3D cadastre, 3D property formation and BIM in Sweden. In: Proceedings of the 4th International Workshop on 3D Cadastres, pp. 17–34 (2014). https://www.researchgate.net/publication/277005533_Integration_of_3D_Cadastre_3D_Property_Formation_and_BIM_in_Sweden_Integration_of_3D_Cadastre_3D_Property_Formation_and_BIM_in_Sweden
29. Khoo, V.: 3D cadastre in Singapore. In: Proceedings of the 2nd International Workshop on 3D Cadastres, pp. 508–520 (2011). https://www.fig.net/resources/proceedings/2011/2011_3dcadastre/3Dcad_2011_40.pdf
30. Shojaei, D., Olfat, H., Briffa, M., Rajabifard, A.: 3D digital cadastre journey in Victoria, Australia. ISPRS Ann. Photogramm. Remote Sens. Spat. Inf. Sci. **IV-4/W5** (2017). https://doi.org/10.5194/isprs-annals-IV-4-W5-117-2017
31. Lemmen, C., van Oosterom, P., Kara, A., Kalogianni, E., Alattas, A., Indrajit, A.: Overview of developments of edition II of the Land Administration Domain Model.

Paper presented at the FIG Working Week 2023, Orlando, Florida, 27 May 2023 (2023) https://www.fig.net/resources/proceedings/fig_proceedings/fig2023/ppt/ts03i/TS03I_lemmen_van_oosterom_et_al_12245_ppt.pdf
32. Kara, A., Lemmen, C., van Oosterom, P., Kalogianni, E., Alattas, A., Indrajit, A.: Design of the new structure and capabilities of LADM edition II including 3D aspects. Land Use Pol. **137** (2024). https://doi.org/10.1016/j.landusepol.2023.107003
33. Zamzuri, A., Abdul-Rahman, A., Hassan, M.I., van Oosterom, P.: BIM-LADM amalgamation—review. Int. Arch. Photogramm. Remote Sens. Spat. Inf. Sci. **XLVIII-4/W9** (2024). https://doi.org/10.5194/isprs-archives-XLVIII-4-W9-2024-391-2024
34. Liu, C., Zhu, H., Li, L., Ma, J., Li, F.: BIM/IFC based 3D spatial model for condominium ownership: a case study of China. Geospatial Inf. Sci. (2023). https://doi.org/10.1080/10095020.2023.2246518
35. Sun, J., Paasch, J., Paulsson, J., Tarandi, V., Harrie, L.: A BIM-based approach to design a lifecycle 3D property formation process: a Swedish case study. Land Use Pol. **131**, 106712 (2023). https://doi.org/10.1016/j.landusepol.2023.106712
36. Gursoy, H., Alkan, M., Koeva, M.: Towards investigation of integrating LADM, BIM, and CityGML of 3D condominium rights for cadastral purposes: the case of Turkish cadastral system. Paper presented at the 10th International FIG Workshop on the Land Administration Domain Model (2022). https://www.researchgate.net/publication/356112301_TOWARDS_INTEGRATION_OF_LADM_AND_CITYGML_FOR_THE_CADASTRAL_SYSTEM_OF_TURKEY
37. Lemmen, C., Thompson, R., van Oosterom, P.: ISO 19152:2012, Land administration—an international 2D and 3D representation standard. GEOInform. (2013). https://www.fig.net/resources/articles_about_fig/geoinformatics/2013_08_geoinformatics.pdf
38. Griffith-Charles, E.: Proposal for taking the current cadastre to a 3D, LADM based cadastre in Trinidad and Tobago. In van Oosterom, P., Fendel, E. (eds.) Proceedings of the 4th International Workshop on 3D Cadastres, pp. 363–378 (2014). https://www.oicrf.org/-/proposal-for-taking-the-current-cadastre-to-a-3d-ladm-based-cadastre-in-trinidad-and-tobago
39. Griffth-Charles, Sutherland, M., Lalloo, S.: Extensions to the LADM Trinidad and Tobago toward a juridical, fiscal and marine cadastre. Paper presented at the 7th International FIG Workshop on the Land Administration Domain Model (2018)
40. Góźdź, K., Pachelski, W., van Oosterom, P.J.M., Coors, V.: The possibilities of using CityGML for 3D representation of buildings in the cadaster. In: van Oosterom, P., Fendel, E. (eds.) Proceedings of the 4th International Workshop on 3D Cadastres, pp. 339–362 (2014). https://repository.tudelft.nl/record/uuid:8b83669c-cfc9-4d6b-aae3-137924f9a8aa
41. Góźdź, K., van Oosterom, P.: Developing the information infrastructure based on LADM—the case of Poland. Surv. Rev. **48**(348), 168–180 (2016). https://doi.org/10.1179/1752270615Y.0000000018
42. Bydlosz, J., Bieda, A.: Developing a UML model for the 3D cadastre in Poland. Land **9**(11), 466 (2020). https://www.mdpi.com/2073-445X/9/11/466
43. Zulkifli, A., van Oosterom, P., Abdul-Rahman, A., Jamil, H.: Towards Malaysian LADM country profile for 2D and 3D cadastral registration system. In Rahmen, A.R.A., Boguslawski, P., Said, M.N. (eds.) Proceedings of the FIG 20144 Research Initiative Seminar, pp. 189–212. International Federation of Surveyors (FIG), Copenhagen (2014)
44. Rajabifard, A., Agunbiade, M., Kalantari, M., Yip, K. M., Atazadeh, B., Badiee, F., Isa, M. N., Adimin, M. K., Chan, K. L., Aien, A., Olfat, H., Shojaei, D. Anaraki, M.R.: An LADM-based approach for developing and implementing a national 3D cadastre—A case study of Malaysia. Paper presented at the 7th International FIG Workshop on the Land Administration Domain Model (2018). https://doi.org/10.4233/uuid:6a337aa8-38bc-4505-855e-fcea28377294

45. Zulkifli, A., Abdul-Rahman, A., Siew, C.B.: Database design and development of 3D cadastral registration based on LADM. Paper presented at the 8th International FIG Workshop on the Land Administration Domain Model (2019). https://doi.org/10.4233/uuid:7280c246-2e8f-4429-b263-2b58107ca0cb
46. Mohd-Hanafi, F., Hassan, M.I.: The integration of 3D spatial and non-spatial component for strata management. Int. Arch. Photogramm. Remote Sens. Spat. Inf. Sci. **XLII-4/W16** (2019). https://doi.org/10.5194/isprs-archives-XLII-4-W16-417-2019
47. Zamzuri, A., Hassan, M.I., Abdul-Rahman, A.: Development of 3D marine cadastre data model based on land administration domain model. Int. Arch. Photogramm., Remote Sens. Spat. Inf. Sci. **XLVI-4/W3** (2021). https://doi.org/10.5194/isprs-archives-XLVI-4-W3-2021-337-2022
48. Atulukwu, Y.H., Hassan, M.I., Abdul-Rahman, A.: Developing 4D Malaysian marine cadastre data model based on LADM—preliminary works. Int. Arch. Photogramm. Remote Sens. Spat. Inform. Sci. **XLVIII-4/W9** (2024). https://doi.org/10.5194/isprs-archives-XLVIII-4-W9-2024-237-2024
49. Zamzuri, A., Hassan, M.I., Abdul-Rahman, A.: Incorporating adjacent free space (AFS) for marine spatial unit in 3D marine cadastre data model based on LADM. Land Use Pol. **139**, 107084 (2024). https://doi.org/10.1016/j.landusepol.2024.107084
50. Kara, A., Kathmann, R., van Oosterom, P., Lemmen, C., Isikdag, U.: Towards the Netherlands LADM valuation information model country profile. Paper presented at FIG Working Week 2019: Geospatial Information for a Smarter Life and Environmental Resilience. (2019). https://www.researchgate.net/publication/335105316_Towards_the_Netherlands_LADM_Valuation_Information_Model_Country_Profile
51. Hagemas, E., Unger, E., Soffers, P., Wortel, T., Lemmen, C.: The new, LADM inspired, data model of the Dutch cadastral map. Land Use Pol. **117**, 106074 (2022). https://doi.org/10.1016/j.landusepol.2022.106074
52. Elia, E., Zevenbergen, J., Lemmen, C., van Oosterom, P.: The land administration domain model (LADM) as the reference model for the Cyprus Land Information System (CLIS). Surv. Rev. **45**(329), 100–110 (2013). https://doi.org/10.1179/1752270611Y.0000000034
53. Demetriades, P., Kalogianni, E., Dimopoulou, E.: Leveraging BIM for the LADM part 4—valuation information model: The case study of Cyprus. In: Proceedings of the 11th International FIG Land Administration Domain Model/3D Land Administration Workshop, pp. 261–288 (2023). https://www.diva-portal.org/smash/get/diva2:1807579/FULLTEXT01.pdf

Enhancing Simple Geometrical Holes in Fused 3D Building Model Using the Laplacian Method

Wahyu Marta Mutiarasari(✉) and Alias Abdul Rahman

3D GIS Research Lab, Department of Geoinformation, Faculty of Built Environment and Surveying, Universiti Teknologi Malaysia, Johor Bahru, Malaysia
wahyu@graduate.utm.my, alias@utm.my

Abstract. Data fusion has been employed to combine multiple point cloud datasets to generate more comprehensive 3D building models. However, discrepancies among datasets can result in unresolved holes, reducing model accuracy. This study aims to enhance data fusion techniques using the advanced Laplacian method, known for its multi-view 3D hole extraction capability, to address simple holes with an additional dataset. The fused 3D building model is derived from three-point cloud datasets, and the automated procedure identifies and restores holes through two main phases: 3D hole extraction and Laplacian data fusion. Python scripting facilitates the application of the Laplacian technique. Experiments on a small part of the 3D model showed that simple holes were covered more accurately by substitute points, with CloudCompare evaluations indicating a significant reduction in mean distance to the target: from 0.45 m to 0.29 m for ALS points and from 0.53 m to 0.31 m for image-based points. Future research will address complex holes, incorporating additional constraints and line extraction from image data to accurately reshape these intricate gaps.

Keywords: Fused 3D Building Model · Point Cloud · Laplacian

1 Introduction

Data fusion has been utilised to integrate diverse datasets, including point cloud data, for the creation of comprehensive 3D building models. Specifically, for generating 3D building objects with complex designs, it is difficult to rely on a single dataset. This integration is driven by the recognition of the limitations inherent in individual datasets and the obstructions that form holes in the 3D models. However, the fused 3D building models reveal discrepancies in distances between datasets, resulting in decreased accuracy. As indicated in previous studies [1–3], the existing distances of integrated datasets after integration were calculated using Multiscale Model-to-Model Cloud Comparison (M3C2) tools. These studies prove that the holes in the 3D models remain unresolved and cannot be closed despite data fusion efforts.

The holes inherent in 3D building models could be simple or complex geometric shapes. Simple holes are small, regular, and easy to identify and fill because they have straightforward geometric shapes. This type of hole typically occurs on planar surfaces,

A. Abdul-Rahman et al. (Eds.): GeoWeek 2024, LNGC, pp. 17–24, 2025.
https://doi.org/10.1007/978-3-031-86654-8_2

such as the roofs and walls of buildings. In contrast, holes with complex geometric forms are sensitive to the quality of the point cloud. They are larger, irregular, often interconnected gaps in the point cloud, and challenging to fill accurately. They can be found in building corners and ornamental parts, which often have complicated patterns. Thus, the method to complete holes could vary based on the shape of the holes, and this issue presents a challenge in terms of increasing the completeness and accuracy of fused 3D building models. In this case, the repair of holes focuses on utilising available datasets through data fusion rather than simulating or interpreting the hole filling. Two methods of data fusion enhancement have been applied, including volumetric data fusion [4] and the Laplacian method [5], with the latter claimed to surpass the former in performance. The enhancement work by the latter utilised two datasets and achieved a completion rate of up to 82% without requiring human intervention, exhibiting errors of less than 4 cm.

Considering a goal of achieving greater completeness than the previous work, this paper addresses simple holes in the 3D building model and subsequently adapts the Laplacian method to focus on filling the simple holes using fused three datasets. The greater completeness of 3D building models is beneficial for further analysis and applications, such as urban planning, architecture, and geographic information systems (GIS). Finally, the improved model can lead to better decision-making and more precise simulations in these fields. The following section elaborates on the methodology, including the data and the Laplacian method. Section 3 contains the results and discussions, and the conclusion is presented in Sect. 4.

2 Methodology

2.1 The Fused 3D Building Model

Data in this study is obtained from previous data fusion work, which is an integration of three point cloud datasets, as seen in Fig. 1, from terrestrial laser scanning (TLS), airborne laser scanning (ALS), and drone photogrammetry.

Fig. 1. The fused 3D building model (*Source* [1]).

The registration errors are measured at 0.927655 m for the TLS-ALS points and 1.14988 m for the TLS-image-based points. Evaluation of the fused 3D model was conducted using the M3C2 distance tool in CloudCompare to calculate the standard deviation, indicating distances of 0.8978 m for TLS-ALS points and 0.7758 m for TLS-image-based points.

2.2 The Laplacian Method

Li et al. [5] introduced a workflow for data fusion and enhancement details using the Laplacian method for 3D building models. The workflow of this method is presented in Fig. 2. This automated method is designed to detect and repair holes through two main stages: 3D hole extraction and Laplacian data fusion.

In the first phase, the integrated model is analysed from multiple viewpoints within the minimum bounding sphere. The method utilises 37 viewpoints located using uniform sampling on the upper hemisphere and the equator. Subsequently, from each viewpoint, the integrated models are projected from 3D to 2D to extract holes in the target data and obtain substitute points from the source data. The holes are identified when the number of points in the target falls below a specified threshold within the search window (L_i), where $L_i = 3.\rho_i$ and ρ_i represents the point density which is the average distance between points. Then, points from the source data that overlap with the holes are extracted and reprojected into 3D.

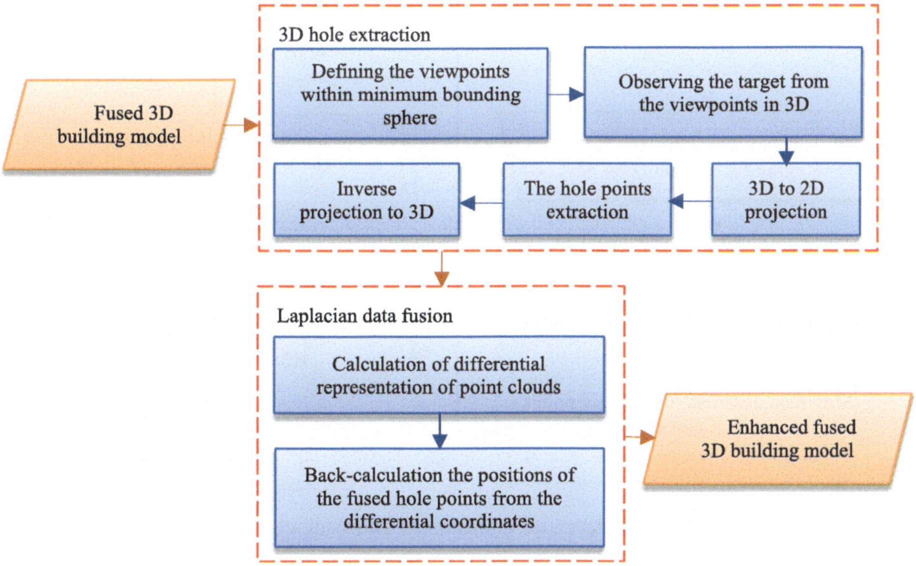

Fig. 2. The workflow of Laplacian method.

In the subsequent phase, the differential coordinates of the substitute points are computed and utilised to seamlessly fill the identified holes (details in Fig. 3). The differential coordinates capture variations in local geometric shapes, supporting accurate point cloud repairs. Each point's variation is explicitly expressed in the corresponding

Laplacian matrix and investigated together with its neighbourhood (N_{vi}) for a radius of $r_i = 5\rho_i$. The Laplacian matrix operator (L) then converts Cartesian coordinates to differential coordinates. Subsequently, back-calculation of the Cartesian coordinates from differential coordinates is conducted using a unique solution set based on boundary conditions. The boundary is extracted from the condition where minimum distances (dis^i_{min}) are less than a distance threshold (dis_{thd}) and moved in the Cartesian system. An additional design matrix is added to set a zero matrix, allowing the x-, y-, and z-components to be derived.

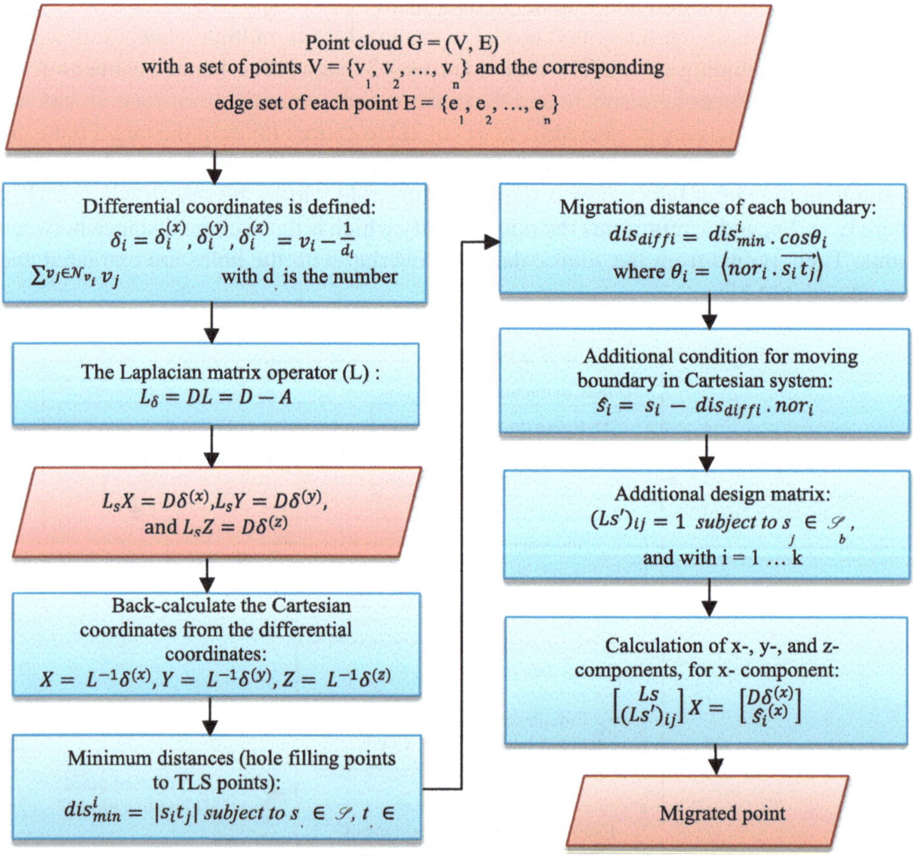

Fig. 3. Details of calculating differential coordinates and back-calculations in Laplacian data fusion.

This study applies the Laplacian method by incorporating an additional dataset to achieve a completeness level of over 82% for the 3D model. Additionally, both simple and complex holes are identified, with the repair focused on the simple geometric holes. This method is implemented and replicated using Python programming, considering the additional data and the identification of the type of holes. In this study, functions that are crucial for the methodology include hole extraction and substitute point migration. The following Figs. 4 and 5 represent the algorithms of the functions (the pseudocode).

```
Algorithm 1: Holes Extraction
procedure HOLES_EXTRACTION(points, density_thd, step_size_multiplier, sample_size)
    num_points ← shape(points, 0)
    sample_size ← min(num_points, sample_size)
    indices_1, indices_2 ← random_samples(num_points, sample_size)
    avg_distance ← mean(norm(points[indices_1] - points[indices_2], axis=1))
    point_density ← 1 / avg_distance
    step_size ← step_size_multiplier * point_density
    min_x, max_x ← min(points[:, 0]), max(points[:, 0])
    min_y, max_y ← min(points[:, 1]), max(points[:, 1])
    num_x_steps, num_y_steps ← calculate_steps(min_x, max_x, min_y, max_y, step_size)
    low_density_mask ← create_mask(points, min_x, min_y, step_size, num_x_steps, num_y_steps, density_thd)
    labeled_holes, num_features ← label(low_density_mask)
    return point_density, labeled_holes, (num_x_steps, num_y_steps), min_x, min_y, step_size, num_features
end procedure
```

Fig. 4. Holes extraction algorithm applied to the target in 2D.

```
Algorithm 2: Coordinate Components
procedure COORDINATE_COMPONENTS(laplacian_mtrx, zero_mtrx, differential, boundary, component)
    degree_matrix ← to_array(laplacian_mtrx['degree_matrix'])
    laplacian_matrix ← to_array(laplacian_mtrx['laplacian_matrix'])
    L_S_prime_ij ← to_array(zero_mtrx)
    if L_S_prime_ij is empty then
        raise ValueError("Invalid zero matrix")
    end if
    delta_component ← extract_component(differential, component)
    S_i_component ← extract_component(boundary, component)
    lhs_matrix ← stack_matrices(laplacian_matrix, L_S_prime_ij)
    rhs_vector ← calculate_rhs_vector(degree_matrix, delta_component, S_i_component)
    solution ← solve_least_squares(lhs_matrix, rhs_vector)
    return solution
end procedure
```

Fig. 5. Algorithm for the calculation of coordinates components.

At the end of the enhancement process, an evaluation is carried out to investigate the improvement of the hole coverage. A distance metric is utilised to compare the state of the filling points before and after enhancement. The subsequent section elaborates on the results and discussion of the experiments.

3 Results and Discussion

The results present the extraction of the holes, the enhanced holes in the 3D model, and the evaluation of the distance metric. This experiment is limited to a small section of a 3D building model's roof (Fig. 6), where the Laplacian method was applied using a single viewpoint (p). In the 3D model, different types of data are represented by various colours: TLS data is shown in blue, ALS data in yellow, image-based data in green, and viewpoints are depicted in red.

3.1 Extraction of the Holes

The TLS-based model, used as the target, was observed from a viewpoint p on a planar surface (2D). Figure 7 (a) and (b) illustrate some holes from the extraction, where blue

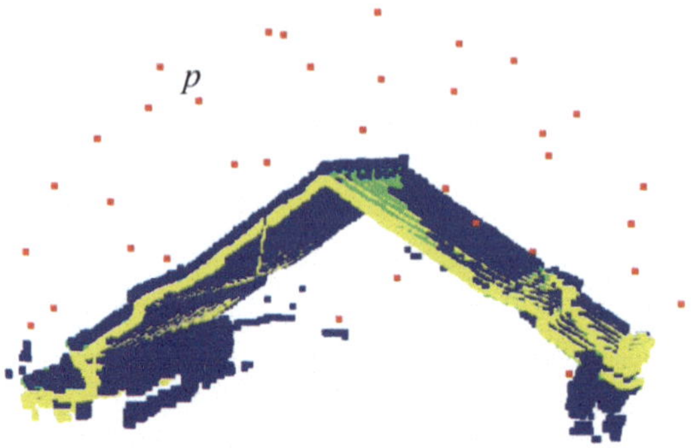

Fig. 6. 3D building model's roof and 37 viewpoints, represented as red points.

points represent TLS points and red polygons indicate the detected holes. A specific parameter, a threshold of 20 points, was set to identify these holes. This parameter facilitated the detection of 22 holes in the model.

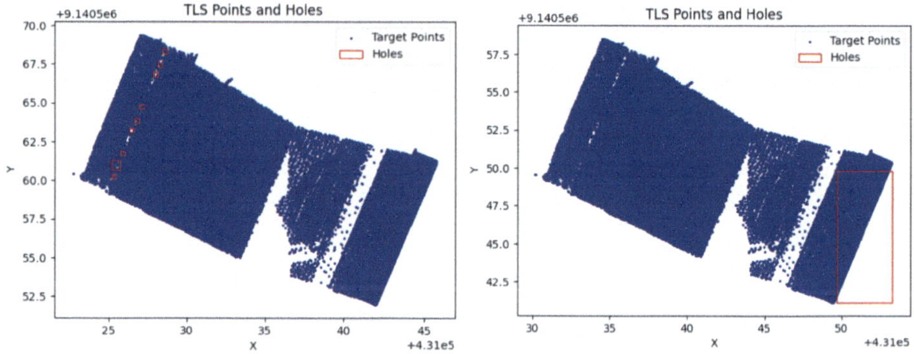

(a) (b)

Fig. 7. Holes (a) and (b) are identified in red polygons.

In this initial identification, it was not possible to categorise the holes as simple or complex due to the limitations of 2D visualisation in representing their actual hole shapes. Therefore, they were reidentified after the improvement process. Based on the extraction, some holes, such as those shown in Fig. 7 (b), were not considered true holes because they exist outside the model. These holes were eliminated during the overlapping process. In this phase, two data sources (ALS points and image-based points) were overlapped with the target to extract substitute points. Subsequently, the substitute points, shown in Fig. 8, were reprojected into 3D.

3.2 Enhanced Holes of the 3D Model

The substitute points were migrated closer to the target model, effectively covering the holes, as indicated in Fig. 9. In this section of the roof, the holes are identified on the

Fig. 8. Substitute points from ALS points (green points), and image-based points (yellow points).

planar surface, e.g., the covered holes inside the yellow polygon in Fig. 9. Consequently, the holes are categorised as simple as no complex holes were found in this section. This classification is based on the straightforward nature of the holes observed, which do not exhibit the intricate shapes or features that would qualify them as complex.

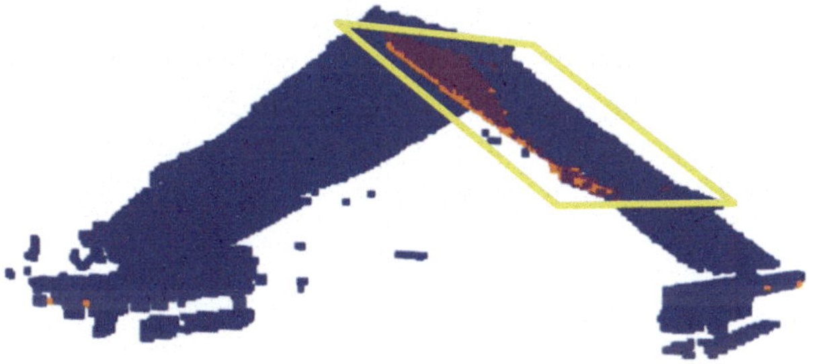

Fig. 9. Migrated points from ALS data (purple points) and image-based data (orange points) cover the 3D model's hole.

From the visualisation, the migrated points appear to cover the holes in the model. However, this visual observation alone is not sufficient to prove that the migrated points cover the simple holes more accurately after the enhancement. To verify the improvement in hole coverage, the subsequent subsection evaluates the distance of the migrated points from the target. This quantitative evaluation will provide concrete evidence of the enhancement's effectiveness, ensuring that the migrated points are indeed filling the holes more accurately and improving the overall quality of the model.

3.3 Evaluation of Distance Metric

Evaluation of the substitute points was conducted using cloud-to-cloud distance tools in CloudCompare, allowing for precise calculation of the distances between the substitute points and the target both before and after the enhancement process. Since the substitute points are derived from two sources, they were evaluated separately. Initially, the distances to the target were measured, resulting in a mean distance of 0.45 m for ALS points and 0.53 m for image-based points. After the enhancement process, the distances were recalculated, yielding a reduced mean distance of 0.29 m for ALS points and 0.31 m for image-based points. This reduction in distance demonstrates the effectiveness of the enhancement process, as the migrated points moved closer to the target, thereby filling the holes more accurately. The decrease in distance signifies that the migrated points are better aligned with the intended target, improving the overall coverage and accuracy of the model.

4 Conclusions

This paper focused on covering simple holes on the target by minimising distances between holes and substituting points from two sources using the Laplacian method. Based on the results from a small part model utilising a single viewpoint, this ongoing research into detail enhancement for the entire building could lead to a complete 3D building model. The utilisation of additional data in this experiment could reduce the need for multiple viewpoints, thus potentially decreasing the number of required viewpoints to fewer than 37. It minimises the need for generating and managing numerous substitute points, streamlining the overall process. In future research, addressing the issue of complex holes will be crucial for further refinement of the model. Complex holes, due to their intricate shapes, present unique challenges that simple holes do not pose. To tackle these challenges, the incorporation of additional constraints could be addressed. One promising approach involves the use of line extraction from image data to accurately reshape these complex holes.

References

1. Mutiarasari, W.M., Rahman, M.Z.A., Rahman, A.A.: Investigating data fusion from three different point cloud datasets using iterative closest point (ICP) registration. In: International 3D GeoInfo Conference, pp. 299–308. Springer Nature Switzerland, Cham (2023)
2. Maset, E.: Integration of photogrammetry and portable mobile mapping technology for 3D modeling of cultural heritage sites: the case study of the Bziza temple, **XLIII**, 6–11 (2022)
3. Jaber, A.S., Abed, F.M.: Revealing the potentials of 3D modelling techniques: a comparison study towards data fusion from hybrid sensors. IOP Conf. Series: Mat. Sci. Eng. **737**(1), 012230. IOP Publishing (2020)
4. Li, W., et al.: A volumetric fusing method for TLS and SfM point clouds. IEEE J. Select. Topics in Appl. Earth Obs. Remote Sens. **11**(9), 3349–3357 (2018)
5. Li, S., Ge, X., Hu, H., Zhu, Q.: Laplacian fusion approach of multi-source point clouds for detail enhancement. ISPRS J. Photogramm. Remote. Sens. **171**, 385–396 (2021)

Designing 3D Triangulated Irregular Network Data Structure for Surface and Subsurface Unified Spatial Data Model – Preliminary Work

Michael Moses Apeh[1,2](✉), Ivin Amri Musliman[1], and Alias Abdul Rahman[1]

[1] Department of Geoinformation, Faculty of Built Environment and Surveying, Universiti Teknologi Malaysia, 81310 UTM Skudai, Johor Bahru, Malaysia
ojonumi@gmail.com, {ivinamri,alias}@utm.my

[2] Department of Surveying and Geoinformatics, Federal Polytechnic, Idah, Nigeria

Abstract. The relationship between surface terrain and subsurface features has revealed a critical gap in current spatial modeling techniques. This research introduces a unified spatial data model utilizing 2D and 3D Triangulated Irregular Network (TIN) data structures. Current methods focus primarily on surface modeling, limiting the analysis of subsurface elements. To address this, we propose extending TIN-based principles into three dimensions for a comprehensive representation and analysis of both surface and subsurface objects. The goal is to develop a flexible unified spatial data model that seamlessly integrates surface and subsurface data through 2D and 3D TINs. The study outlines methodologies for creating and structuring integrated TIN-based models, encompassing algorithms for TIN generation and topology in both 2D and 3D. Initial progress includes their data structure based on 3D triangulation. Further research aims to develop accurate 3D integrated topology, topology tables, and a unified spatial data model. The research highlights the flexibility of TIN in constructing topology and emphasizes its usefulness in establishing a unified spatial data model.

Keywords: Data Structure · Triangulated Irregular Network (TIN) · Surface and Subsurface Objects · 3D Topology · Unified Spatial Data Model

1 Introduction

Integrating surface and subsurface spatial data into a unified model poses a significant challenge in GIS and spatial analysis. While 2D TIN structures are commonly used for surface modeling, there is a need for comprehensive methodologies for 3D TIN structures that integrate both surface and subsurface features. Existing models have not been proven to provide a unified framework for efficient spatial analysis and visualization in both domains. [1] highlight the Triangular Irregular Network (TIN) as a standout method due to its adaptability and large storage capacity, which makes it well-suited for unified spatial data models representing structures, terrains, and sub-surface objects in both 2D and 3D. Al Kalbani et al. [2] stress the importance of integrating surface and subsurface data structures for a comprehensive viewpoint.

A. Abdul-Rahman et al. (Eds.): GeoWeek 2024, LNGC, pp. 25–40, 2025.
https://doi.org/10.1007/978-3-031-86654-8_3

The Triangulated Irregular Network data structure plays a vital role in spatial data modeling, allowing for the 3D portrayal of surfaces, terrains, and subsurface blending [3]. It is a valuable data structure in spatial data modeling, providing benefits such as accurately representing terrain features for cross country mobility (CCM) [4], generating continuous surfaces using interpolation in GIS platforms, and facilitating spatial skyline queries based on TIN distances for more precise calculation of travel distances [5]. TIN represents geographic surfaces using triangular faces and visibility graphs, which are essential for determining visibility among TIN vertices [6]. Moreover, TIN-based digital elevation models (DEMs) are widely utilized for calculating terrain derivatives, highlighting the broad application and significance of TIN in spatial analysis and modeling [7]. The advantages of the TIN approach, including enhanced visualization, routing capabilities, and detailed feature representation, establish it as a versatile and powerful tool in spatial data modeling.

The TIN acts as a linking interface in an Object-Oriented 3D Integrated Spatial Data Model (OO3D-ISDM), aiding the merger of surface structures, terrain, and geological features by representing them through triangles and generalized tri-prisms [8]. It serves as a standardized geometrical model to seamlessly connect aboveground structures, belowground layers, and terrain. This enables the creation of interconnected triangular sets that establish a unified topological relationship between aboveground and belowground components.

This study aims to create a unified spatial data model that combines 2D and 3D spatial objects, emphasizing topological data structure for 3D TIN. Section 2 of this document discusses related works while Sect. 3 outlines the approach for designing the 2D and 3D TIN data structures. Following this is Sect. 4, which considers the initial results and Sect. 5 reflects the overall conclusions.

2 Related Works

In a unified spatial data model, integrating 2D and 3D TIN data structures is essential. Previous research, such as [3], has utilized TIN for terrain visualization, enhancing algorithms with entity structures and attributes. This is needed for accurately representing spatial connections and interactions among surface and subsurface elements, ensuring seamless connectivity between different dimensions. The use of TIN data structures in a unified spatial data model offers numerous advantages, including precise surface representation, efficient handling of variable density data, enhanced topological relationships, improved terrain visualization, robust interpolation methods, seamless integration of 2D and 3D data, scalability, flexibility, and alignment with open standards.

The 3D TIN data structure is justified for a unified spatial data model due to its ability to represent both natural and man-made features above, on, and below the earth's surface effectively [1, 10–12].

This paper explores research related to TIN data structures in spatial data modeling, the integration of surface and subsurface objects, topological data structures for 2D and 3D, and unified spatial data models.

2.1 Integration of Surface and Subsurface Objects

Surface spatial entities indicate the depiction of above-ground features in a three-dimensional spatial data model, essential for urban planning, infrastructure development, and spatial analysis [1, 2]. They help depict real-world traits in spatial analysis, modeling, and visualization. Examples include buildings, bus stops, train stations, airports, bike lanes, parks, bridges, and railways. Terrain represents a land area's physical attributes like topography, elevation, and landforms such as hills, valleys, mountains, and plains. Varying in size and shape, these objects significantly influence a location's physical features. While numerous initiatives concentrate on surface features, integrating subsurface spatial elements is just as vital for a complete 3D spatial data infrastructure (3D SDI) [2]. Subsurface spatial objects, on the other hand, refer to features and structures below the Earth's surface, vital in disciplines like geology, hydrology, mining, and environmental science. A variety of 3D spatial models depicting the integration between surface, terrain, and subsurface objects are increasingly crucial because of the complexity of urban environments and infrastructure advancements in recent years [13].

The relationship between the TIN data structure and surface and subsurface spatial objects is found in their function of representing three-dimensional space characteristics. Integration of Surface and Subsurface Objects.

2.2 Topological Data Structure for 2D and 3D

Topological data structures are essential in Geographic Information Systems (GIS) as they handle spatial relationships between objects in both 2D and 3D spaces. While many current GIS systems lack robust 3D topological support, recent research has proposed innovative solutions. One approach includes implementing 3D topology rules based on models like the 36-Intersection Model (36IM) to establish relationships between 3D objects stored in databases [14]. Additionally, a generic 3D vector model has been introduced to merge limited 3D spatial data types with 2D topological relations to effectively implement 3D topological relations [15]. Moreover, the development of Compact Abstract Cell Complexes (CACC) data structure offers a concise way to represent the topology of 3D models without the need for discrete or geometric topological spaces, enabling efficient storage and representation of topological information for spatial analysis [16]. These advancements in topological data structures enhance the capabilities of GIS systems to handle complex spatial relationships in both 2D and 3D environments, enabling more comprehensive spatial analysis and visualization [17]. Computational topology algorithms have been developed to compute incidences, adjacencies, and ordering of cells in 2D and 3D spaces using sparse vectors and matrices. These algorithms have applications in geometric design, computer graphics, and solid modeling [18]. Furthermore, the implementation of 3D topological relations in GIS systems involves combining 2D topological relations with limited 3D spatial data types and operations, highlighting their effective use in 3D applications [15]. Overall, topological data structures are essential for analyzing complex data and providing insights into the shape of data in various physical systems.

2D and 3D topological data structures primarily differ in dimensionality and complexity of maintaining connectivity information. While 2D structures emphasize relationships within a plane, 3D structures broaden this to include relationships in three-dimensional space, enabling more intricate connectivity analysis [19]. The implementation of 3D topological relations often requires combining restricted 3D spatial data types with 2D topological relations to fulfill desired properties between objects [15].

2.3 3D Unified Spatial Data Model

A 3D Unified Spatial Data Model comprises key components that establish a unified data model for data types, create mapping relations between entity data and virtual data objects, define data presentation views, and utilize object-oriented methods for spatial data modeling [20, 21]. Additionally, utilized an approach that combines surface and subsurface models [22] to create a unified spatial integrated model. This unified spatial data model is essential for efficiently integrating various spatial data types. Several research papers have proposed novel approaches to meet this requirement. Li et al. have concentrated on Precinct Information Modeling (PIM) and spatial data management within a Spatial Database Management System (SDBMS) to integrate diverse data sources in a semantic PIM [23]. Zhang et al. have suggested a 3D cadastral data model using conformal geometric algebra (CGA) to harmonize geometry and topological relationships for 3D spatial units, thereby enhancing data organization and representation efficiency [24]. These studies collectively advance the creation of comprehensive and efficient 3D unified spatial data models, facilitating the seamless integration and analysis of spatial information across multiple domains.

To create a 3D unified spatial data model, the process involves generating TINs through the creation of topological structures, integrating surface and subsurface models, and improving spatial connectivity through Delaunay triangulation. This approach guarantees accurate representation and seamless integration of 2D and 3D spatial objects.

3 Methodology

The creation of a unified 2D and 3D TIN data structure for a spatial data model involves several crucial stages. Majorly, the methods include TIN generation through topology generation and integration of the topology data structure of building, terrain, and subsurface spatial objects representing both 2D and 3D. The steps for the design and the connection between them are highlighted at this stage.

3.1 Generating 2D and 3D TIN Data Structure

In the design phase, attention is given to creating data structures for 2D and 3D TIN to represent surface and subsurface objects. The methodology begins with the acquisition of three-dimensional structure data from OBJ files.

These files contain the vertices and faces of the model's representing terrain, surface (building), and subsurface objects. Figure 1. Illustrates a typical OBJ file indicating

Fig. 1. A typical OBJ file of a model indicating vertices and face.

the vertices (black) and face (red and grey) of a model. The pseudocode's function for parsing the file is illustrated in Fig. 2.

This code is a function that reads a given.obj file, which typically contains information about vertices and faces for 3D models. It reads and parses the vertices and faces from an OBJ file. Vertices are points in 3D space, defined by 'v' lines in the OBJ file while faces are polygons formed by connecting vertices, defined by 'f' lines. The vertices and faces are extracted from the OBJ files using a function that reads the file line by line, identifying vertex (v) and face (f) definitions. In summary, this code parses a.obj file to extract vertex coordinates and face indices, handling exceptions and errors along the way. This data is stored in arrays for further processing.

3.1.1 Triangulation of Faces

The faces of the structures are triangulated to ensure all faces are composed of triangles. This is done by converting faces with more than three vertices into multiple triangles, The input mesh data is analyzed to identify faces with more than three vertices. For instance, Fig. 3, illustrates triangulated faces of the library building which is the subsurface object utilized as part of integration.

This is obtainable for the terrain and the subsurface also. Those faces are then split into triangles by creating new edges between adjacent vertices, resulting in a triangulated mesh that can be further processed or displayed efficiently.

The code illustrated in Fig. 4, defines a function called "triangulate_faces" that accepts a list of faces as input. It processes each face in the input list according to the following criteria: If a face has exactly 3 vertices, it is added to a new list called "triangulated_faces". If a face has more than 3 vertices, it is subdivided into triangles by connecting the first vertex with each pair of adjacent vertices. These resulting triangles are then included in the "triangulated_faces" list. Upon completion, the function returns

```
# Initialize empty lists to store vertices and faces
Initialize empty list vertices
Initialize empty list faces

# Read and parse each line in the .OBJ file
For each line in obj_file:
   Decode the line to utf-8 and strip whitespace
   Split the line into parts

   # Check if the line defines a vertex
   If parts is not empty and the first part is 'v':
      Append the next three parts as floats to vertices

   # Check if the line defines a face
   Else if parts is not empty and the first part is 'f':
      Initialize empty list face
      For each part after the first part in parts:
         Split part by '/'
         If the first element in indices is not empty:
            Append the first element as an integer (minus 1) to face
      Append face to faces

# Error handling
Catch Exception as e:
   Display error message

# Return the parsed vertices and faces
Return vertices as an array and faces
```

Fig. 2. Function to read vertices and faces from the.OBJ file.

the list of triangulated faces. The purpose of the function is to convert faces with more than 3 vertices into triangles while leaving faces with exactly 3 vertices unchanged.

3.1.2 Translation of Vertices

To align the models correctly in the visualization space, translation vectors are applied to the vertices of each model. This ensures that the terrain, surface, and subsurface structures are positioned accurately relative to each other about their geographical locations and spatial relationships.

The translation process as illustrated in Fig. 5, reflect that by applying these translation vectors, the different components within the visualization can accurately represent the real-world features and topographical details, enhancing the overall quality and precision of the modeling process.

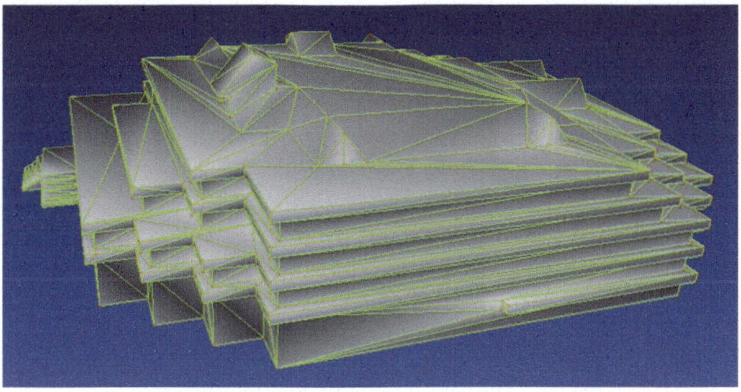

Fig. 3. Triangulation of faces of the Library Building

```
# Initialize empty list to store triangulated faces
Initialize empty list triangulated_faces

# Process each face to ensure it is triangulated
For each face in faces:
   # If the face is already a triangle, add it directly
   If face has exactly 3 vertices:
      Append face to triangulated_faces

   # If the face has more than 3 vertices, subdivide it into
triangles
   Else if face has more than 3 vertices:
      For i from 1 to length of face - 1:
         Append a triangle formed by face[0], face[i], and
face[i + 1] to triangulated_faces

# Return the list of triangulated faces
Return triangulated_faces
```

Fig. 4. Function to triangulate faces with more than 3 vertices

Figure 6, illustrates this function. The above code applies a translation vector to vertices to align models correctly in the visualization space. It also calculates the new positions of the vertices based on the provided translation vector to ensure proper alignment of the models within the visualization space. This process maintains the relative positions of the vertices while effectively moving the entire model to its desired location.

3.1.3 Creating TIN

Delaunay triangulation is used to create TINs from the vertices of the models. This technique generates a network of non-overlapping triangles that cover the surfaces of the

Fig. 5. Translation among spatial objects.

```
# Function to apply translation to a list of vertices
Function translate_vertices(vertices, translation):
   # Add the translation vector to each vertex and return the new list of vertices
   Return vertices plus translation as an array
```

Fig. 6. Function to apply translation to vertices.

structures, providing a basis for further topological arrangement. In this way, the Delaunay triangulation process ensures that there are no empty spaces or overlaps between the triangles, resulting in a precise division of the surface and subsurface into interconnected geometric elements. The create TIN code is as shown in Fig. 7.

```
# Function to create TIN structures using Delaunay triangulation
If vertices has at least 2 columns:
   # Perform Delaunay triangulation on the first two columns of vertices
   Perform Delaunay triangulation on the first two columns of vertices
   # Return the resulting simplices (triangles)
   Return the simplices (triangles)
```

Fig. 7. Function to create TIN structures using Delaunay triangulation.

The following code defines a function named 'create_tin' that requires a parameter called 'vertices'. The function checks if the input 'vertices' has a minimum of 2 columns. If it does, the code conducts a Delaunay triangulation on the first two columns of 'vertices' and then returns the resulting simplices (triangles) that represent the triangulation. If the input 'vertices' does not have at least 2 columns, the function will return an empty list. In summary, this function will create a Delaunay triangulation only if the input data has enough dimensions; otherwise, it will return an empty list.

3.2 Topology for 2D and 3D TIN

Topology in TIN refers to the spatial connections among the elements (points, edges, and triangles) composing the framework. Managing these connections effectively is vital for spatial data modeling. For instance, topology plays a crucial role in ensuring correct geometric relationships and connectivity between different features represented in a TIN model.

```
# Initialize empty adjacency table
Initialize empty adjacency table as a defaultdict of lists

# Process each face to build the adjacency table
For each face in faces:
    For i from 0 to length of face:
        # Define the edge as a tuple of consecutive vertices
        Set edge as a tuple of face[i] and face[(i + 1) % length of face]
        # Update adjacency table with the edge connections
        Append edge[1] to adjacency_table[edge[0]]
        Append edge[0] to adjacency_table[edge[1]]
```

Fig. 8. Function to create topology tables.

This is a summary of the code as illustrated in Fig. 8. This code contains a function called "create_topology_table" that accepts a list of faces as input. The function is responsible for creating an adjacency table that represents the connections between vertices in a network of faces. The code initializes an empty adjacency table using defaultdict from the collections module. Figure 9 illustrates a visual representation showing how the vertices are interconnected across the various models. It stores lists of adjacent vertices for each vertex. It then goes through each face in the input list of faces. For each face, the code iterates through each vertex and creates an edge using each pair of consecutive vertices on the face.

After creating the edge, the code updates the adjacency table by adding the second vertex of the edge to the list of adjacent vertices for the first vertex, and vice versa. Finally, the code returns the completed adjacency table representing the topology of the input faces. This code essentially builds a data structure that captures the connections between vertices in a network of faces.

As illustrated in Fig. 10, the workflow includes uploading terrain, surface, and subsurface OBJ files through a Streamlit interface. The system checks if all the files have been uploaded successfully. If any file is missing, an error is shown, and the process stops. Once the upload is successful, the vertices and faces from each OBJ file are read and decoded. The system makes sure that the vertices arrays are not empty; if they are, an error is displayed, and the process stops. Next, the vertices are adjusted to align the models correctly. We use Delaunay triangulation to create TIN (Triangulated Irregular Network) structures for each set of vertices. Then, topology tables are generated to show

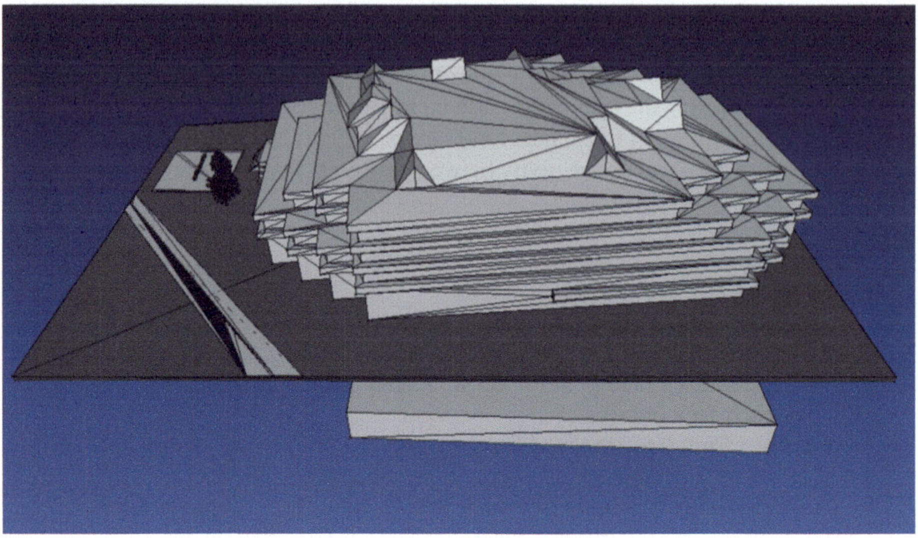

Fig. 9. Visual representation of Building, terrain, and subsurface topology

the connections between vertices in each model. These topology tables are combined to create a unified topology table that includes adjacency information from the surface, subsurface, and terrain models. The 3D models are visualized using Plotly, with each model's faces triangulated if necessary. The visualization is updated and displayed with a suitable layout. The process is completed once the integrated 3D model is successfully visualized.

3.3 Integration of 2D and 3D TIN Topology

The topology tables from the different models are integrated into a unified topology table. This merging process allows for the identification of interconnected triangles, providing a comprehensive overview of the relationships between the terrain, surface, and subsurface elements as illustrated in Table 1. Additionally, the unified topology table facilitates the visualization of potential hotspots or areas of interest where multiple elements converge. The interconnected triangles offer insights into the complex interactions within the environment, allowing for a more holistic understanding of the system. Further development regarding this is ongoing in this research.

The function (Fig. 11) integrate_topology_tables take two tables as input: surface_table and subsurface_table. It merges the data from both tables into a new table called unified_table. The process is as follows:

1. Copy all data from surface_table to unified_table.
2. For each key and value in subsurface_table:

   - If the key already exists in unified_table, add the value to the existing values for that key.

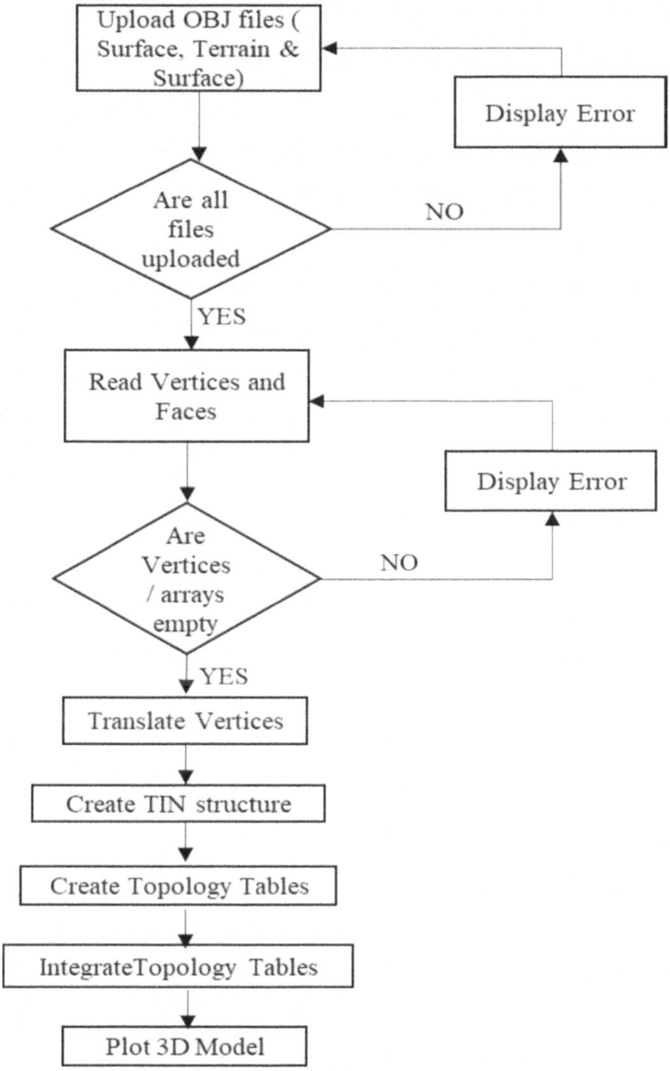

Fig. 10. Flow chart of the methodology

- If the key does not exist in unified_table, create a new entry for that key with the corresponding value.

3. Return the final unified_table that now contains integrated data from both input tables.

In summary, this code integrates data from two tables by merging them based on common keys, thus creating a unified table that combines the information from both the surface and subsurface tables.

As illustrated in the integrated topology function, it merges topology tables from different models into a unified table, identifying interconnected triangles. Additionally,

```
# Function to integrate two topology tables
Copy surface_table to unified_table

# Merge data from subsurface_table into unified_table
For each key and value in subsurface_table:
   If key exists in unified_table:
      Extend unified_table[key] with value
   Else:
      Set unified_table[key] to value

# Return the integrated topology table
Return unified_table
```

Fig. 11. Function for integrating topology tables.

Table 1. Showing adjacent (topology) vertices for surface, terrain and subsurface

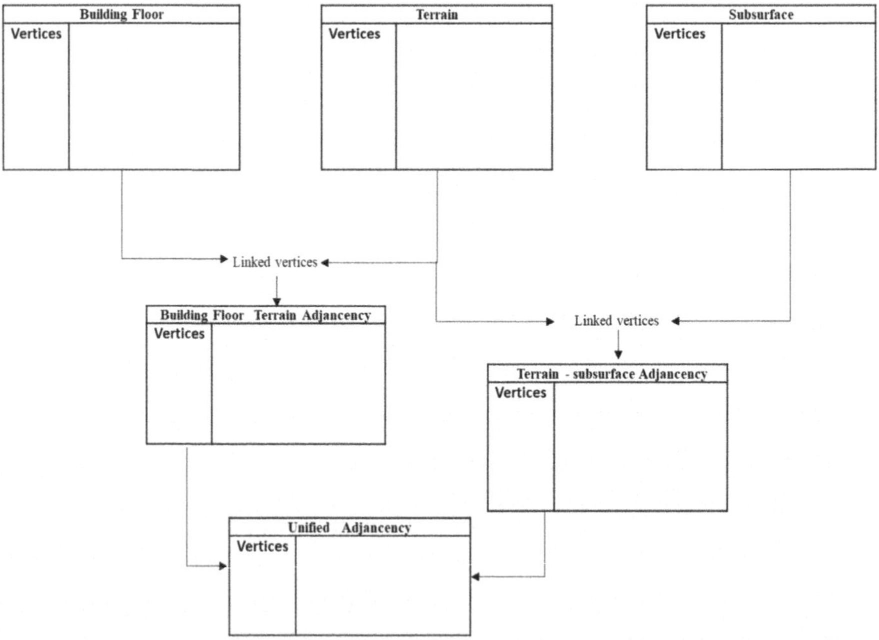

it enables the identification of any potential issues or bottlenecks that may arise due to interconnections between triangles in the topology. These visualizations provide insights into the spatial relationships within each structure. The integrated topology presents a holistic view, showcasing the interconnected nature of building, terrain, and subsurface elements. Further discussions will follow in the subsequent sections.

4 Initial Result

The initial outcome of the process above is the creation of a 3D TIN topological data structure and a unified 3D spatial data model that depicts buildings, terrain, and subsurface objects. This process combines individual models, ranging from objects to characters, within a single file or scene. Figure 6 shows the conceptual view of the unified spatial data model.

4.1 Integration Process

The integration process combines the topology tables of the terrain, surface (building), and subsurface models to create a unified representation. The key steps are as follows:

Building-Terrain Integration: The topology table of the surface (building) and terrain is created using adjacency relationships between the triangles of the two structures. The create_topology_table function is used to generate these tables, capturing the connections between adjacent triangles.

Terrain-Subsurface Integration: The topology table of the terrain and subsurface is similarly generated. The adjacency relationships between the terrain and subsurface structures are established using the create_topology_table function.

Building-Subsurface Integration: The topology table of the surface (building) and subsurface is created by identifying adjacent triangles between these two structures. Again, the create_topology_table function is utilized to record these spatial relationships.

4.2 Final Integration

The three topology tables from the previous steps (surface-terrain, terrain-subsurface, and surface-subsurface) are integrated into a single unified topology table. The integrate_topology_tables function is used to merge these tables, ensuring that all relationships between adjacent triangles in the terrain, surface, and subsurface structures are captured. This final integration provides a comprehensive overview of the spatial relationships between all elements, resulting in an "Integrated Topology" table.

4.3 Data Structure for a Unified Spatial Data Model

The process of creating a 2D and 3D TIN data structure for a combined spatial data model involves essential stages from the creation of TIN data structure to a unified spatial data model. Surface and subsurface elements are projected onto the TIN Data Structure, capturing complexities.

The Unified Spatial Data Model combines surface, subsurface, and terrain spatial data to capture their complex interconnections in a cohesive framework. As illustrated in Fig. 12, the process starts by aligning vertices to coordinate terrain, surface, and subsurface models. Next, Delaunay triangulation creates Triangulated Irregular Networks (TINs) for each model, accurately representing the 3D surface. Adjacency tables are then generated to establish relationships between vertices, forming the topology of each

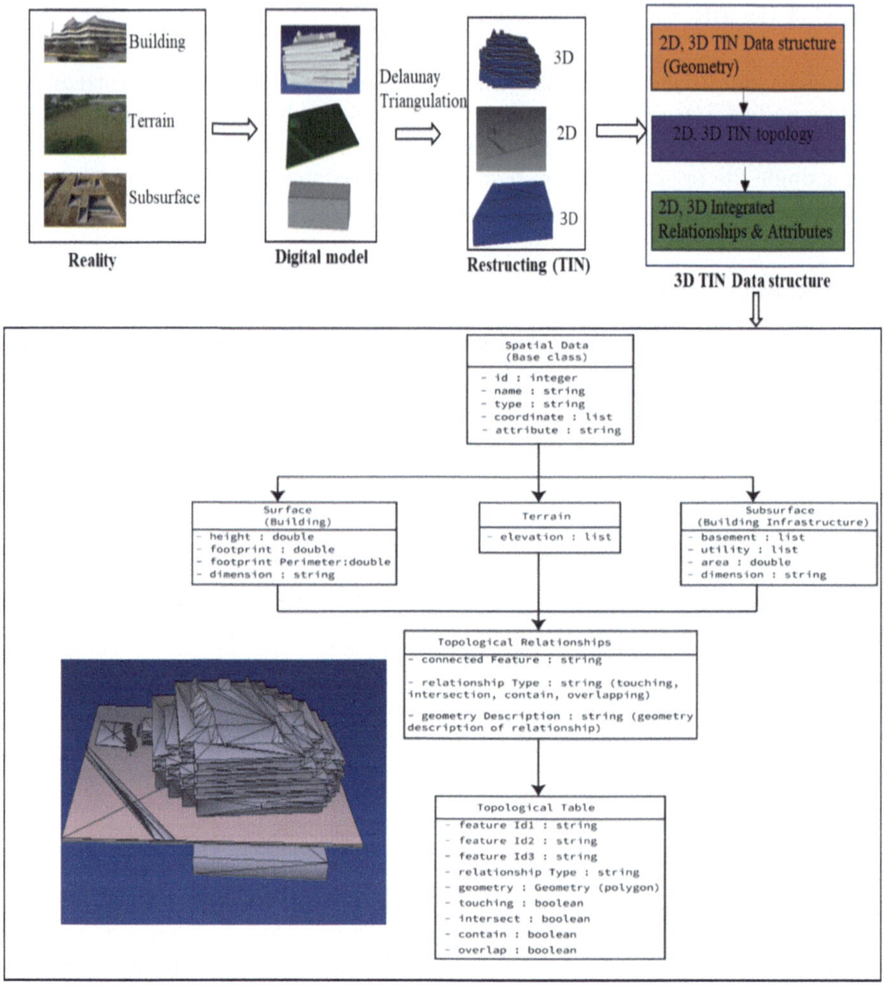

Fig. 12. A conceptual unified spatial data model for surface and subsurface

model. These topology tables are integrated to merge surface, subsurface, and terrain data into a unified table that reflects physical connections, such as between building floors and terrain, and terrain and basement roofs. This unified table comprehensively lists vertex connections across all elements, ensuring spatial continuity. Finally, the integrated model is visualized using Plotly, providing a 3D representation that enhances the understanding of spatial relationships and structures.

4.4 Challenges

Maintaining consistent topological relationships among surface, subsurface, and terrain features during integration is a significant challenge. Correctly representing complex geometries within a unified Triangulated Irregular Network (TIN) requires a thorough understanding of object relationships, which has to be translated into programming for real-world visualization. Although triangulation of individual objects such as buildings, terrain, and subsurface was conducted, detailed design aspects of these faces have not been produced. Furthermore, the correct integration of objects within the TIN data structure to achieve precise triangulation continues to pose challenges. The connections and topology between faces, especially concerning intricate geometries and their corresponding topology tables, are critical issues in this research phase. Additionally, the extensive topological tables that describe the relationships among various faces of surfaces, terrain, and subsurface features have not yet been addressed in this study.

5 Conclusion

In the current research, there is a significant challenge in correctly representing complex geometries by maintaining consistent topological relationships among surface, subsurface, and terrain features within a unified Triangulated Irregular Network (TIN). Successfully integrating these varied features requires a deep understanding of their interrelationships. While progress has been made in triangulating individual objects such as buildings, terrain, and subsurface features and defining their data structure, the detailed design and integration of these faces within the TIN data structure remains incomplete. Correctly connecting and maintaining topology between faces, particularly in intricate geometries, underscores the critical importance of developing comprehensive topology tables. These tables describe the relationships among various faces, surfaces, terrain, and subsurface features.

Addressing these challenges is vital for advancing research and development. The future work will focus on refining the integration processes, enhancing the detail and accuracy of topology tables, and ensuring that the TIN data structure faithfully represents the complex interrelationships among all involved features. Through continued research and development, the study aims to overcome these obstacles and achieve a robust, unified spatial data model for real-world visualization.

References

1. Apeh, M.M., Rahman, A.A.: A review of 3D spatial data models for surface and subsurface integration of spatial objects. Int. Arch. Photogramm. Remote Sens. Spat. Inf. Sci.—ISPRS Arch. **48**(4/W6), 15–23 (2023). https://doi.org/10.5194/isprs-archives-XLVIII-4-W6-2022-15-2023
2. Al Kalbani, K., Rahman, A.A.: Integration between surface and subsurface spatial objects for developing Oman 3D SDI based on the CityGML standard. Int. Arch. Photogramm. Remote Sens. Spat. Inf. Sci.—ISPRS Arch. **42**(4/W16), 79–84 (2019). https://doi.org/10.5194/isprs-archives-XLII-4-W16-79-2019
3. Apeh, M.M., Musliman, I.A., Rahman, A.A.: Experimenting tin data structure for representing plane surface and subsurface spatial objects—preliminary work. Int. Arch. Photogramm. Remote Sens. Spat. Inf. Sci.—ISPRS Arch. **48**(4/W9), 45–51 (2024). https://doi.org/10.5194/isprs-archives-XLVIII-4-W9-2024-45-2024
4. Laboratory, G.R., Rehrer, S., Griffin, A., Renner, M.: Tactical geospatial information capabilities cross country mobility (CCM) modeling using triangulated irregular networks (TIN) (2022) [Online]. Available: https://erdclibrary.on.worldcat.org/discovery

5. Kasai, Y., Sugiura, K., Ishikawa, Y.: Spatial skyline queries on triangulated irregular networks. ACM Int. Conf. Proceeding Ser., 64–73 (2021). https://doi.org/10.1145/3469830.3470901
6. Boomari, H., Ostovari, M., Zarei, A.: Recognizing visibility graphs of triangulated irregular networks. Fundam. Informaticae **179**(4), 345–360 (2021). https://doi.org/10.3233/FI-2021-2027
7. Hu, G., et al.: Using vertices of a triangular irregular network to calculate slope and aspect. Int. J. Geogr. Inf. Sci. **36**(2), 382–404 (2022). https://doi.org/10.1080/13658816.2021.1933493
8. Wang, Y.: 3D GIS spatial modeling for city surface and subsurface integration. Int. Geosci. Remote Sens. Symp., 1511–1518 (2006). https://doi.org/10.1109/IGARSS.2006.391
9. Li, P.: Design and implementation of 3D visualization system. In: 2023 IEEE International Conference on Sensors, Electronics and Computer Engineering (ICSECE), pp. 693–697 (2023). https://doi.org/10.1109/ICSECE58870.2023.10263437
10. Zhang, R.C., Li, H., Wu, M.F., Wang, Y.Q.: An automatic unified modeling method of geological object and engineering object based on tri-prism (TP). J. Cent. South Univ. **22**(4), 1419–1426 (2015). https://doi.org/10.1007/s11771-015-2659-7
11. Duncan, E.E., Rahman, A.A., Siew, C.B., Baig, S.U.: An amalgamated 3D spatial data model for city modelling, 3075(5), 45–49 (2012)
12. Wu, M., Shen, J., Wen, Y.: A hybrid data structure for integrated modeling of geographic 3D data. In: First International Conference on Information Science and Engineering, pp. 1943–1946. https://doi.org/10.1109/ICISE.2009.48
13. Fadli, F., et al.: Extending indoor open street mapping environments to navigable 3D CityGML building models: emergency response assessment, Volume XLI, pp. 161–168 (2018)
14. Salleh, S., Ujang, U., Azri, S.: Topology models and rules: a 3D spatial database approach. https://doi.org/10.5194/isprs-archives-XLVIII-1-W2-2023-117-2023
15. Belussi, A., Migliorini, S., Negri, M.: A framework for evaluating 3D topological relations based on a vector data model. GeoInform. **24**(4), 915–950 (2020). https://doi.org/10.1007/s10707-020-00413-y
16. Ujang, U., Castro, F.A., Azri, S.: Abstract topological data structure for 3D spatial objects. ISPRS Int. J. Geo-Inform. **8**(3) (2019). https://doi.org/10.3390/ijgi8030102
17. Lu, M., et al.: A topological framework for real-time 3D weather radar data processing. Int. J. Digit. Earth **15**(1), 813–831 (2022). https://doi.org/10.1080/17538947.2022.2061616
18. Paoluzzi, A., Shapiro, V., Dicarlo, A., Furiani, F., Martella, G., Scorzelli, G.: Topological computing of arrangements with (Co)chains. ACM Trans. Spat. Algorithms Syst. **7**(1) (2021). https://doi.org/10.1145/3401988
19. Salleh, S., Ujang, U., Azri, S., Choon, T.L.: Spatial adjacency analysis of CityGML buildings via 3D topological data structure. Int. Arch. Photogramm. Remote Sens. Spat. Inf. Sci.—ISPRS Arch. **42**(4/W16), 573–579 (2019). https://doi.org/10.5194/isprs-archives-XLII-4-W16-573-2019
20. Li, Y., Wang, Z., Zhou, P., Yang, J.G.: Data integration method based on unified data model (2015)
21. Ma, D., Liu, X., Ma, L.: Integrated 3D spatial data model of object-oriented digital landslide. https://doi.org/10.4028/www.scientific.net/AMM.204-208.4872
22. Che, D., et al.: Spatial integrated modeling method and device
23. Li, W., Zlatanova, S., Diakite, A.A., Aleksandrov, M., Yan, J.: Towards integrating heterogeneous data: A spatial DBMS solution from a CRC-LCL project in Australia. ISPRS Int. J. Geo-Inf. **9**(2) (2020). https://doi.org/10.3390/ijgi9020063
24. Zhang, J.Y., Yin, P.C., Li, G., Gu, H.H., Zhao, H., Fu, J.C.: 3D cadastral data model based on conformal geometry algebra. ISPRS Int. J. Geo-Information **5**(2) (2016). https://doi.org/10.3390/ijgi5020020

Enhancing 3D Building Models Segmentation in CityGML

Hanis Rashidan[1](✉), Ivin Amri Musliman[1], Gurcan Buyuksalih[2], Alias Abdul Rahman[1], and Ismail Rakip Karas[3]

[1] GIS Research Lab, Faculty of Built Environment and Surveying, Universiti Teknologi Malaysia, Johor Bahru, Malaysia
mhanis27@gmail.com, {ivinamri,alias}@utm.my

[2] Institute of Marine Sciences and Management, Istanbul University, Istanbul, Türkiye
gurcanb@istanbul.edu.tr

[3] Department of Computer Engineering, Karabuk University, Karabuk, Türkiye

Abstract. This paper presents a methodology for enhancing the segmentation of 3D building models using the BuildingGNN method and converting these segmented outputs to CityGML, the standardised format for 3D city models. The process involves leveraging a pre-trained BuildingGNN model to segment various components of 3D building models, such as walls, roofs, windows, and doors. Key challenges addressed include georeferencing the segmented models using affine transformations and ground control points (GCPs) to ensure real-world positioning and geometry validation to comply with CityGML specifications. The segmented models were converted into 3D tiles to visualise their integration within a geographic context, demonstrating their location and orientation on a map. This approach provides a practical means for urban planning and geographic information systems by delivering precise and semantically rich 3D building models. The paper discusses the methods used, the challenges encountered, and the solutions implemented to enhance the conversion process.

Keywords: 3D Building Models · Semantic Segmentation · CityGML

1 Introduction

Building on our previous work using the BuildingGNN method for semantic segmentation of building models with deep learning, this paper focuses on enhancing the process of converting segmentation outputs to CityGML, a standardised format for 3D city models. Figure 1 shows the basic structural components of a CityGML building, such as walls and roofs. Therefore, accurate CityGML models are essential for applications in urban planning, management, and analysis. This paper aims to detail the conversion process, highlight the challenges of georeferencing and geometry validation, and discuss the methods used to enhance the conversion process.

The BuildingGNN model was chosen for its demonstrated effectiveness in segmenting 3D building models due to its accuracy and reliability [2]. Utilising a pre-trained

A. Abdul-Rahman et al. (Eds.): GeoWeek 2024, LNGC, pp. 41–49, 2025.
https://doi.org/10.1007/978-3-031-86654-8_4

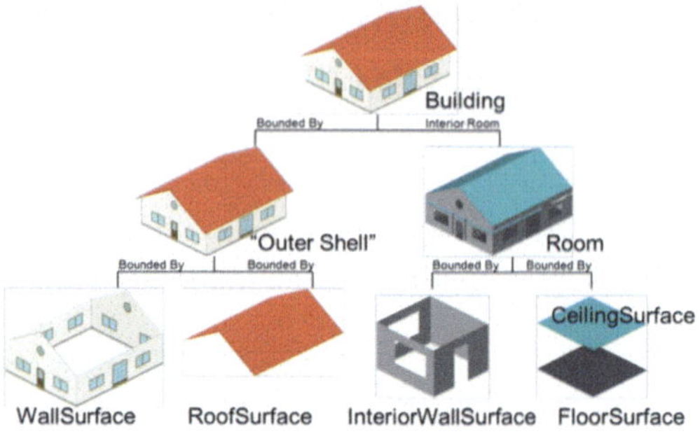

Fig. 1. Basic structure of CityGML building components [1]

BuildingGNN model, the dataset was processed to generate segmented outputs. This involved preparing the input data to be compatible with the model, running the model to perform segmentation, and interpreting the resulting segmented components.

Georeferencing is crucial for aligning 3D building models with real-world coordinates, ensuring their utilisation in urban planning and geographic information systems [3]. In this study, the affine transformation method was employed to georeferenced the segmented 3D models accurately. To meet CityGML specifications, the geometry of the 3D models must be watertight and free of anomalies such as overlapping faces and gaps [4]. The approach involved geometry validation methods, which allowed us to detect and correct geometric issues. Utilising the FME Workbench, the custom workflows were developed to automate the conversion of segmented 3D models to CityGML. This involved creating scripts to handle data transformation, attribute assignment, and format conversion (Fig. 2).

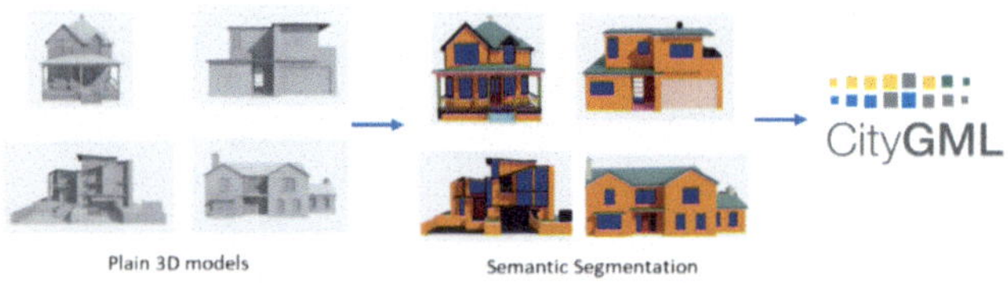

Fig. 2. The overall concept of the building segmentation within CityGML [5].

2 Methodology

BuildingGNN is a deep learning framework specifically designed for the semantic segmentation of 3D building models. This method leverages graph neural networks (GNNs) to process the intricate geometric and topological information inherent in 3D structures. In this study, we utilised a pre-trained BuildingGNN model to segment various components of 3D building models, such as walls, roofs, windows, and doors. The following sections detail the steps taken to apply the BuildingGNN method to our dataset and the subsequent conversion to CityGML.

2.1 Segmentation Process

The segmentation process begins with data preparation and preprocessing. The 3D building models, initially available in various formats, were converted into a uniform format compatible with the input requirements of BuildingGNN. Key steps included: first, the 3D building models were collected from publicly available datasets and proprietary sources. Next, the models underwent preprocessing, which involved normalising the scale, orientation, and positioning of the models to ensure consistency. Following this, the 3D geometry of the building models was converted into graph representations where nodes represented geometric primitives (e.g., vertices, edges) and edges represented spatial relationships. Once the data was preprocessed, the BuildingGNN model was employed to perform semantic segmentation. The model assigned a semantic label to each component of the building model, resulting in a segmented output that identified the various parts of the building accurately.

Figure 3 illustrates the BuildingGNN approach used for the segmentation process. The pipeline begins with the extraction of point descriptors from the 3D building model using MinkNet [6]. These point descriptors capture various features and groupings within the model, such as polygon groups. The descriptors are then used to construct a structure graph, where nodes represent geometric elements like columns, walls, and domes, and edges represent relationships such as support, similarity, and proximity. The graph is processed by a multi-layer perceptron (MLP) to predict the semantic labels for each component of the building model.

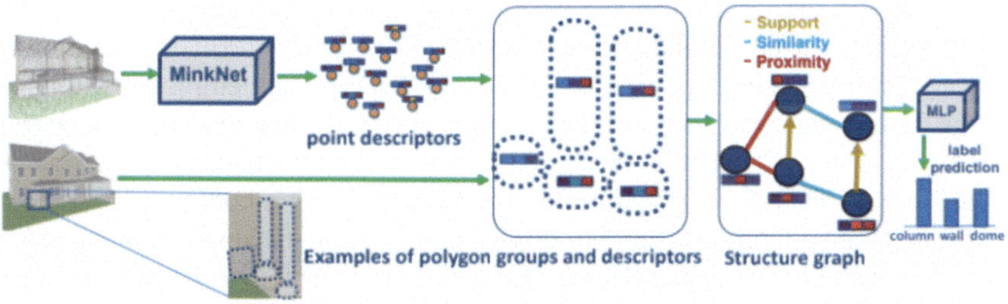

Fig. 3. The BuildingGNN approach [2].

Figure 4 provides a visual representation of the building model at different stages of the segmentation process. The first image shows the building model in a 3D plane, depicting its geometric structure without any additional information. The second image presents the building as a point cloud, highlighting the density and distribution of data points that define the model's shape. The third image displays the final semantic model, where each part of the building is colour-coded based on the semantic labels assigned by the BuildingGNN model, clearly identifying different components such as walls, windows, roofs, doors, and ground surfaces.

Fig. 4. The building model in 3D plane, point cloud, and semantic model.

2.2 Georeferencing Techniques

Accurate georeferencing of the segmented building components was essential to ensuring their correct positioning in real-world coordinates. Affine transformations were employed to achieve this. Initially, coordinate transformation was applied to convert the local coordinate system of the segmented models to a global coordinate system, such as WGS84. This transformation was based on ground control points (GCPs) to enhance accuracy. Additionally, affine transformations were applied to adjust the scale, rotation, and translation of the models based on reference data. Challenges in georeferencing included differences between local and global coordinate systems and alignment errors. These issues were addressed through refinement and validation against reference datasets, ensuring the accuracy of the georeferencing process.

To achieve the georeferencing of the converted 3D model in CityGML, it is important to understand the differences between planar surfaces and geographic coordinate systems. Models created on planar surfaces lack geographic information and are often positioned using the Local Coordinate System (LCS), typically employed in computer 3D design tools, where the origin of the model is at (0,0,0). In contrast, the Geographic Coordinate System (GCS) defines the location of objects on the Earth, considering its 3D spherical surface. Therefore, transforming the LCS building model to a GCS model is crucial for accurate georeferencing. For this transformation, the absolute projected coordinate system EPSG:4326 was selected. The local coordinates of the objects' vertices were computed using the following Eq. (1):

$$\begin{bmatrix} x\prime \\ y\prime \\ z\prime \end{bmatrix} = Ax \begin{bmatrix} Wx \\ Wy \\ Wz \end{bmatrix} + \begin{bmatrix} x \\ y \\ z \end{bmatrix} \tag{1}$$

where:

(Wx, Wy, Wz): direction vector of sweeping.

A: sweeping distance.

(x', y', z'): LCS coordinates.

(x, y, z): GCS coordinates.

Finally, the computation and transformation of the LCS to GCS are performed using a transformation matrix equation [7] in Eq. (2).

$$\begin{bmatrix} x \\ y \\ z \end{bmatrix} = \begin{bmatrix} 1 & 0 & 0 \\ 0 & cos\varphi x & sin\varphi x \\ 0 & -sin\varphi x & cos\varphi x \end{bmatrix} \begin{bmatrix} cos\varphi y & 0 & -sin\varphi y \\ 0 & 1 & 0 \\ sin\varphi y & 0 & cos\varphi y \end{bmatrix} \begin{bmatrix} cos\varphi z & sin\varphi z & 0 \\ -sin\varphi z & cos\varphi z & 0 \\ 0 & 0 & 1 \end{bmatrix} \begin{bmatrix} x\prime \\ y\prime \\ z\prime \end{bmatrix} + \begin{bmatrix} \Delta x \\ \Delta y \\ \Delta z \end{bmatrix} \tag{2}$$

where:

(x,y,z): GCS coordinates.
(x',y',z'): LCS coordinates.
(Δx,Δy,Δz): translation (change) from LCS to GCS from the origin.
(φx,φy,φz): rotation angles with respect to x-axis, y-axis, and z-axis.

The georeferencing techniques ensured the accurate positioning of segmented building models in a global coordinate system, overcoming challenges related to coordinate system differences and alignment errors through iterative refinement and validation. Future work will focus on improving the robustness of these techniques and addressing remaining issues related to large-scale georeferencing.

2.3 Geometry Validation

Ensuring the geometric validity of the segmented models was critical for compliance with CityGML specifications. The validation process was carried out using the geometry validator function in FME, which provided a solution for geometry validation, including the basic integrity check as shown in Fig. 5—the figure is a snapshot of the FME's 'Transformer' tools for geometry validation.

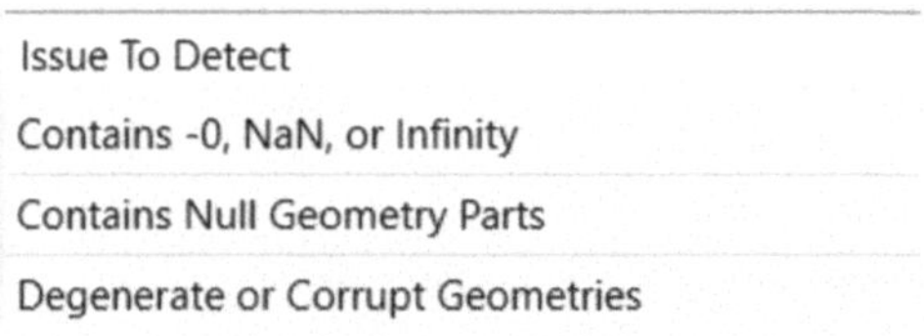

Fig. 5. Basic integrity geometry validation implemented.

Several specific issues were detected and addressed: the presence of undefined or infinite values within the geometry data, such as -0, NaN, or Infinity, was identified and corrected; instances, where parts of the geometry were null or missing, were detected and rectified; and any geometries that were degenerate or corrupt were flagged and corrected. These basic integrity checks were essential to maintain the accuracy and reliability of the 3D building models. By automating the detection and correction of these issues, the Geometry Validator ensured that all segmented models adhered to the required geometric standards, thereby ensuring their compliance with CityGML specifications.

3 Results and Discussion

The results of the conversion process from BuildingGNN-segmented outputs to CityGML demonstrate the effectiveness of our methodology. The segmented 3D building components, including walls, roofs, windows, and doors, were successfully converted into CityGML format, preserving their semantic labels and geometric integrity (Fig. 6).

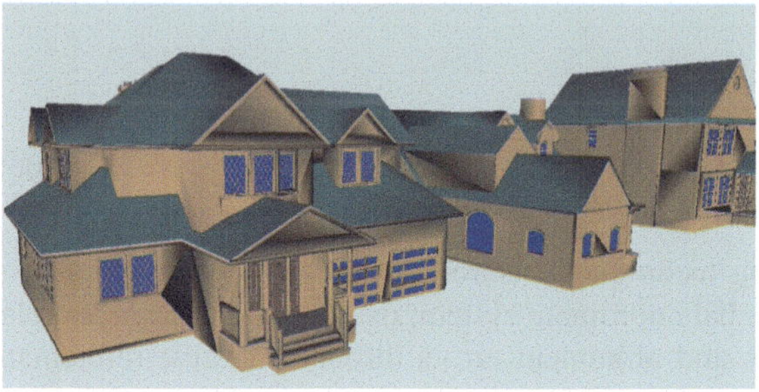

Fig. 6. The overall appearance of the CityGML models with embedded semantic information.

To visualise the 3D model within a geographic context, the CityGML models were converted into 3D tiles. This process allows for the placement and visualisation of 3D building models on a map. Figure 7 illustrates the resulting 3D model positioned within its real-world location and orientation. This visualisation demonstrates the integration of the 3D model into a map-based environment, providing insights into the spatial relationships and positioning of buildings. The conversion to 3D tiles facilitates a practical way to view and analyse the models in their geographic context, which can be useful for various applications, including urban planning and geographic information systems.

Fig. 7. Visualization of the 3D building model placed on a map: (left) the 3D model integrated within its geographical context, and (right) a zoomed-in view of the building.

Geometry validation played a crucial role in ensuring the compliance of the converted models with CityGML specifications. The geometry validator ensured that the segmented models adhered to the required geometric standards, thereby ensuring their compliance with CityGML specifications. Several specific issues were detected and addressed: the presence of undefined or infinite values within the geometry data, such as -0, NaN, or Infinity, was identified and corrected; instances where parts of the geometry were null or missing were detected and rectified; and any geometries that were degenerate or corrupt were flagged and corrected. The automated tools and scripts used in this process led to CityGML models that were both geometrically and semantically rich (Fig. 8).

Figure 8 provides an example of the attribute table showing the metadata and properties of CityGML objects. This table includes several key fields:

- **gml_id**: This is a unique identifier for each CityGML object. It ensures that each element within the CityGML model can be distinctly referenced.
- **gml_parent_id**: This indicates the parent object, showing hierarchical relationships between objects. It helps in understanding the structure and organisation of the 3D building model.
- **citygml_target_uri**: This field links to the URI (Uniform Resource Identifier) that identifies the object in the CityGML schema. It provides a reference to the standard definitions and properties of the object.
- **citygml_feature_role**: This indicates the role of the object within the CityGML model, such as "*cityObjectMember*," which signifies that the object is a member of the city model.

Building ✕ Window ✕ Door ✕ RoofSurface ✕ WallSurface ✕

	gml_id	gml_parent_id	citygml_target_uri	citygml_feature_role
1	fme-gen-fefe2e...	fme-gen-5f55550a...	http://www.opengis.ne...	cityObjectMember
2	fme-gen-ed52b...	fme-gen-5f55550a...	http://www.opengis.ne...	cityObjectMember
3	fme-gen-22350...	fme-gen-5f55550a...	http://www.opengis.ne...	cityObjectMember
4	fme-gen-ab066...	fme-gen-5f55550a...	http://www.opengis.ne...	cityObjectMember
5	fme-gen-38af3...	fme-gen-5f55550a...	http://www.opengis.ne...	cityObjectMember
6	fme-gen-df0f7e...	fme-gen-5f55550a...	http://www.opengis.ne...	cityObjectMember
7	fme-gen-fe885...	fme-gen-5f55550a...	http://www.opengis.ne...	cityObjectMember

Fig. 8. An example of the attribute table that accompanies these models, highlighting the metadata and properties assigned to each CityGML object.

While Fig. 6 shows the overall appearance of the CityGML models with embedded semantic information, Fig. 8 provides an example of the attribute table that accompanies these models, highlighting the metadata and properties assigned to each CityGML object. Figure 9 offers a close-up view of the building models, showcasing the detailed and accurate representation of building components.

Figure 10 presents a detailed view of the properties and attributes assigned to the 3D building models in CityGML. This figure showcases the various exposed attributes, such as GML IDs, parent IDs, target URIs, and feature roles, along with other relevant metadata. The figure also highlights unexposed attributes and specific details such as the number of vertices, the coordinate system used, and the dimensions of the models.

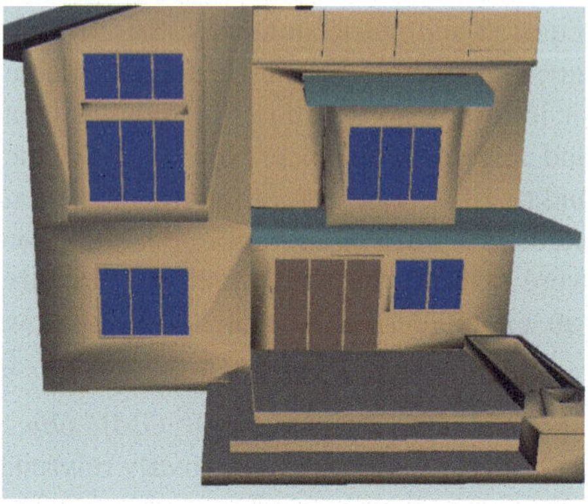

Fig. 9. A close-up view of the building models, showcasing the detailed representation of building components.

Property	Value
Exposed Attributes (6)	
gml_id	fme-gen-9bcf8db2-0a43-4363-bb69-215c97fb4fcb
gml_parent_id	fme-gen-5f55550a-6469-44b9-9607-90f732f1bcce
citygml_target_uri	http://www.opengis.net/citygml/bridge/2.0
citygml_feature_role	cityObjectMember
label	2
citygml_level_of_detail{} (1)	
citygml_level_of_detail{0}	3
Unexposed Attributes (4)	
xml_type	xml_surface
FME Attributes (3)	
fme_feature_type	Window
fme_geometry	fme_aggregate
fme_type	fme_surface
Geometry	
Coordinate System	Unknown
Dimension	3D
Number of Vertices	143868

Fig. 10. Detailed properties and attributes of the 3D building models in CityGML.

Potential work may focus on improving the robustness of these techniques and addressing remaining issues related to large-scale georeferencing. Enhancing the automation of the conversion and validation processes could further streamline the workflow, making it more efficient and scalable. Additionally, extending the methodology to incorporate other types of 3D models and integrating additional semantic information could provide more comprehensive solutions for urban planning and management.

4 Conclusion

This study explored the conversion of BuildingGNN-segmented 3D building models into CityGML format, emphasising the challenges and solutions related to georeferencing and geometry validation. The methodology involved using a pre-trained BuildingGNN model for semantic segmentation, followed by the application of affine transformations and ground control points to achieve correct georeferencing. Geometry validation ensured the compliance of the models with CityGML specifications by addressing issues such as undefined values and degenerate geometries.

The conversion process was successfully demonstrated, and the resulting CityGML models were visualised within a geographic context by converting them into 3D tiles. This provided a practical means to view and analyse the spatial relationships and positioning of the buildings. The approach detailed in this study has potential applications in urban planning, geographic information systems, and other domains that require precise and semantically rich 3D building models.

Future work will focus on enhancing the robustness of the georeferencing techniques and automating the conversion process further. Additionally, extending the methodology to include other types of 3D models and integrating additional semantic information will be investigated to provide more comprehensive solutions for various applications.

References

1. Häfele, K.H., Benner, J.: STREAMER deliverable 6.5: advance mapping structures and standards. Streamer Consort. **2015**, 2–80 (2015)
2. Selvaraju, P., Nabail, M., Loizou, M., Maslioukova, M., Averkiou, M., Andreou, A., Chaudhuri, S., Kalogerakis, E.: BuildingNet: learning to label 3D buildings. In: Proceedings of the IEEE/CVF International Conference on Computer Vision, pp. 10397–10407 (2021)
3. Jaud, Š, Donaubauer, A., Heunecke, O., Borrmann, A.: Georeferencing in the context of building information modelling. Automat. Construct. **118**, 103211 (2020)
4. Zhao, J., Stoter, J., Ledoux, H.: A framework for the automatic geometric repair of CityGML models. In: Cartography from Pole to Pole: Selected Contributions to the XXVIth International Conference of the ICA, Dresden 2013, pp. 187–202. Springer Berlin Heidelberg (2014)
5. Rashidan, H., Musliman, I.A., Abdul Rahman, A., Coors, V., Buyuksalih, G.: Semantic segmentation of building models with deep learning in CityGML. Int. Arch. Photogramm. Remote. Sens. Spat. Inf. Sci. **48**, 97–102 (2024)
6. Choy, C., Gwak, J., Savarese, S.: 4D spatio-temporal convnets: Minkowski convolutional neural networks. In: Proceedings of the IEEE/CVF conference on computer vision and pattern recognition, pp. 3075–3084 (2019)
7. Wu, I.C., Hsieh, S.H.: Transformation from IFC data model to GML data model: methodology and tool development. J. Chinese Institute Eng. **30**(6), 1085–1090. https://doi.org/10.1080/02533839.2007.9671335

Geo-based Artificial Intelligence / Deep Learning

Geospatial Analysis of Electric Vehicle Charging Infrastructure Deployment: Evaluation of Technical Factors via Artificial Intelligence

Mohamed Benayad[1,2](✉), Abdelilah Rochd[2], Nouriddine Houran[2], Mohamed Rabii Simou[1], Mehdi Maanan[1], and Hassan Rhinane[1]

[1] Geosciences Laboratory, Faculty of Sciences-Ain Chock, Hassan II University, Casablanca, Morocco
med.benayad98@gmail.com
[2] Green Energy Park (GEP), Benguerir, Morocco

Abstract. This study addresses the challenge of optimizing charging infrastructure for electric vehicles (EVs) in urban areas of Morocco, focusing on the cities of Casablanca and Marrakech. The aim is to develop an effective planning model that integrates various technical factors, such as bus stations, gas stations, hotels, as well as main and secondary roads. Using ArcGIS software, we first calculated the Euclidean distance of each factor across the city of Casablanca. Then, by applying the Analytic Hierarchy Process (AHP) method, we assigned weights to each factor. These weights were used to evaluate the potential of each pixel in our study area, by multiplying the Euclidean distance by the corresponding weight. For predictive modeling, we compared two popular machine learning models: Random Forest (RF) and XGBoost. Our results showed that the RF model outperforms XGBoost in terms of accuracy and reliability for this specific application case. Finally, we applied our optimized model to a new city, Marrakech, thus demonstrating its ability to be adapted and implemented in other urban contexts in Morocco. This research offers an innovative and practical approach to guide the strategic deployment of EV charging infrastructure, thereby supporting the growing adoption of electric vehicles in Moroccan urban areas.

Keywords: EV · IA · GIS · AHP · RF · XGBoost

1 Introduction

In the current context of global energy transition, the adoption of electric vehicles (EVs) is emerging as a central pillar of efforts to reduce greenhouse gas emissions and combat climate change [1]. At the heart of this transition, the adoption of electric vehicles (EVs) is emerging as a key strategy, promising a significant reduction in air pollution and less dependence on fossil fuels [2]. However, the success of this transformation largely depends on the provision of an accessible and efficient recharging infrastructure, capable of meeting the needs of EV users [3]. This requirement is particularly pressing in urban areas, where high population density and complex transport systems exacerbate the challenges of integrating EVs into the daily fabric of city life [4].

A. Abdul-Rahman et al. (Eds.): GeoWeek 2024, LNGC, pp. 53–68, 2025.
https://doi.org/10.1007/978-3-031-86654-8_5

Morocco, with its growing commitment to renewable energies and sustainability, is at the forefront of this transition in Africa [5]. Moroccan cities, and Casablanca and Marrakech in particular, offer fertile ground for exploring the dynamics of EV adoption [2]. Casablanca, as the country's economic centre and most populous metropolis [6], presents a complex urban setting characterized by a growing demand for sustainable mobility solutions. Marrakech, with its status as a leading tourist destination and its commitment to environmental conservation [7], underlines the importance of adopting environmentally-friendly transportation practices tailored to both residents and visitors. In these contexts, optimizing the location and distribution of EV charging infrastructures becomes a crucial issue, requiring a methodical and innovative approach to mapping and evaluating the most strategic locations for the deployment of these essential infrastructures.

This paper proposes an advanced planning model that incorporates advanced geospatial analysis to optimally assess the locations of EV charging infrastructure. Leveraging the capabilities of ArcGIS software and applying sophisticated analytical methods such as the Hierarchical Analysis Process (AHP) [2], this study aims to map deployment potential across the cities of Casablanca and Marrakech. We focus on integrating various technical factors, such as proximity to bus stations, service stations, hotels, as well as the presence of main and secondary roads, to establish a hierarchy in the criteria for selecting optimal sites for charging infrastructures.

In addition, the predictive approach adopted via the comparison of machine learning models, notably Random Forest and XGBoost, illustrates our quest for a robust and reliable methodology capable of guiding the strategic deployment of charging infrastructures. The results obtained not only validate the effectiveness of our model in the Casablanca context, but also pave the way for its application and adaptation to other urban contexts in Morocco, notably Marrakech. Through this research, we aspire to make a significant contribution to the body of knowledge on the deployment of EV charging infrastructure, offering an innovative and practical approach that supports the growing adoption of electric vehicles in Moroccan urban areas [8]. This initiative is part of a broader vision of sustainable development and ecological urban mobility, responding to the challenges of urbanization and energy transition in developing countries.

2 Materials and Methods

2.1 Study Area

Casablanca, Morocco's bustling metropolis, is an ideal study area for the deployment of electric vehicle charging infrastructure, given its status as the country's largest city and economic hub (Fig. 1). Home to over 3.5 million people [9], it embodies rapid urbanization and socio-economic diversity, hosting Morocco's largest commercial port and serving as a critical junction for national and regional economies. The city's urban fabric, blending dense residential zones, modern business centers, and industrial spaces, mirrors the complexities of challenges and opportunities in sustainable mobility. Major roadways, diverse activity areas ranging from historic spaces to modern sectors like the Morocco Mall and Casablanca Finance City, along with a concentration of hotels and tourist centers, underscore the need for well-integrated EV charging infrastructure. This urban setting, combined with an expanding middle class and increased environmental awareness, positions Casablanca as a living laboratory to explore mechanisms supporting EV adoption, marking the city as a potential model for urban energy transition in Morocco.

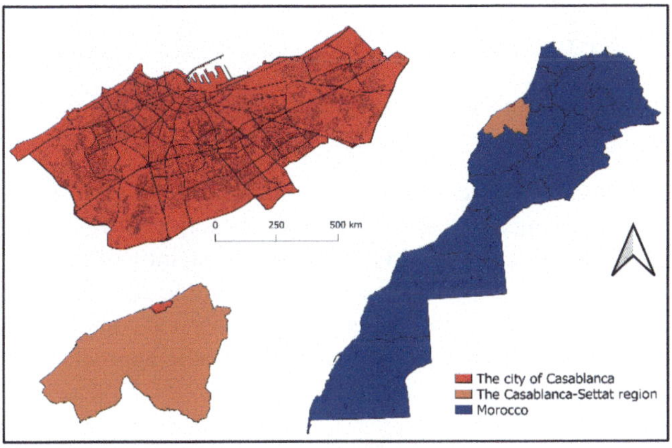

Fig. 1. Location of the study area

2.2 Explicative Factors

Deploying electric vehicle (EV) charging infrastructure in an urban environment requires a comprehensive understanding of the various urban and technical factors that influence both the demand for and accessibility of charging stations. In Casablanca, a series of factors were identified as essential elements of the planning model, each selected for its direct or indirect impact on the electric vehicle charging ecosystem. Cafes and restaurants are essential for deployment, as they are places where people spend time every day. Their integration as charging points offers unrivalled convenience. These locations, often situated in high-traffic areas, guarantee high visibility and good accessibility to charging infrastructures, encouraging EV use by becoming a natural part of the routine [10]. The inclusion of service stations as potential sites for the installation of EV charging stations plays on familiarity and availability. Installing charging stations in these locations helps to overcome EV range anxiety by offering a visible and accessible charging network in places already recognized for energy refueling [11]. This facilitates drivers' transition to EVs, by integrating charging infrastructures into the existing urban landscape. These strategic choices encourage the adoption of EVs by combining utility and pleasure, thus contributing to a more sustainable urban environment (Fig. 2).

The installation of EV charging stations in hotels meets a dual objective: to ensure the comfort of travelers with EVs, and to encourage the adoption of EVs among visitors. Hotels, as places to stay par excellence, offer a unique opportunity to recharge vehicles during rest periods or overnight stays, guaranteeing users a hassle-free experience with their EV away from home [12]. This strategic choice not only enhances the hotel offer in

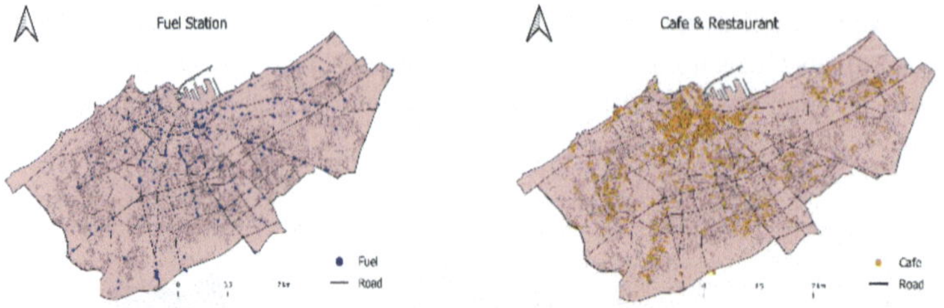

Fig. 2. Factors determining technical Potential (Fuel Station Café & Restaurant)

terms of sustainability, but also encourages eco-tourism in the city. As for bus stations, their selection as recharging sites capitalizes on the synergy between public and private transport. Installing charging stations in the immediate vicinity of these public transport hubs facilitates access to charging for EV users who combine different modes of transport in their urban journeys [13]. This strengthens connectivity and complementarity between mobility services, supporting an integrated and sustainable approach to urban mobility in Casablanca (Fig. 3).

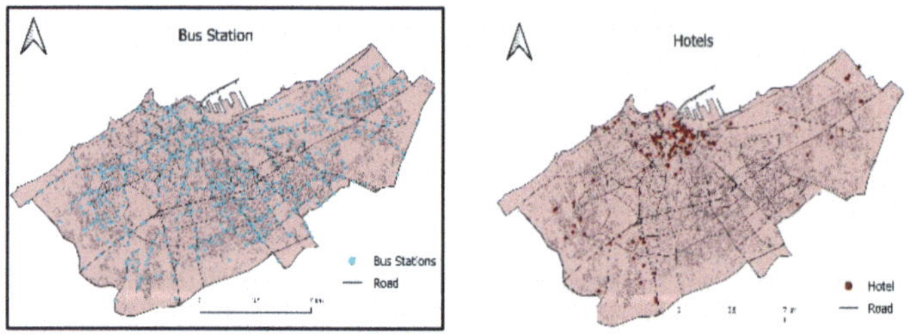

Fig. 3. Factors determining technical Potential (Bus Station/Hotels)

The presence of EV charging stations around hospitals is essential to support the constant flow of patients, visitors and staff. Offering charging solutions in these critical locations ensures greater accessibility and convenience for those who rely on electric mobility for their essential journeys. This not only enhances the experience of EV users, but also supports the efforts of healthcare facilities in their transition to a greener, more sustainable infrastructure [10]. As far as sports facilities and leisure centers are concerned, integrating EV charging infrastructure meets the growing demand for green services among citizens engaged in sports and recreational activities. These venues, often visited for extended periods, offer the necessary recharging time while promoting a healthy, environmentally-friendly lifestyle [14]. The installation of charging stations in these areas underlines the city's commitment to supporting a sustainable mobility ecosystem, encouraging the use of EVs among an active, environmentally conscious population (Fig. 4).

The installation of EV charging stations along Casablanca's main and secondary roads ensures widespread and equitable accessibility, making it easier for EV drivers to get around the city and its residential areas, while aiming to eliminate areas without access

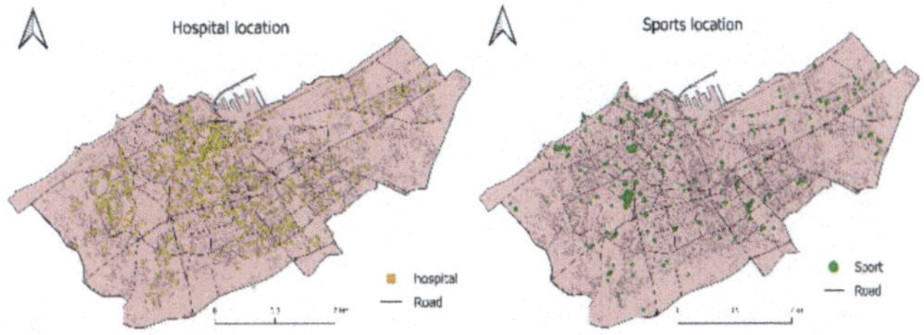

Fig. 4. Factors determining technical Potential (Hospital/Sport)

to charging [15]. At the same time, shopping malls and shopping areas, as high-traffic areas, are identified as strategic locations for the deployment of charging infrastructures, enabling EV users to conveniently combine charging their vehicles with their commercial and leisure activities [2]. This comprehensive approach encourages the adoption of electric vehicles by integrating charging into daily commuting and consumption routines (Fig. 5).

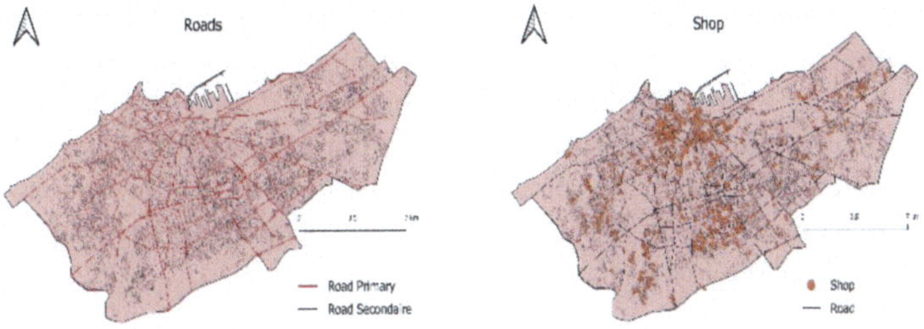

Fig. 5. Factors determining technical potential (Roads/Shop)

Including EV charging stations in parking lots meets a fundamental need for convenience and safety for EV drivers, by enabling them to charge their vehicles in places already designated for parking [10]. This strategy maximizes the use of existing spaces while encouraging an easier transition to electric mobility. At the same time, the integration of charging infrastructure into buildings, whether residential or commercial, is essential to support the adoption of EVs among the urban population. This ensures that residents and professionals have access to charging points on a daily basis, directly at their home or workplace, reinforcing the appeal and practicality of electric vehicles [16]. Together, these initiatives create a coherent and accessible recharging network, essential to support Casablanca in its transition to more sustainable mobility (Fig. 6).

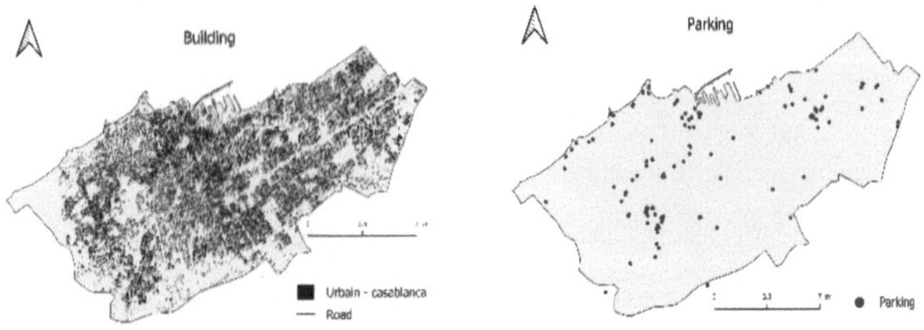

Fig. 6. Factors determining technical potential (Building/Parking)

3 Models Theoretical Background

3.1 Fuzzy Analytic Hierarchy Process (Fuzzy AHP)

The Fuzzy AHP method combines the principles of Analytic Hierarchy Process (AHP) with Fuzzy Logic to deal with the uncertainties and imprecision inherent in human evaluation when making multi-criteria decisions. Traditionally [17], AHP allows a complex problem to be broken down into a hierarchy of simpler sub-problems, facilitating the analysis and comparison of different options by assigning weights to each criterion according to their relative importance. By integrating fuzzy logic, Fuzzy AHP goes a step further, enabling the manipulation of data that are not precise, but rather expressed as degrees of membership of fuzzy sets [18]. This approach is particularly relevant to the evaluation of potential sites for EV charging infrastructures, where evaluation criteria, such as a site's accessibility or attractiveness, can be subjective and vary according to the perceptions of stakeholders.

3.2 Machine Learning (ML)

Artificial intelligence (AI) in the context of EV charging infrastructure deployment refers to the use of machines to simulate human intelligence, particularly for problem-solving and decision-making [19]. In this study, two predominant AI models are highlighted: Random Forest (RF) and XGBoost. Random Forest is a supervised learning algorithm that works by constructing a multitude of decision [20] trees during training and producing the class mode (classification) or average of predictions (regression) of the individual trees. XGBoost, or Extreme Gradient Boosting, is an optimized implementation of the gradient boosting algorithm that aims to be efficient, flexible and portable. It is particularly well known for its speed and performance [21]. In this study, these AI models are used to predict the effectiveness of charging station locations based on criteria evaluated by Fuzzy AHP, enabling accurate and reliable predictive analysis of optimal sites for charging infrastructure deployment.

4 Methodology for Mapping the Technical Potential

The systematic approach to mapping the electric vehicle (EV) charging infrastructure in Casablanca is outlined (Fig. 2), including the collection and pre-processing of geographical, technical and infrastructural data. The Fuzzy AHP method is applied to evaluate and weight the factors influencing the location of charging stations, with the use of fuzzy logic to deal with uncertainties. AI techniques such as Random Forest and XGBoost are then used for predictive modeling, determining potential sites for the installation of charging stations. After validating this model in Casablanca, we projected our model onto the city of Marrakech, using GIS integration and spatial analysis to visualize sites with high technical potential for the implementation of EV charging infrastructure. This extension to Marrakech illustrates the model's ability to adapt and be relevant in different Moroccan urban contexts, demonstrating a robust methodological approach for the strategic deployment of EV charging infrastructure in Morocco (Fig. 7).

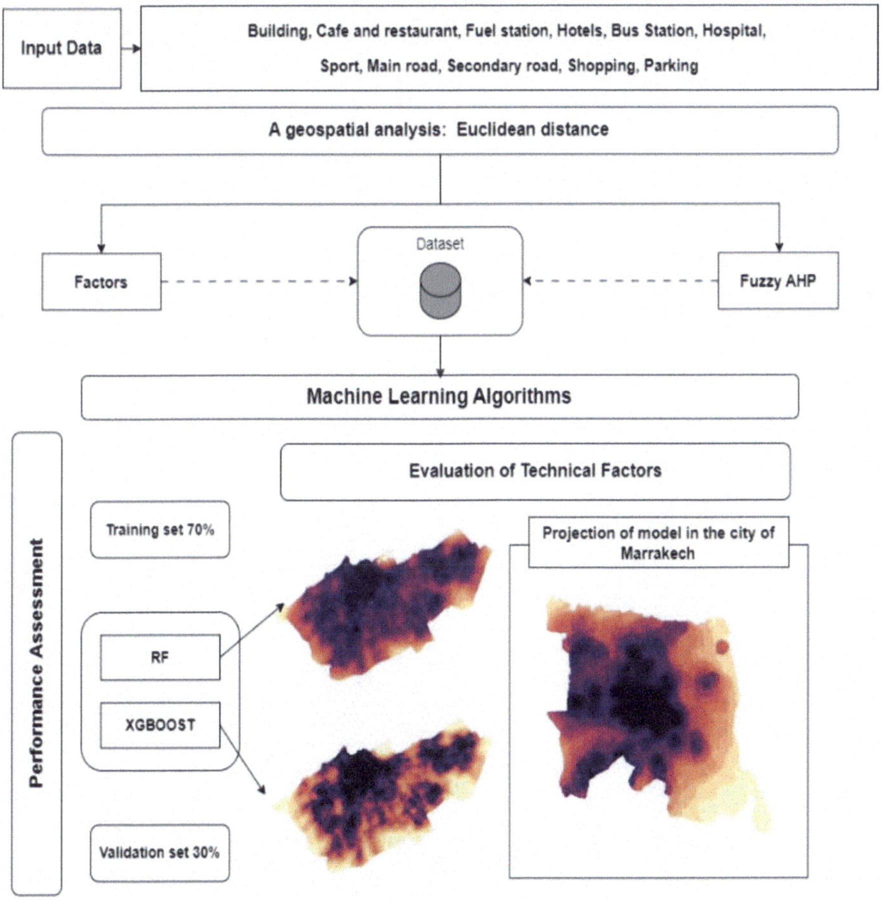

Fig. 7. Methodology for mapping technical potential for EV charging

Table 1. The factors and according to weights using fuzzy AHP

Factors													Weighted sum value	Criteria weight	Factor weight, %
Factors	Class	*Fuel Station*	*Café & restaurant*	*Bus Station*	*Hotels*	*Hospital*	*Sport*	*Main Roads*	*Road Secondary*	*Shop*	*Building*	*Parking*			
Fuel Station	[1–9]	1	2	4	2	4	7	1	7	2	4	1	2,0208	0,179	18
Café & restaurant	[1–7]	1/2	1	2	1	2	5	1/2	5	1	2	1/2	1,0853	0,0954	10
Bus Station	[1–5]	1/4	1/2	1	1/2	1	3	1/4	3	1/2	1	1/4	0,5676	0,0496	5
Hotels	[1–7]	1/2	1	2	1	2	5	1/2	5	1	2	1/2	1,0853	0,0954	10
Hospital	[1–5]	1/4	1/2	1	1/2	1	3	1/4	3	1/2	1	1/4	0,5385	0,0471	5
Sport	[1, 2]	1/7	1/5	1/3	1/5	1/3	1	1/7	1	1/5	1/3	1/7	0,2376	0,0212	2
Main Road	[1–9]	1	2	4	2	4	7	1	7	2	4	1	1,8397	0,1597	16
Road Secondary	[1, 2]	1/7	1/5	1/3	1/5	1/3	1	1/7	1	1/5	1/3	1/7	0,3091	0,0288	3
Shop	[1–7]	1/2	1	2	1	2	5	1/2	5	1	2	1/2	1,1407	0,0973	10
Building	[1–5]	1/4	1/2	1	1/2	1	3	1/4	3	½	1	1/4	0,5465	0,0477	5
Parking	[1–9]	1	2	4	2	4	7	1	7	2	4	1	2,0208	0,1790	18

4.1 Creation of a Technical Database Using Fuzzy AHP

The first crucial step towards the efficient deployment of EV charging infrastructure involves the creation of a robust technical database. This database incorporates a multi-criteria evaluation of potential sites for the positioning of charging stations, using the Fuzzy Analytic Hierarchy Process (Fuzzy AHP) method. The choice of Fuzzy AHP addresses the often subjective and uncertain nature of expert judgements when weighting different criteria, such as accessibility, visibility, expected demand, and proximity to existing infrastructure. The process of applying Fuzzy AHP to create a technical database for the deployment of EV charging infrastructures begins with the definition of criteria and sub-criteria influencing the choice of locations. Experts from various fields then assess the relative importance of these criteria, using fuzzy logic to translate their judgments into fuzzy values, thus managing the uncertainty of human assessments. The fuzzy weights calculated for each criterion are synthesized to establish a hierarchy, identifying the key factors for site selection. This procedure results in the creation of an accurate technical database, based on a multi-criteria analysis and reflecting the nuances specific to Casablanca and Marrakech, effectively guiding the selection of high-potential sites for the installation of charging stations.

Complementing the Fuzzy AHP application process, the results obtained (Table 1), in the form of weights assigned to each factor, are used for further spatial analysis. Each fuzzy weight corresponding to a specific factor is multiplied by the Euclidean distance of that factor in each pixel on the map of the city of Casablanca. This mathematical operation makes it possible to calculate the technical potential of each location, taking into account both the relative importance of the criteria (as assessed by the experts via Fuzzy AHP) and their concrete spatial distribution in the urban areas studied. In other words, this method refines the selection of potential sites for EV charging stations by incorporating a spatial dimension into the multi-criteria evaluations, enabling the precise identification of locations that maximize coverage and accessibility for EV users. This approach ensures that the technical database reflects not only the preferences and judgments of experts, but also the specific geographical realities of the city, leading to accurate and optimized mapping of potential charging sites.

5 Machine Learning Algorithm

5.1 Creation of Dataset on the Technical Potential

After establishing a rich and detailed technical database of ten critical factors identified for the deployment of EV charging infrastructure in the city of Casablanca and calculating a "Target" for each pixel via a mathematical calculation method based on Fuzzy AHP, our study progressed to the application of machine learning algorithms. Our database, integrated into ArcGIS 10.8, comprises over 20,000 pixels (Fig. 8), each characterized by factor weights and a target value representing the calculated VE recharge technical potential. The "target" of our model is a multiclass variable, representing different categories of technical potential for the installation of EV charging stations. Given the multiclass nature of our target variable and the continuous values derived from the mathematical calculations performed with Fuzzy AHP, we opted for a regression approach in our ML model to predict the recharging potential of each pixel.

To train our model, the database was divided into two sets: 70% of the data was used for training and 30% for testing. This division ensures both complete training of the model on a wide range of data and reliable evaluation of its performance on unseen data. The aim is to develop a model capable of accurately predicting the potential of each location to host an EV charging infrastructure, based on the multi-criteria analysis previously carried out.

This methodological approach combines spatial expertise with advanced machine learning techniques, illustrating the effectiveness of integrating geospatial analysis and AI to address the complex challenges of deploying EV charging infrastructure in urban environments.

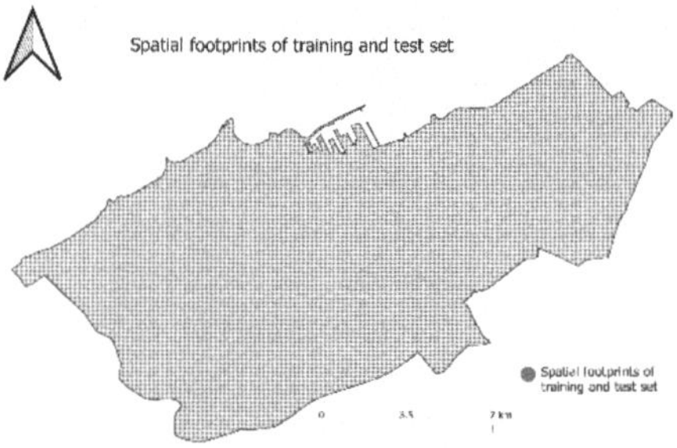

Fig. 8. Spatial footprints of training and test set

6 Configuration of the Potential Models

To analyze the potential of VE recharge sites in Casablanca, the Random Forest (RF) and XGBoost regression models were carefully configured. RF was adjusted with parameters such as the number of trees (n_estimators) around 200, and a controlled maximum depth (max_depth) to avoid overlearning. XGBoost, renowned for its speed, was set with a similar number of trees, a learning rate (learning_rate) between 0.01 and 0.3 for a balanced contribution from each tree, and a maximum depth (max_depth) of 10 for tree complexity. Both models also explored strategies to manage overlearning and improve robustness, such as downsampling samples and features. The dataset was divided into 70% for training and 30% for testing, enabling precise evaluation using metrics such as R^2 and RMSE, to optimize the accuracy of recharge site potential predictions.

7 Metrics Validation

To evaluate and validate the performance of the Random Forest and XGBoost regression models configured in our study, several validation metrics were used. These metrics quantify the accuracy of the models' predictions in comparison with actual test data [22]. The main metrics include the Coefficient of Determination (R^2), assessing the proportion of variance in the target variable explained by the model; Mean Squared Error (MSE) and its root (RMSE), measuring the average difference between predicted and actual values; and Mean Absolute Error (MAE), calculating the average of the absolute differences between predictions and observations, providing a direct perspective on model accuracy without overemphasizing large errors. These indicators allow for a comprehensive and reliable evaluation of the models' performance.

8 Results and Discussion

8.1 Mapping of Technical Potential and Performance of Models

Random Forest (RF)

The performance of the Random Forest (RF) model demonstrates outstanding predictive accuracy and reliability (Fig. 9). The RF model achieved a Coefficient of Determination (R^2) of 0.971, indicating that approximately 97.1% of the variance in the potential for EV charging sites is explained by the model, showcasing its exceptional fit to the data. The Mean Squared Error (MSE) and the Root Mean Squared Error (RMSE) values stand at 0.0383 and 0.1959, respectively, suggesting that the model's predictions are very close to the actual data, with minimal error. The Mean Absolute Error (MAE) of 0.0954 (Fig. 11) further confirms the model's precision in predicting the potential of charging sites (Table 2). Moreover, the Pearson correlation coefficient of 0.9985 nearly approaches perfection, illustrating a near-perfect linear relationship between the predicted and actual values. The Adjusted Coefficient of Determination at 0.9881, adjusted for the number of predictors in the model, reaffirms the model's robustness and its capability to generalize well to new data. These results highlight the Random Forest model's efficacy in accurately mapping out potential sites for EV charging infrastructure deployment, providing a reliable tool for urban planning and development strategies in Casablanca.

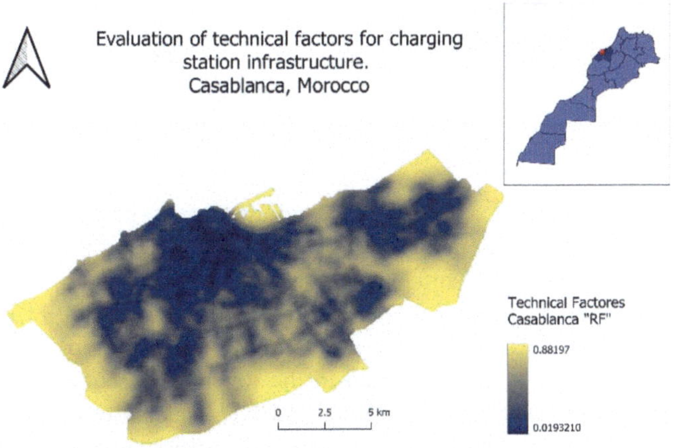

Fig. 9. Prediction of Technical potential using the RF models

eXtreme Gradient Boosting (XGBOOST)

The results of the XGBOOST model reveal significant performance but with room for optimization compared to the Random Forest model (Fig. 10). The Coefficient of Determination (R^2) obtained by XGBoost is 0.827, indicating that the model explains approximately 82.7% of the variance in the potential of EV charging sites, which demonstrates a good fit of the model to the data although lower than that of RF. The Mean Square Error (MSE) and Root Mean Square Error (RMSE) are 0.3010 and 0.5486, respectively, indicating that the model predictions have a larger error dispersion compared to those of RF. The Mean Absolute Error (MAE) of 0.4032 (Fig. 11) shows that the model predictions can deviate from the actual values with some regularity (Table 2). The Pearson correlation coefficient of 0.8588, although strong, suggests a less direct linear correlation between predicted and actual values compared to RF. The Adjusted Determination Coefficient of 0.8177 confirms the effectiveness of the model but also highlights the space for potential improvements. These results demonstrate that, although the XGBoost model performs well in predicting the potential of EV charging sites, it presents optimization opportunities to improve its accuracy and reliability in the specific context of Casablanca.

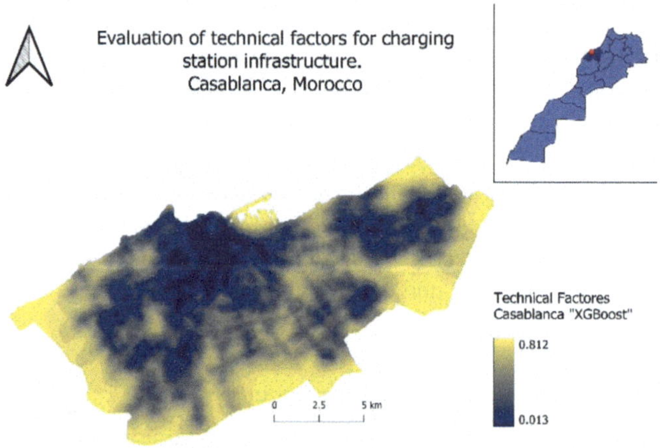

Fig. 10. Prediction of Technical potential using the XGBoost mode

In comparing the performance of the Random Forest (RF) and XGBoost models (Table 2) for the deployment of electric vehicle (EV) charging infrastructure in Casablanca, it is clear that the RF model outperforms XGBoost in terms of accuracy and reliability. The RF model displays an exceptionally high Coefficient of Determination (R^2) of 0.971 versus 0.827 for XGBoost, indicating RF's superior ability to explain data variance. This superiority is also reflected in the lower values of Mean Square Error (MSE) and Root Mean Square Error (RMSE) for RF, suggesting predictions closer to actual values. Furthermore, the Pearson correlation coefficient near perfection for RF (0.9985) versus that for XGBoost (0.8588) highlights a closer match between predictions and actual observations with RF. These differences highlight the superior robustness and

Table 2. Model Performances with Validating Samples

Evaluation Metrics	RF	XGBoost
R^2	0.971	0.827
MSE	0.0383	0.3010
RMSE	0.1959	0.5486
MAE	0.0954	0.4032
Pearson correlation coefficient	0.9985	0.8588
Adjusted coefficient of determination	0.9881	0.8177

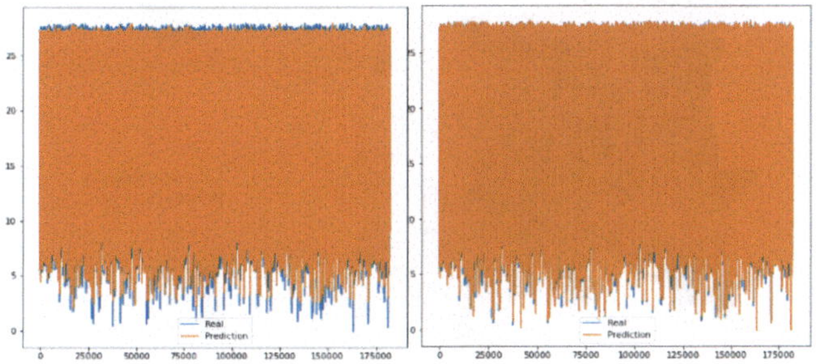

Fig. 11. MAE visualization (RF and XGBoost)

efficiency of the RF model in the context of the study, making it a more reliable tool for guiding the strategic deployment of EV charging infrastructure in Casablanca.

8.2 Projection of the RF Model Optimized for the Prediction of a New City "Marrakech"

After demonstrating exceptional performance in predicting potential sites for EV charging infrastructure in Casablanca, the Random Forest (RF) model has been projected onto Marrakech, a city with a distinct profile and urban challenges. This extension is designed to test the flexibility and robustness of the RF model in the face of new geographical and technical data, illustrating its ability to adapt to different urban environments.

To apply the RF model to Marrakech (Fig. 12), a series of preparatory steps were necessary, including the collection and processing of Marrakech-specific data, similar to the variables used for Casablanca. This involved re-evaluating the weights of the factors according to Marrakech's local peculiarities, while retaining the structure and parameters of the RF model optimized for Casablanca.

Preliminary results from the projection of the Random Forest (RF) model on Marrakech have revealed an encouraging ability to map areas of high potential for the deployment of electric vehicle (EV) charging stations, despite the clear urban and infrastructural differences between Casablanca and Marrakech. To visualize these potentials, high-potential areas are shown in black, gradually fading to yellow to indicate areas of lesser potential. This gradation of colors provides an intuitive reading of the potential of each area, making it easier to identify the most promising locations for the installation of new charging infrastructures. These results underline not only the versatility of the RF model, but also its potential as an essential urban planning tool to support the development of electric mobility in Morocco, illustrating its ability to adapt and provide valuable guidance for the strategic deployment of EV charging infrastructures across different urban configurations.

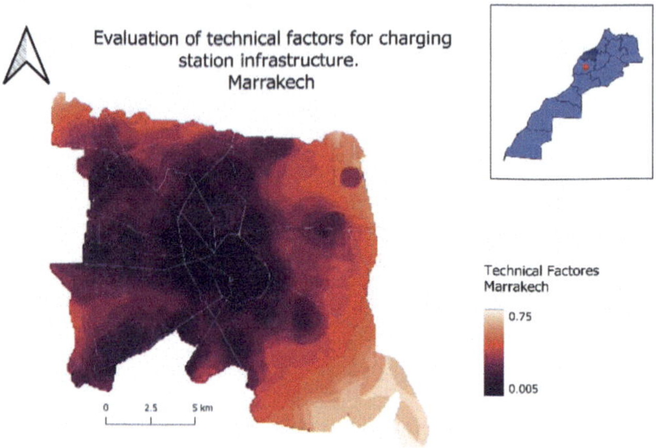

Fig. 12. Prediction of technical potential "Marrakech"

The successful projection of the RF model onto Marrakech paves the way for wider applications of the model for EV charging infrastructure planning in other M11oroccan cities and, potentially, in similar urban contexts. This step underlines the importance of a modular and adaptable approach in the development of predictive models for the deployment of critical infrastructure, enabling informed and strategic urban planning that can easily be adjusted to the unique characteristics of different cities.

9 Conclusion

This study has highlighted the significant potential of applying machine learning models, particularly Random Forest, to optimize the deployment of electric vehicle (EV) charging infrastructure within Moroccan urban contexts, exemplified by Casablanca and extended to Marrakech. The results not only demonstrate the accuracy and reliability of these models in identifying the most suitable sites for charging stations but also their adaptability to the unique characteristics of different cities. This research significantly contributes to the literature on urban electric mobility and offers a replicable methodological framework for other regions aiming to support EV adoption. Ultimately, it underscores the importance of integrating advanced technologies into urban planning for a successful energy transition, marking a step towards achieving a coherent and efficient EV charging infrastructure, essential for the sustainable future of Moroccan cities.

References

1. Rochd, A., Laamim, M., Benazzouz, A., Kissaoui, M., Raihani, A., Sun, H.: Public charging infrastructure for EVs: a comprehensive analysis of charging patterns & real-world insights—case study of Rabat City, Morocco. Energy Rep. **9**, 216–234 (2023). https://doi.org/10.1016/j.egyr.2023.05.259
2. Benayad, M., et al.: Assessing the socio-economic potential of electric vehicle charging infrastructure: a machine learning based approach for Marrakech-Safi Region, Morocco. In: 2023 12th International Conference on Renewable Energy Research and Applications (ICRERA), pp. 166–174. https://doi.org/10.1109/ICRERA59003.2023.10269379
3. Al-Hanahi, B., Ahmad, I., Habibi, D., Masoum, M.A.S.: Charging infrastructure for commercial electric vehicles: challenges and future works. IEEE Access **9**, 121476–121492 (2021). https://doi.org/10.1109/ACCESS.2021.3108817
4. Carrese, F., Colombaroni, C., Fusco, G.: Accessibility analysis for urban freight transport with electric vehicles. Transport. Res. Proced. **52**, 3 (2021). https://doi.org/10.1016/j.trpro.2021.01.002
5. Okpanachi, E., Ambe-Uva, T., Fassih, A.: Energy regime reconfiguration and just transitions in the Global South: lessons for West Africa from Morocco's comparative experience. Futures **139**, 102934 (2022). https://doi.org/10.1016/j.futures.2022.102934
6. Alaya, M.B., Sadqi, A.: Le développement des places financières en Afrique : Cas de Casablanca finance city. Int. J. Account. Fin. Audit. Manag. Econ. **3**(2–1), Art. no. 2–1 (2022). https://doi.org/10.5281/zenodo.6388094
7. Gray, E.: Morocco's green mosque program in the City of Marrakesh (2023). Accessed: Feb. 09, 2024. [Online]. Available: https://urn.kb.se/resolve?urn=urn:nbn:se:uu:diva-495587
8. WEVJ | Free Full-Text | Inhibitors of Battery Electric Vehicle Adoption in Morocco. Accessed: Feb. 09, 2024. [Online]. Available: https://www.mdpi.com/2032-6653/15/1/6
9. "Monographies de la Région," Site de la Direction Régionale de Casablanca-Settat. Accessed: Feb. 28, 2024. [Online]. Available: https://www.hcp.ma/reg-casablanca/Monographies-de-la-Region_a2.html
10. Zhou, X., Ji, Y., Lv, M.: The impact of the parking spot' surroundings on charging decision: a data-driven approach. KSCE J. Civ. Eng. (2024). https://doi.org/10.1007/s12205-024-0960-4
11. Brown, A., Cappellucci, J., Heinrich, A., Cost, E.: Electric vehicle charging infrastructure trends from the alternative fueling station locator: third quarter 2023. National Renewable Energy Laboratory (NREL), Golden, CO (United States), NREL/TP-5400-88223 (2024). https://doi.org/10.2172/2293493
12. Zhou, Y., Liu, X., Zhao, Q.: A stochastic vehicle schedule model for demand response and grid flexibility in a renewable-building-e-transportation-microgrid. Renew. Energy **221**, 119738 (2024). https://doi.org/10.1016/j.renene.2023.119738
13. Ramasamy, G., Vedi, A., Madasamy, M.: Multi-objective optimization for optimal energy transporting path and energy distribution in electric vehicles energy internet. Int. J. Comm. Syst. https://doi.org/10.1002/dac.5700
14. Zhao, B., Li, H., Yang, C., Wang, W., Sun, T., Chen, R.: A novel bilevel electromechanical compound braking coordinated control strategy for electric vehicles. Energy Technol. 2300835. https://doi.org/10.1002/ente.202300835
15. Piedra-de-la-Cuadra, R., Ortega, F.A.: Bilevel optimization for the deployment of refuelling stations for electric vehicles on road networks. Comput. Oper. Res. **162**, 106460 (2024). https://doi.org/10.1016/j.cor.2023.106460
16. Sheng, Z., Liu, Y., Wang, X., Chiu, A.S.F.: The impact of charging infrastructure on coupling coordination of urbanization and ecological environment: evidence from China. Technol. Forecast. Soc. Chang. **199**, 122992 (2024). https://doi.org/10.1016/j.techfore.2023.122992

17. Prapanchan, V.N., Subramani, T., Karunanidhi, D.: GIS and fuzzy analytical hierarchy process to delineate groundwater potential zones in southern parts of India. Groundw. Sustain. Dev. **25**, 101110 (2024). https://doi.org/10.1016/j.gsd.2024.101110
18. Metegam, I.F.F., Bomeni, I.Y., Chara-Dackou, V.S., Njomo, D., Tchinda, R.: Assessment of the geographic and technical potential for large-scale wind energy production using a GIS-based MCDM-AHP multi-criteria analysis method and sensitivity analysis: case of Cameroon. Preprints, Feb. 02 (2024). https://doi.org/10.20944/preprints202402.0122.v1
19. Krichen, M.: Amélioration des performances de l'Intelligence Artificielle, Feb. (2024) Accessed: Feb. 28, 2024. [Online]. Available: https://hal.science/hal-04446438
20. Siqueira, R.G., et al.: Modelling and prediction of major soil chemical properties with random forest: machine learning as tool to understand soil-environment relationships in Antarctica. CATENA **235**, 107677 (2024). https://doi.org/10.1016/j.catena.2023.107677
21. Machine learning for skin permeability prediction: random forest and XG boost regression. J. Drug Target. **32**(1). Accessed: Feb. 28, 2024 [Online]. Available: https://www.tandfonline.com/doi/abs/https://doi.org/10.1080/1061186X.2023.2284096
22. Liu, J., Liu, W., Allechy, F.B., Zheng, Z., Liu, R., Kouadio, K.L.: Machine learning-based techniques for land subsidence simulation in an urban area. J. Environ. Manage. **352**, 120078 (2024). https://doi.org/10.1016/j.jenvman.2024.120078

Predictive Analysis of Road Accidents in Casablanca: A Machine Learning Approach

Laaziza Hammoumi(✉), Mohamed Benayad, Saad Farah, Chaymae Khaloua, and Hassan Rhinane

Geoscience Laboratory, Department of Geology, Faculty of Science Aïn Chock, University Hassan II, Casablanca, Morocco
laaziza.hammoumi-etu@etu.univh2c.ma

Abstract. Road accidents significantly contribute to traffic congestion, causing severe impacts on public health, the environment, and the economy. This study focuses on mapping road accident susceptibility in Casablanca using machine learning algorithms integrated with Geographic Information Systems (GIS). By analysing spatial relationships between various causality factors and accident pixels, models such as Random Forest (RF), Support Vector Machine (SVM), K-nearest neighbours (KNN), and Artificial Neural Network (ANN) were employed. The resulting predictive maps, particularly from the RF model with an AUC of 0.884, demonstrate high accuracy and are crucial tools for road safety management. These susceptibility maps are essential for optimizing planning, management, and safety measures on roadways, offering proactive solutions to mitigate traffic accidents in urban environments.

Keywords: Accidents · Predictive modelling · Machine learning algorithms · Geographic Information Systems (GIS)

1 Introduction

Road accidents represent a significant global public health and economic challenge, resulting in approximately 1.3 million fatalities each year [1]. These incidents impose substantial economic burdens due to medical costs and productivity losses, impacting victims, their families, and broader communities. The economic impact is estimated to account for approximately 3% of the gross domestic product (GDP) in many countries [2]. Research indicates that road accidents are not random events; they tend to cluster in specific areas, known as hotspots, characterized by unique road, traffic, and operational features [3]. Identifying these hotspots is crucial for improving road safety and ensuring the sustainability of transportation systems [4].

Urbanization and increased motorization have contributed to a notable rise in road accidents, with significant disparities between developed and developing nations. In developing countries, the rate of accidents, fatalities, and injuries is 10 to 60 times higher than in developed countries, with over 80% of road accidents occurring in these regions [5, 6]. A study by Berhanu, Alemayehu, and Schröder [6] highlights these disparities,

A. Abdul-Rahman et al. (Eds.): GeoWeek 2024, LNGC, pp. 69–80, 2025.
https://doi.org/10.1007/978-3-031-86654-8_6

noting the higher incidence of road accidents in low-income countries compared to high-income countries [7]. Furthermore, the World Health Organization (WHO) has projected that road traffic deaths will become the sixth leading cause of death globally by 2020 [8].

In Morocco, road accidents result in approximately 3,500 deaths and 12,000 severe injuries annually, equating to around 10 deaths and 33 severe injuries per day [9]. Despite various efforts to mitigate these incidents, the situation remains critical. For example, in September 2021, Morocco reported 10,028 bodily traffic accidents, marking an increase of 18.6% compared to September 2020 and 13.8% compared to September 2019, according to data from the National Road Safety Agency (NARSA) [10]. Analysing road safety indicators between 2008 and 2018, a study by Chamseddine and Ait Boubkr [10] revealed a 10.78% increase in accidents and a 7.16% rise in severe injuries, despite a 1.30% decrease in fatal accidents [9].

The primary objective of this research is to assess, predict, and classify the susceptibility to road accidents in Casablanca using machine learning algorithms, Geographic Information Systems (GIS), and remote sensing. By evaluating the effectiveness of various road accident prediction models, we aim to generate spatial sensitivity maps that identify critical areas requiring intervention. This approach not only contributes valuable insights for managing the traffic system in Casablanca but also aligns with global efforts to create safer, more efficient urban environments.

While there has been considerable research on road safety and accident prediction, many studies have primarily focused on descriptive analyses, identifying risk factors and the spatial distribution of accidents [11]. Few studies have utilized predictive modelling to forecast road accidents, particularly using machine learning techniques combined with GIS [12]. This gap in the literature presents a unique opportunity to enhance road safety management by employing advanced predictive models that can provide real-time insights and proactive measures. Additionally, the concept of smart cities, characterized by integrating information and communication technologies (ICT) to manage urban assets, has become increasingly relevant in addressing urbanization challenges, including traffic congestion and road safety [13, 14].

2 Materials and Methods

2.1 Study Area

Casablanca, known as Morocco's economic powerhouse, is the largest urban agglomeration in the country, covering an area of 1,140.54 km^2 (Fig. 1). It comprises 16 administrative prefectures, which collectively form a diverse and densely populated metropolitan area. Strategically located on the Atlantic coast, approximately 80 km south of Rabat, Casablanca serves as both the economic and administrative capital of the Casablanca-Settat region. This positioning along the coast not only facilitates international trade but also underscores the city's importance in national and regional economic activities.

In 2014, the city's population was recorded at 3,359,818, making it one of the most densely populated cities in the Maghreb region. The Greater Casablanca area, encompassing the municipalities of Mohammedia, Aïn Harrouda, and the provinces of Médiouna and Nouaceur, spans 1,615 km^2 and housed an estimated 4.3 million people

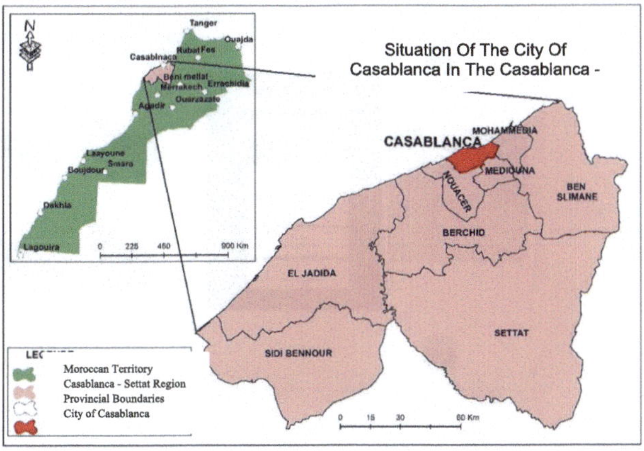

Fig. 1. Study Area Map Covering Casablanca and Neighbouring Provinces.

in the same year. This population further expanded in 2015 with the administrative integration of Settat, Benslimane, and Berrechid provinces, forming the larger Casablanca-Settat region, home to approximately 6.9 million inhabitants as per the 2014 census [15, 16].

The city's infrastructure includes a critical urban expressway network that stretches for 22 km, enhancing east-west urban connectivity. Additionally, Casablanca's 33.5 km bypass motorway is integral to the nation's transportation infrastructure, connecting three major highways: the A3 motorway, which links Casablanca to Rabat; the A5 motorway, connecting the city to El Jadida; and the A7 motorway, which provides a route to Agadir via Marrakech. This extensive network not only supports intra-city transportation but also reinforces Casablanca's role as a central hub in Morocco's national motorway system. Between 2001 and 2014, the number of vehicles in the city increased dramatically from 400,000 to 1.4 million, illustrating the rapid urbanization and rising vehicular density that pose significant challenges to traffic management and urban planning [17].

2.2 Data Collection and Analysis

The study on road accident susceptibility in Casablanca utilizes a comprehensive dataset encompassing 4289 accident pixels, covering the period from July 1, 2022, to August 9, 2022. This dataset serves as a crucial component in constructing an accident inventory map, representing the target variable for the analysis. The accident pixels were collected through a navigation platform Waze, an active participant in the Connected Citizens Program (CCP). This program facilitates data exchange between the platform and its partners, including the city of Casablanca, enabling data-driven infrastructure decisions and optimizing incident response efficiency [13].

The collaboration between the navigation platform and Casablanca provides valuable insights into the spatial and temporal distribution of road accidents. This partnership has facilitated the generation of hotspot traffic maps, highlighting areas with a high frequency of accidents.

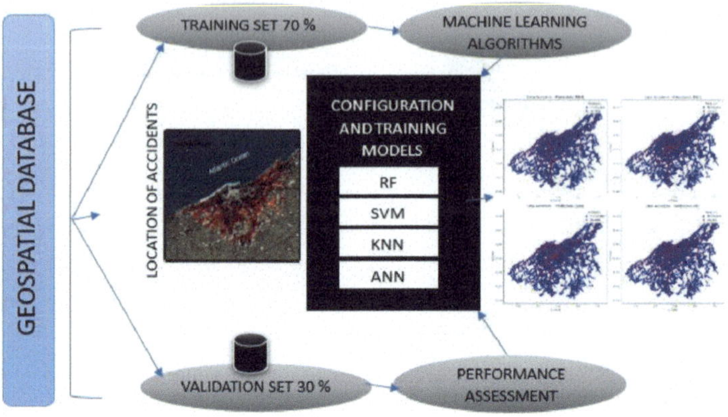

Fig. 2. Workflow of the methodology used in this study.

To ensure robust model development and validation, as depicted in Fig. 2, the dataset was divided into two subsets: 70% (3002 pixels) were allocated for model training, while the remaining 30% (1287 pixels) were reserved for model validation. A binary classification was implemented to distinguish between accident and non-accident classes, identifying pixels with accident risk indexes. Additionally, 5558 points without accidents were randomly selected, with 3891 for training and 1667 for validation. This resulted in a training dataset comprising 2954 samples and a validation dataset comprising 6893 samples.

2.3 Identification of Explanatory Variables

A crucial phase in road crash mapping and modelling involves identifying a set of independent explanatory variables, or crash causal factors, that may influence the likelihood of accidents. These variables are selected based on their potential correlation with the local characteristics of the study area, historical crash patterns, and the availability of data. Drawing on prior research and the specific context of Casablanca, we identified 10 significant factors: road type, day of the week, district, population density, traffic congestion level, proximity to schools, train stations, roundabouts, road intersections, and traffic signs (Table 1). Among these, the differentiation in road characteristics, particularly the road type and district delineation, is crucial as accidents tend to cluster in areas with high population density.

To accurately capture these factors, a systematic sampling procedure was employed. This procedure extracted values for each of the influential factors, ensuring comprehensive data collection. The use of spatial analysis tools within ArcGIS 10.3 enabled the detailed mapping of topographic features and compliance with standards for road construction, including attributes like slope, curvature, speed limits, and road width. These attributes are critical in influencing driver behaviour and alertness and were carefully considered in the modelling approach. Additionally, variables such as traffic volume and weather conditions were included to enhance the understanding of road accident dynamics.

Table 1. Conditioning Factors Used in Road Accident Susceptibility Analysis

Conditioning factor	Unit	Source
Road	–	Roads map
Day	Day	Data Waze (programme Connected Citizens)
Borough	–	https://siredd.environnement.gov.ma/
Population	–	http://www.equipement.gov.ma/
Traffic jam level	–	Data Waze (programme Connected Citizens)
Distance to intersection	m	Open street map
Distance to school	m	Open street map
Distance to train station	m	Open street map
Distance to Roundabout	m	Open street map
Distance to traffic sign	m	Open street map

The comprehensive analysis of these factors forms the basis for developing predictive models for road accident susceptibility, allowing for targeted interventions and improved traffic safety strategies.

2.4 Machine Learning Algorithms

To predict road accident susceptibility in Casablanca, we employed four machine learning algorithms: Random Forest (RF), Support Vector Machine (SVM), K-Nearest Neighbours (KNN), and Artificial Neural Network (ANN). Each algorithm was chosen for its unique strengths in handling the complexity of traffic data and the spatial aspects integrated with GIS.

Random Forest (RF), an ensemble learning method, constructs multiple decision trees and aggregates their outputs to determine the final prediction. This algorithm was chosen for its robustness against overfitting and its ability to handle large datasets with numerous features, making it ideal for identifying key factors contributing to road accidents [18]. RF also provides feature importance scores, aiding in pinpointing the most critical variables influencing accident occurrence.

Support Vector Machine (SVM) is a powerful classification technique that identifies the optimal hyperplane to separate data points into different classes. SVM was selected for its high accuracy and effectiveness in binary classification, distinguishing between accident and non-accident cases. This method's ability to use kernel functions enables it to handle non-linear relationships in the data, which is essential for modelling the complexities of road traffic incidents [19].

K-Nearest Neighbours (KNN) operates by classifying a data point based on the majority class of its nearest neighbours. KNN is particularly effective for pattern recognition tasks, making it useful for detecting local data clusters. This algorithm was chosen for its simplicity and its ability to classify accident-prone areas based on proximity to similar features [20].

Artificial Neural Network (ANN) mimics the neural structure of the human brain, allowing it to learn complex patterns from data. ANNs were selected for their capability to model non-linear interactions and relationships between variables. This flexibility is crucial for understanding the multifaceted nature of road accidents, where multiple factors interact in complex ways [21].

The integration of GIS with these machine learning algorithms was critical for extracting and visualizing spatial features related to road accidents. GIS tools enabled the mapping of topographic features, road networks, and traffic conditions, which were essential inputs for the models. This combination allowed for a detailed spatial analysis, highlighting accident hotspots and providing valuable insights for targeted interventions. By leveraging GIS, we were able to create a comprehensive dataset that enriched the machine learning models and enhanced the accuracy of the road accident susceptibility predictions.

3 Results

3.1 Key Findings

The analysis revealed significant insights into the relationship between various crash causal factors and the occurrence of road accidents. Roundabouts, intersections, and traffic lights emerged as critical components within the road network, with intersections, in particular, identified as hazardous points due to their complexity and the increased likelihood of accidents. Statistical analysis indicated an overrepresentation of accidents at intersections, attributed to the challenges drivers face in detection and estimation during the driving task.

The presence of traffic lights, intended to enhance safety by regulating traffic flow and protecting pedestrians, was paradoxically associated with a notable number of accidents. This phenomenon was linked to the non-compliance with traffic regulations by a significant segment of the population. In our analysis, these elements were categorized using a binary framework: a value of 1 indicated accidents occurring within approximately 100 m of these elements, while a value of 0 denoted accidents beyond this threshold.

Furthermore, proximity to educational institutions and train stations also showed a correlation with increased accident frequency. The classification for these elements similarly followed a dichotomous approach, with a value of 1 assigned to accidents occurring within 150 m and a value of 0 beyond this distance.

Human factors, particularly population density within districts, played a pivotal role in influencing accident rates. The data showed that areas with higher population densities experienced more accidents, underscoring the impact of human factors on road safety.

3.2 Model Performance and Confusion Matrix Analysis

The performance of the four machine learning models—Random Forest (RF), K-Nearest Neighbours (KNN), Support Vector Machine (SVM), and Multilayer Perceptron (MLP)—was assessed using a validation dataset distinct from the training set. Key evaluation metrics include Accuracy, Precision, Sensitivity, Specificity, Positive Predictive Value (PPV), and Negative Predictive Value (NPV), as shown in Table 2.

Table 2. Evaluation Metrics for Machine Learning Models.

Heading level	RF	SVM	KNN	ANN
True Positive	1544	1296	1389	1404
True Negative	1069	969	954	985
False Positive	127	375	282	267
False Negative	215	315	330	299
Accuracy	0.884	0.766	0.792	0.808
Precision	0.893	0.720	0.771	0.786
F1-Score	0.90	0.86	0.86	0.83
recall	0.832	0.754	0.742	0.767
Positive predictive value %	0.893	0.720	0.771	0.786
Negative predictive value %	0.877	0.804	0.808	0.824
Specificity	0.923	0.775	0.831	0.840
AUC	0.88	0.77	0.79	0.80

The Random Forest model demonstrated superior performance (Fig. 3), achieving an Accuracy of 88.43%, Precision of 89.38%, Sensitivity of 83.26%, and Specificity of 92.40%. The confusion matrix (Fig. 4) highlights RF's strong performance, with 1544 true negatives and 1209 true positives, leading to 1759 correctly classified instances. The model's ability to minimize false positives (127) and false negatives (215) further underscores its robustness in classifying accident-prone areas.

K-Nearest Neighbours achieved an Accuracy of 79.29%, with Specificity at 83.12% and Sensitivity at 74.30%. The model's Precision was 77.18%, with 1389 true negatives and 954 true positives. Despite its lower performance compared to RF, KNN's confusion matrix analysis shows a reasonable balance between precision and recall, indicating its usefulness in specific contexts.

The SVM model, with an Accuracy of 76.65%, demonstrated a balanced performance across various metrics, including Specificity (77.56%) and Sensitivity (75.47%). The confusion matrix (Fig. 4) revealed 1296 true negatives and 969 true positives, highlighting its capacity to manage both classes effectively.

MLP showed competitive performance with an Accuracy of 80.85%, Specificity of 84.02%, and Sensitivity of 76.71%. The model recorded 1404 true negatives and 985 true positives, with precision and recall values reflecting a balanced approach to classification. Despite being the lowest among the four, MLP's confusion matrix still indicates a viable predictive capacity, particularly in recognizing true negatives.

Overall, the confusion matrix analysis underscores RF's superior performance across all metrics, followed by KNN, SVM, and MLP. The detailed breakdown provides insights into the models' strengths and weaknesses, particularly in handling false positives and false negatives. This analysis is crucial for optimizing the model selection process for road accident prediction tasks, as it highlights the nuances in each model's performance.

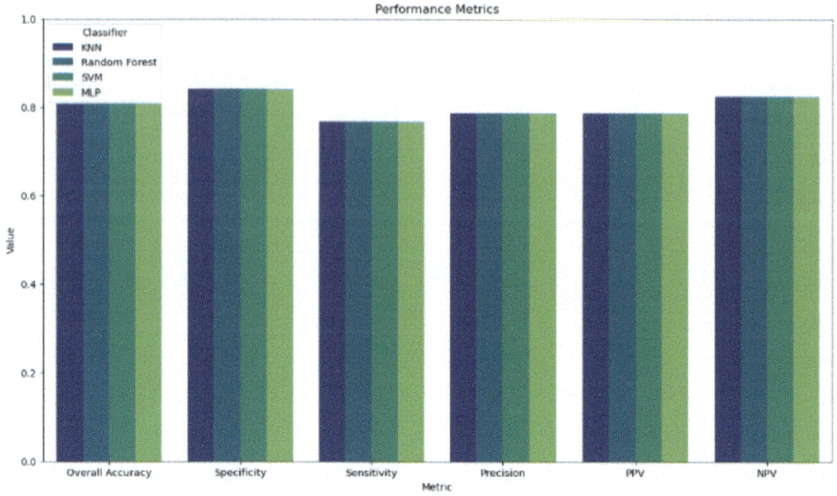

Fig. 3. Comparative Performance Metrics Across Classification Algorithms.

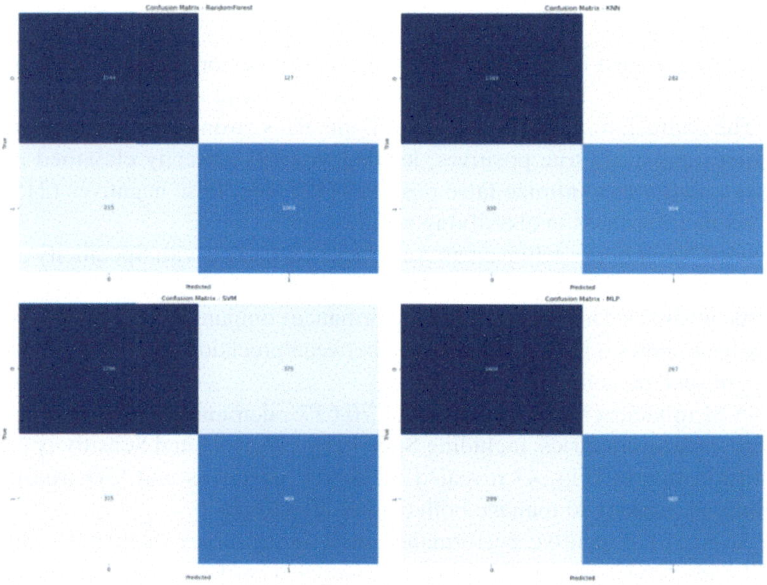

Fig. 4. Confusion Matrices for Classification Algorithms: Evaluating Predictive Performance in Road Accident Risk Analysis.

The ROC curves and AUC values provide a comprehensive view of the models' classification capabilities, illustrating the trade-off between true positive and false positive rates (Fig. 5). The RF model achieved the highest AUC value at 0.98, followed by SVM at 0.97, MLP at 0.96, and KNN at 0.95. These results highlight the RF model's superior

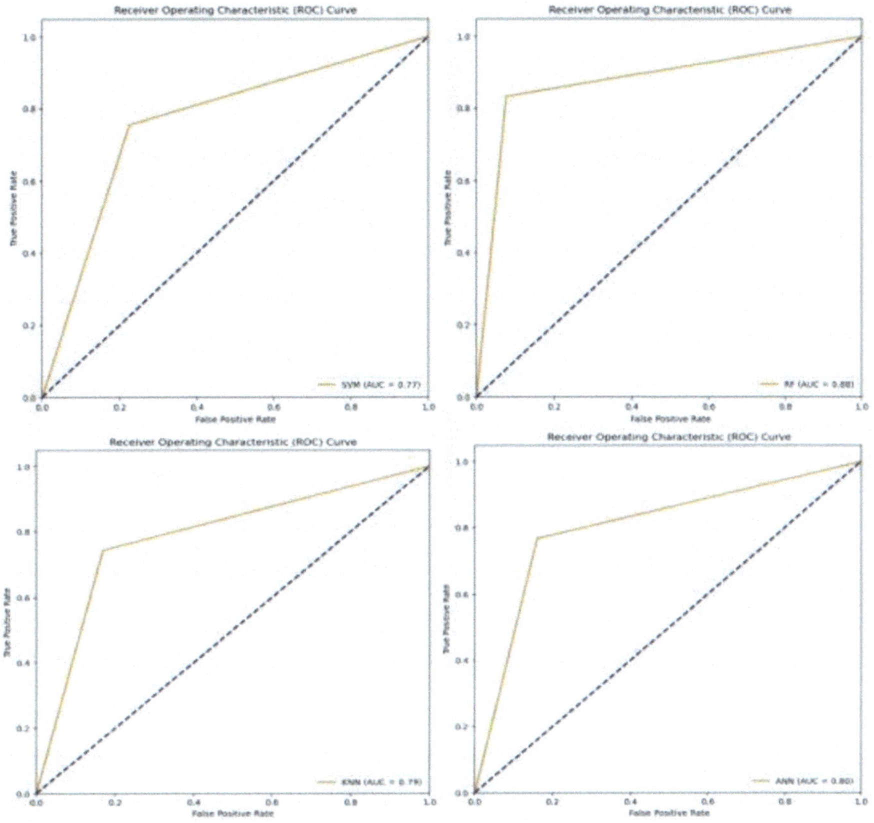

Fig. 5. ROC Curves and AUC Values: Comparative Analysis of Classification Models for Accident Risk Prediction.

ability to distinguish between accident-prone and non-accident-prone areas, making it the most effective among the models studied.

The risk maps generated using the four models offer a practical application for local and regional authorities to identify areas with a high risk of accidents. Each map, developed using data on physical characteristics such as traffic volume, population density, and proximity to schools and train stations, provides a visual representation of the predicted susceptibility to road accidents (Fig. 6). The maps highlight areas for targeted interventions aimed at improving road safety.

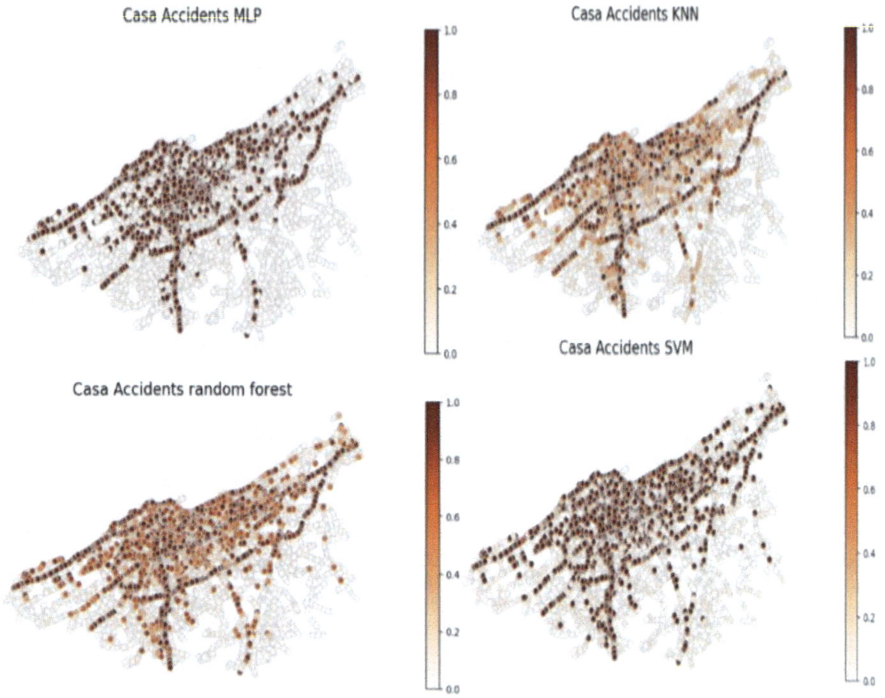

Fig. 6. Road Accident Susceptibility Maps Generated by Different Machine Learning Models.

4 Conclusion

Road accidents continue to pose a significant public health challenge, contributing to millions of deaths and injuries annually. Despite various governmental efforts, many regions still struggle to reduce accident rates effectively. This study aimed to address this issue by developing predictive models for road accident risk and identifying high-risk areas that require urgent management action.

Using a dataset comprising 4289 accident locations and ten conditioning factors, four machine learning algorithms—Random Forest (RF), Support Vector Machine (SVM), K-Nearest Neighbours (KNN), and Multilayer Perceptron (MLP)—were employed to model road accidents. The results indicate that all four algorithms performed well, with accuracy scores ranging from 76.65% to 88.43%. Among them, the Random Forest algorithm achieved the highest accuracy of 88.43%, followed by MLP (80.85%), KNN (79.29%), and SVM (76.65%).

These findings highlight the potential of machine learning as a valuable tool for predicting areas with a high risk of road accidents. The generated risk maps serve as practical instruments for road safety managers and authorities, enabling them to prioritize interventions and improve road safety more effectively.

Furthermore, the methodology demonstrated in this study can be adapted to other domains with similar predictive modelling requirements. For instance, it could be applied to forecast the risk of workplace accidents or medical errors, thereby enhancing safety measures in various sectors.

In summary, this research underscores the efficacy of integrating machine learning and GIS in road accident analysis. By providing accurate predictions and actionable insights, it offers a robust framework for enhancing urban safety measures and guiding policymakers in their efforts to reduce road accident rates. Future studies could expand on this work by incorporating additional variables and exploring real-time data to further refine the predictive models.

References

1. World Health Organization Global Road Safety Status Report.: Available online: https://www.who.int/violence_injury_prevention/road_safety_status/2018/en/ (2018). Accessed 28 Jan 2019
2. Ali, M.S., Adnan, M., Noman, S.M., Baqueri, S.F.A.: Estimation of traffic congestion cost—A case study of a major arterial in Karachi. Procedia Eng. **77** (2014). https://doi.org/10.1016/j.proeng.2014.07.030
3. Akhtar, M., Moridpour, S.: A review of traffic congestion prediction using artificial intelligence. J. Adv. Transp. **2021** (2021). https://doi.org/10.1155/2021/8878011
4. Banerjee, K., Bali, V., Sharma, A., et al.: Traffic accident risk prediction using machine learning. In: 2022 International Mobile and Embedded Technology Conference, MECON 2022 (2022). https://doi.org/10.1109/MECON53876.2022.9752273
5. Augustine, T., Shukla, S.: Road accident prediction using machine learning approaches. In: 2022 2nd International Conference on Advance Computing and Innovative Technologies in Engineering, ICACITE 2022 (2022). https://doi.org/10.1109/ICACITE53722.2022.9823499
6. Berhanu, Y., Alemayehu, E., Schröder, D.: Examining car accident prediction techniques and road traffic congestion: a comparative analysis of road safety and prevention of world challenges in low-income and high-income countries. J. Adv. Transp. **2023** (2023). https://doi.org/10.1155/2023/6643412
7. Road Traffic Injuries.: [https://www.who.int/fr/news-room/fact-sheets/detail/road-traffic-injuries] (https://www.who.int/fr/news-room/fact-sheets/detail/road-traffic-injuries)
8. World Health Organization: World report on road trafc injury prevention. WHO, Geneva (2014)
9. Maroc.: les accidents de la route en hausse de près de 19% et les efforts se poursuivent pour stopper l'hémorragie—La Vie éco. https://www.lavieeco.com/actualite-maroc/maroc-les-accidents-de-la-route-en-hausse-de-pres-de-19-et-les-efforts-se-poursuivent-pour-stopper-lhemorragie/
10. Chamseddine, Z., Ait Boubkr, A.: Exploring the place of social impacts in urban transport planning: The case of Casablanca City. Urban, Plan. Transp. Res. **8**(1) (2020). https://doi.org/10.1080/21650020.2020.1752793
11. Tillinghast, D. J., Duffy, V.G.: Crowdsourcing data to improve transportation safety and efficiency: A systematic review. Lecture Notes in Computer Science (Including Subseries Lecture Notes in Artificial Intelligence and Lecture Notes in Bioinformatics), 14057 LNCS (2023). https://doi.org/10.1007/978-3-031-48047-8_28
12. Fiorentini, N., Leandri, P., Losa, M.: Defining machine learning algorithms as accident prediction models for Italian two-lane rural, suburban, and urban roads. Int. J. Inj. Control. Saf. Promot. **29**(4) (2022). https://doi.org/10.1080/17457300.2022.2075397
13. Hernandez-Callejo, L., Nesmachnow, S.: Mobility and IoT for the Smart Cities (n.d.). www.mdpi.com/journal/smartcities

14. Bellini, E., et al.: An IOE and big multimedia data approach for urban transport system resilience management in smart cities. Sensors (Switzerland) **21**(2) (2021). https://doi.org/10.3390/s21020435
15. Haut-Commissariat au Plan.: Recensement Général de la Population et de l'Habitat 2014. Rabat, Morocco: Haut-Commissariat au Plan. http://www.hcp.ma/ (2014)
16. RGPH2014.: Recensement Général de la Population et de l'Habitat 2014. Rabat, Morocco: Haut-Commissariat au Plan. http://rgph2014.hcp.ma/ (2014)
17. Waze.: Connected Citizens Program. Retrieved from Waze CCP. www.waze.com/ccp (2019). Accessed 14 Apr 2019
18. Breiman, L.: Random forests. Mach. Learn. **45**(1), 5–32 (2001)
19. Cortes, C., Vapnik, V.: Support-vector networks. Mach. Learn. **20**(3), 273–297 (1995)
20. Cover, T., Hart, P.: Nearest neighbor pattern classification. IEEE Trans. Inf. Theory **13**(1), 21–27 (1967)
21. Rumelhart, D.E., Hinton, G.E., Williams, R.J.: Learning representations by back-propagating errors. Nature **323**(6088), 533–536 (1986)

Development and Evaluation of Geo-AI for Healthcare Discovery

Myriame Virginie Matius(✉) and Ivin Amri Musliman

University Teknologi Malaysia, 81310 Skudai, Johor, Malaysia
virginiemyriame@gmail.com, ivinamri@utm.my

Abstract. Integration of AI and geo-location technologies can transform healthcare services by providing more personalised and efficient recommendations. Current healthcare facility locators prioritise geographic proximity over specialised medical service. Due to reliance on static databases and needing more personalised, condition-centric matching, individuals need help finding suitable care based on specific conditions. Traditional systems often need to incorporate real-time data and advanced AI, limiting accuracy and usefulness. Recent advancements, particularly in GPT-3.5 and GPT-4, offer new opportunities for enhancing healthcare service recommendations. This study aims to develop a GeoAI-augmented web platform to recommend healthcare facilities for minor illnesses such as toothache, fever, headache, and stomachache. The platform combines AI-enhanced geo-location with precise condition-centric matching to improve user experience and accuracy, utilising a fine-tuned GPT-3.5 model to enhance interaction and generate relevant prompts. The system integrates OSM data to provide tailored facility suggestions. Methodologies include data collection of user prompts, model ne-tuning, expert reviews for relevance and accuracy, user surveys for satisfaction, and performance testing for response times. Comparative analysis between GPT-3.5 and GPT-4 assesses relevance, accuracy, user satisfaction, and response times. Results indicate that GPT-4 achieves higher accuracy and relevance, improved user satisfaction, and more contextually accurate responses compared to GPT-3.5, though with slightly longer response times. The study demonstrates GPT-4's superior performance in generating accurate healthcare facility recommendations. The anticipated outcome is a chatbot-like web tool offering customised healthcare recommendations within specific regions, emphasising ease of use, precision, and contextual relevance. This study highlights the potential of AI and geo-location technologies in providing more personalised and effective healthcare solutions.

Keywords: GeoAI · Healthcare facilities locator · Prompt engineering

1 Introduction

The advent of Artificial Intelligence (AI) has been a cornerstone in revolutionising healthcare since the mid-20th century. AI, fundamentally rooted in machine learning, has significantly advanced disease prediction, diagnosis, and treatment, enhancing the

A. Abdul-Rahman et al. (Eds.): GeoWeek 2024, LNGC, pp. 81–94, 2025.
https://doi.org/10.1007/978-3-031-86654-8_7

overall healthcare experience for patients and providers. This transformative journey of AI in healthcare is a testament to technological advancement and a beacon of hope for improved patient care and healthcare management. This research is poised to tackle the intricate challenges and limitations inherent in healthcare accessibility and information retrieval systems.

Over the past decade, AI has seen unprecedented growth, profoundly influencing various sectors, including Natural Language Processing, computer vision, recommender systems, healthcare, and finance. This growth is primarily fueled by the substantial and high-quality training data that drives the success of contemporary AI models like Generative Pre-trained Transformers (GPT), which have maintained consistent architectural frameworks but have expanded model weights. The efficacy of these models is further enhanced when they are sufficiently powered and fine-tuned with specific engineering prompts and fixed models.

Despite these advancements, the current landscape of healthcare facility locators is predominantly focused on geographical proximity, often overlooking the need for specialised medical services. This gap presents a significant challenge for individuals seeking healthcare tailored to their medical conditions. Addressing this gap, the current study proposes the development of a GeoAI-augmented web platform. This platform is designed to recommend healthcare facilities by aligning with the individual health conditions users report. The study will further explore and identify the application of AI language models in enhancing user interaction within the GeoAI context for healthcare, drawing insights from various studies in this field.

1.1 Problem Background

The influence of big data and machine learning spans numerous sectors of modern life, notably in entertainment, commerce, and healthcare. As Bohr and Memarzadeh [1] state, platforms like Google exemplify AI applications, utilising algorithms to interpret user inquiries about health symptoms and conditions. This demonstrates AI's capacity to perform on par with or better than humans in many tasks. For instance, Google can identify which symptoms and conditions people are searching for, showcasing the effectiveness of AI in processing and analysing vast amounts of data.

The work of Tchango [2] further highlights the potential of AI in healthcare. They designed a system that interacts with patients, collects relevant evidence, and predicts underlying diseases, underscoring the rising interest in Automatic Diagnosis (AD) and Automatic Symptom Detection (ASD) systems within machine learning research. These systems aim to assist doctors in telemedicine services despite challenges like the lack of datasets for training models. This issue has been partially resolved by introducing large-scale synthetic datasets, including diagnoses and ground truth pathology for each patient. Adding to the body of research [3] developed a web-based GIS system for health facilities mapping, monitoring, and reporting in Zambia.

This initiative addressed public health management challenges by creating a user-friendly system using MySQL, Apache web server, and PHP. System requirements were determined through interviews with health officials, leading to an online health facilities web GIS application. The hybrid architecture conducts spatial functions on the client side, minimising data traffic and enabling real-time data collection, visualisation, and reporting. The system integrates geospatial data on health facilities, diseases, and environmental factors, facilitating disease surveillance and planning for new service locations. It also enhances responsiveness to public health emergencies by offering tools for search, drawing, measurement, layer overlay, and data collection functionalities.

The output of this study presented a robust online health facilities web GIS application, enabling users to search for health facilities, draw and measure features, overlay external layers, collect data on outbreaks, perform nearest neighbour searches, generate heat maps, and utilise routing functionalities. This study focuses on applying web-based solutions in healthcare discovery, emphasising the critical nature of timely emergency access to healthcare services. Traditional systems, which rely on static databases to locate healthcare facilities, often lack real-time accuracy and fail to consider dynamic factors such as traffic conditions, potentially leading to dangerous delays in receiving medical care. The core of this study is exploring an AI-driven web-based solution for healthcare discovery.

By integrating the advanced language processing capabilities of the GPT-3.5 model with real-time location data from OpenStreetMap (OSM), the aim is to provide precise and timely recommendations for healthcare facilities. This combination promises to enhance user experience and improve healthcare outcomes by offering condition-specific, real-time, location-based recommendations. However, implementing the GPT-3.5 model presents challenges, particularly regarding the model's language version and token usage limitations, which are critical factors in determining the platform's effectiveness and reliability.

1.2 Problem Statement

The concept of innovative healthcare, as discussed in Zhou's work [4], is fundamentally aimed at enhancing efficiency, accuracy, accessibility, and outcomes for patients. Intelligent healthcare systems facilitate remote monitoring of patients' health conditions through connected devices and telehealth solutions, allowing healthcare providers to manage patient health outside traditional clinical settings. Based on Luo's paper [5] mentions the recent advances in deep reinforcement learning and task-oriented dialogue systems; recent studies have proposed human-computer dialogue systems for automatic diagnosis; the automatic diagnosis system applied a deep Q network (DQN) to decide whether to continue the dialogue by asking about a symptom or conclude the diagnosis by predicting a disease.

Despite these advancements, contemporary healthcare systems still face significant challenges in providing suitable healthcare options, particularly for individuals with unique health conditions. Current healthcare locators tend to focus more on the proximity of facilities rather than on the specific medical services they offer, creating obstacles for patients needing specialised care.

Addressing this gap, the focus shifts to developing a GeoAI-augmented web platform. This platform is envisioned to provide customised recommendations for healthcare facilities based on users' input on health conditions. The goal is to enhance the accessibility of specialised healthcare services, catering specifically to the unique needs of individual patients.

1.3 Objectives

1. To investigate the role of AI language models in improving user engagement in the realm of GeoAI for healthcare.
2. To implement an AI-powered geo-locator tool to identify the nearest healthcare facility accurately.
3. To produce customised recommendations in sync with each user's unique preferences and specific health needs.

This study aims to create a web-based geo-locator application enhanced with Artificial Intelligence support. The prototype will provide detailed, near-real-time data about recommended healthcare facilities tailored to the user's current location and specific health condition.

2 Research Background

It discusses the comprehensiveness of studies application of the AI language model in enhancing user interaction within the context of GeoAI for healthcare. Recap to Sect. 1, this study aims to develop a web-based healthcare discovery platform using an AI model, a GPT 3.5 languages model linked with OpenStreetMap. This section will focus on refining the AI Model to enhance its effectiveness in providing personalised healthcare facility recommendations. As a result, the goal is to optimise the model for real-time usage within the web-based platform. Based on the development phase, the model's performance will be strictly assessed using relevant metrics such as accuracy, precision and F1 score. This evaluation process aims to validate the effectiveness and reliability of the model's recommendations.

Artificial Intelligence represents the development of computer systems that can perform tasks that require human intelligence. This involves various capabilities such as problem-solving, learning from data, understanding natural language, recognising patterns, and making decisions. Within the realm of AI, Large Language Models (LLMs) represent a specific type of AI model that excels in understanding and generating human-like text. There are two types of LLMs: Bidirectional Encoder Representations from Transformers (BERT) and GPT. Google developed BERT. Devlin's paper stated that BERT is designed to pre-train deep bidirectional representations from the unlabeled text by joint conditioning on the left and proper context in all layers. As a result, the pre-trained BERT model can be fine-tuned with just one additional output layer to create state-of-the-art models for a wide range of tasks, such as question answering and language inference, without substantial task-specific architecture modifications [6]. However, this paper will focus more on GPT. GPT was developed by OpenAI, which is another transformer-based language model. GPT is designed for generative tasks and operates based on unidirectional architecture; it also generates human-like text based on a given prompt. However, pre-trained GPT models lack context awareness, which can result in awkward dialogue in specific contexts. Ngyen-Mau's [7] research paper proposed the latest information fusion-based approach to fine-tuning GPT models based on contextual data as the solution [7].

LLM is a specific type of AI Model trained using various amounts of text data. It is called "large" because it contains massive amounts of information and has many layers or components that help it understand and generate human-like text. However, there are several challenges in terms of uncertainty in LLM effectiveness in healthcare, which is uncertain in clinical text mining impact brought more poor performance when directly employing LLM as a task. Meanwhile, Natural Language Processing (NLP) in the AI model refers to the interaction between computers and humans using Natural language and emphasises the computer's ability to understand human language. NLP is necessary for various applications of extensive data analysis within healthcare [1]. AI's future role in medical practice remains an unanswered question. Computers learn to detect patterns that are not decipherable using biostatistics by processing massive datasets (big data) through algorithms. Then, the training where the correcting algorithm mistakes adds to AI predictive model confidence [8].

In NLP fundamentals, the encoder is responsible for processing input sequences and encoding them into a fixed-dimensional vector representation. At the same time, the decoder takes the context vector generated by the encoder and uses it to create an output. In the case of LLMs, the encoder processes the input tokens and then creates embeddings for each token, which passes through multiple layers to generate a high-level representation. At the same time, the decoder focuses on mechanisms that focus on various parts of the input sequence while generating the output tokens.

Overall, this section will elaborate on and discuss the literature review regarding the AI-language models, mainly focusing on applying GPT 3.5 within the scope of GeoAI for healthcare.

2.1 GPT 3.5 vs GPT 4 for Healthcare Facility

GPT models, developed by OpenAI, use a neural network architecture called "Transformer" to generate human-like language. GPT-3.5 and GPT-4 were launched in March 2022 and March 2023, respectively, with GPT-4 currently available under a paid subscription. There is significant interest in applying ChatGPT in the medical field [9]. Integrating iterative questioning, chain-of-thought workflows, domain-specific databases, external tools, and few-shot in-context learning with ChatGPT has improved accuracy and relevance in scientific areas, including medicine [10].

Figure 1 shows the percentage of correct answers for humans, GPT-4, and GPT-3.5 across different medical subspecialties and practice areas. GPT-4 outperformed GPT-3.5 and human participants, scoring 82.4%, 65.9%, and 75.5% respectively. GPT-4 demonstrated greater consistency across various medical practice areas compared to GPT-3.5. Clinical questions were more accessible for both AI models due to the availability of answers online. In contrast, surgical questions involving specialised techniques that needed to be covered better in AI training proved more challenging. GPT-4 performed better than GPT-3.5 in these areas. For complex questions (incorrectly answered by over 50% of humans), both AI models did better than humans, but the difference was insignificant. All models performed similarly for more straightforward questions (answered correctly by 95% or more humans). This study suggests improvements in AI, especially GPT-4, for medical questions, but challenges remain in dealing with complex medical knowledge [9].

Question classification	Questions number (%)	Humans mean ± SD (mean ± SD%)	GPT-4.0 correct answers (%)	GPT-3.5 correct answers (%)	GPT-4.0 vs. humans P value (post-hoc power)	GPT-3.5 vs. Humans P value (post-hoc power)	GPT-4.0 vs. GPT-3.5 P value (post-hoc power)
All	1023 (100.0%)	774 ± 175 (75.7 ± 17.1%)	843 (82.4%)	674 (65.9%)	0.0002* (0.96)	<0.0001* (1.00)	<0.0001* (1.00)
Subspecialty							
Cataract/anterior segment	116 (11.3%)	90 ± 18 (77.6 ± 16.3%)	89 (76.7%)	68 (58.6%)	0.8757 (0.03)	0.0019* **(0.88)**	0.0032* **(0.84)**
Cornea/external disease	98 (9.6%)	75 ± 15 (77.1 ± 16.1%)	86 (87.8%)	72 (73.5%)	0.0508 (0.50)	0.5510 (0.08)	0.0114* (0.72)
Glaucoma	136 (13.3%)	104 ± 22 (77.1 ± 16.7%)	119 (87.5%)	88 (64.7%)	0.0253* (0.61)	0.0237* (0.61)	<0.0001* **(1.00)**
Neuro-ophthalmology/orbit	82 (8.0%)	61 ± 12 (75.5 ± 15%)	63 (76.8%)	44 (53.7%)	0.8431 (0.04)	0.0034* **(0.84)**	0.0018* **(0.88)**
Ocular pathology/oncology	35 (3.4%)	24 ± 6 (71.3 ± 18.6%)	27 (77.1%)	27 (77.1%)	0.5754 (0.08)	0.5754 (0.08)	1.000 (N/A)
Oculoplastics/orbit	116 (11.3%)	88 ± 19 (75.9 ± 17.1%)	100 (86.2%)	81 (69.8%)	0.0444* (0.52)	0.3015 (0.18)	0.0026* **(0.86)**
Pediatric ophth/strabismus	122 (11.9%)	91 ± 20 (74.8 ± 16.7%)	101 (82.8%)	65 (53.3%)	0.1279 (0.33)	0.0004* **(0.94)**	<0.0001* **(0.99)**
Refractive Mgmt/intervention	68 (6.6%)	46 ± 13 (68.3 ± 19.8%)	53 (77.9%)	42 (61.8%)	0.2052 (0.24)	0.4236 (0.12)	0.0398* (0.53)
Retina/vitreous	128 (12.5%)	98 ± 22 (77 ± 17.9%)	101 (78.9%)	97 (75.8%)	0.7141 (0.05)	0.8173 (0.04)	0.5504 (0.09)
Uveitis	122 (11.9%)	92 ± 21 (75.7 ± 17.5%)	104 (85.2%)	90 (73.8%)	0.0591 (0.46)	0.7337 (0.05)	0.0264* (0.60)
Practice area							
Diagnostics/clinics	646 (63.1%)	487 ± 111 (75.4 ± 17.2%)	541 (83.7%)	440 (68.1%)	0.0002* **(0.96)**	0.0035* **(0.83)**	<0.0001* **(1.00)**
Medical treatment	235 (23.0%)	181 ± 40 (76.9 ± 16.9%)	196 (83.4%)	153 (65.1%)	0.079 (0.42)	0.0047* **(0.81)**	<0.0001* **(1.00)**
Surgery	142 (13.9%)	106 ± 24 (74.7 ± 17.2%)	106 (74.6%)	81 (57.0%)	0.9989 (0.03)	0.0017* **(0.89)**	0.0017* **(0.88)**

Fig. 1. The table shows the number and percentage of correct answers for humans, GPT-4.0, and GPT-3.5 across different subspecialties and practice areas. The post-hoc power > 80 was highlighted in bold [9].

2.2 Limitation of the Study

GPT-4 is generally more accurate than GPT-3.5 but has limitations. According to Wu [10], these include the depth of technical analysis, data currency, and subjectivity and bias.

GPT-4 can analyse text using patterns in data, but it needs more depth and detail than a subject matter expert can provide. Its data is only sometimes current, relying on training data that may not include the latest research. This is significant for papers published after 2023 or those introducing new concepts. Furthermore, GPT-4's interpretations can be biased and influenced by the data it was trained on, leading to potential misunderstandings of cultural or contextual subtleties.

3 Methodology

This study consists of five main steps to achieve the objectives, as illustrated in Fig. 2.

The system architecture diagram in Fig. 3 illustrates a web-based GIS application that integrates OpenAI and a mapping service. The user interacts with the system through the device, inputting data sent to the web server. The web server, handling HTML and GeoJSON, receives the input and returns the response to the web server via a Python Flask. Concurrently, the web server communicates with the mapping service to retrieve the relevant geospatial data based on the user's query and the AI's response. The mapping service processes the request and sends the map data back to the web server, relaying the AI response and the map data to the user's device. The device finally displays this information to the user, ensuring seamless integration between user queries, AI processing and geospatial mapping services for a comprehensive interactive experience.

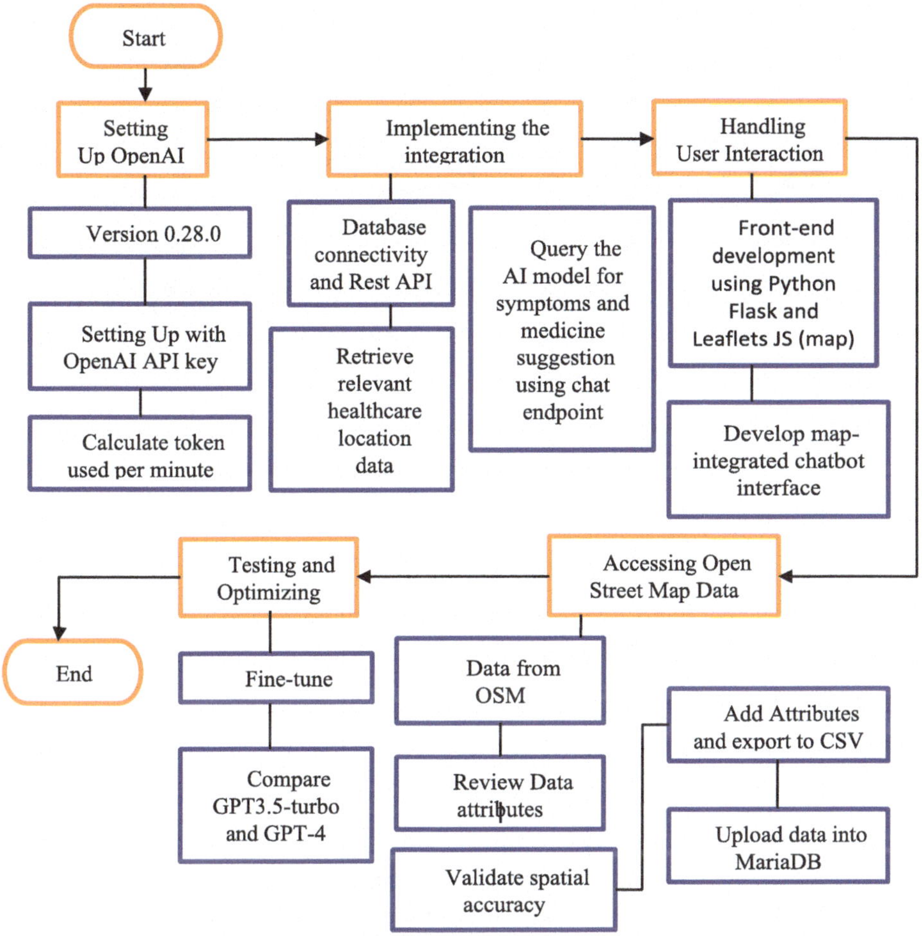

Fig. 2. Shows the flowchart of five main steps, including (i) Setting up OpenAI API Integration, (ii) Implementing the Integration, (iii) Handling User Interaction, (iv) Accessing OSM Data, and (v) Testing and Optimizing.

3.1 Accessing Open Street Map Data

Accessing OSM data: collectingvolunteerdata:volunteerdatafrom https://data.humdata.org/dataset/hotosm_mys_health_facilities?, reviewing data attributes, validating spatial accuracy, exporting to a CSV file and uploading into MariaDB database. In reviewing data attributes, it contains osm_id and osm_type registered in the LocationIQ database. LocationIQ is a geo-location service provider that offers various APIs and location-based services. In this case, one may obtain the coordinates of the healthcare location by calculating using "calculate geometry" tools in ArcGIS Pro (Fig. 3). Before that, add new fields, which are latitude and longitude, on the data table (Fig. 4).

After validating spatial accuracy and reviewing the data attributes, export the table into *.csv and upload it into the MariaDB database. In the phase of uploading the data

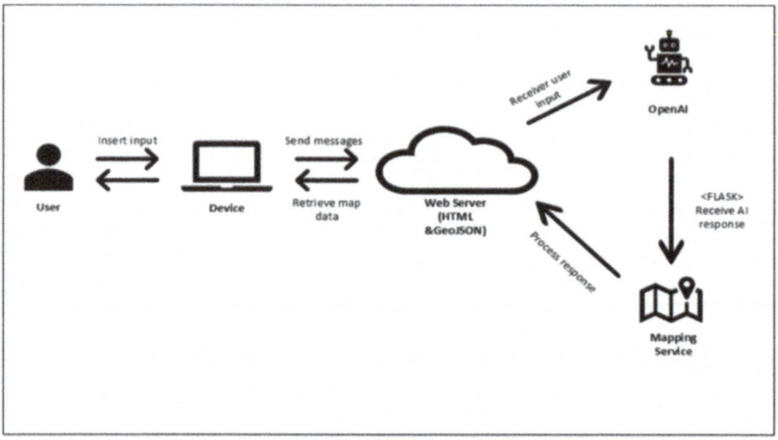

Fig. 3. Illustrates the system architecture of this study.

Fig. 4. Shows the longitude and latitude calculated using "calculate geometry" tools.

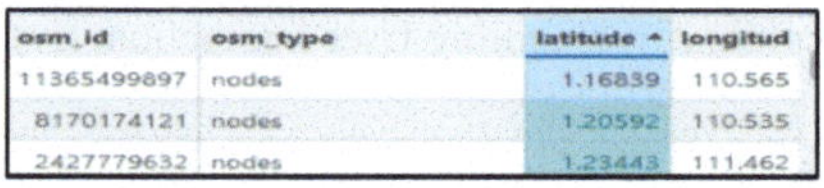

osm_id	osm_type	latitude	longitud
11365499897	nodes	1.16839	110.565
8170174121	nodes	1.20592	110.535
2427779632	nodes	1.23443	111.462

Fig. 5. Shows an example of the results of coordinates calculated using the "Calculate Geometry" tool.

into a database (Fig. 5), a new field "geom" data type "GEOMETRY" are required for spatial query purposes. In this case, perform adding spatial index into the database; below are the SQL commands for adding spatial index.

```
ALTER TABLE healthcare_data_updated_1
ADD SPATIAL INDEX (geom);
```

Finally, the system was tested and optimized through model fine-tuning and a comparative analysis of the performance between GPT-3.5-turbo and GPT-4. The application

osm_id	osm_type	latitude	longitud	geom
11,365,499,897.0	nodes	1.16839	110.5650024	æ□ □□ VŠÓý(¤[@6"o¶¹±ò?
8,170,174,121.0	nodes	1.20592	110.5350037	æ□ □□ L¯(€=¢[@%z□ÅrKó?

Fig. 6. Shows the results after running spatial query in MariaDB.

leverages the OpenAI GPT-3.5- turbo model to provide symptom-based medical suggestions. It integrates with a MariaDB database to retrieve relevant healthcare location data within a specified radius of the user's location. This is accomplished through Python Flask for the web framework, OpenAI for AI-based recommendations, and MariaDB for database management.

3.2 Applications Initialisation and Configuration (Backend)

Based on the Fig. 6 flowchart, the first step of the integration is to set up Python Flask and import the OpenAI library. Setup Python Flask is a micro web framework used to build web applications while the OpenAI library is imported to access OpenAI's GPT-3.5-turbo model for generating medical suggestions. For database connection functions, it establishes a connection to the MariaDB database using the provided credentials, and it also handles connection errors and returns a connection object if successful. Next, the main ingredient recipe in this study, the routing and request handling phase, contains five routes: the index route, get treatment route, database query for healthcare locations, formatting and returning the response and symptom-based filtering (amenity filter function) (Fig. 7).

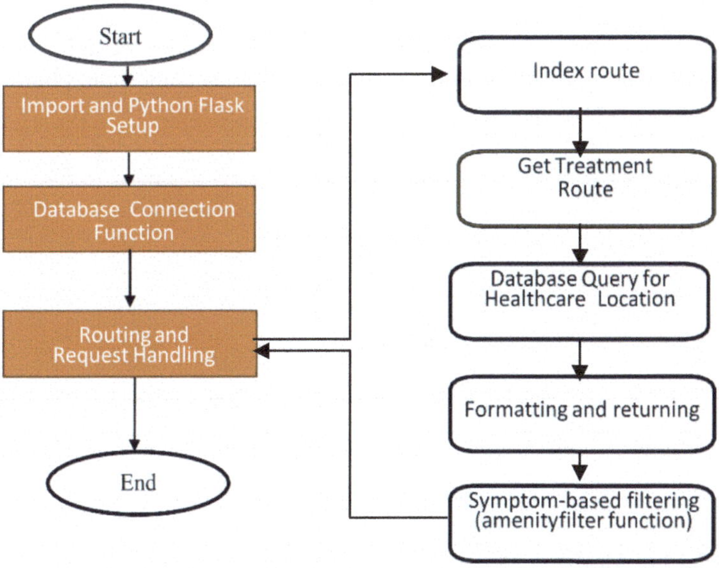

Fig. 7. Illustrates the flowchart of the integration of the application.

4 User Interface Overview

Figure 8 illustrates the user interface of the Discovery Healthcare GeoAI-based platform, which displays a map with healthcare facility makers and a chat interface where users input symptoms to receive AI-generated health advice.

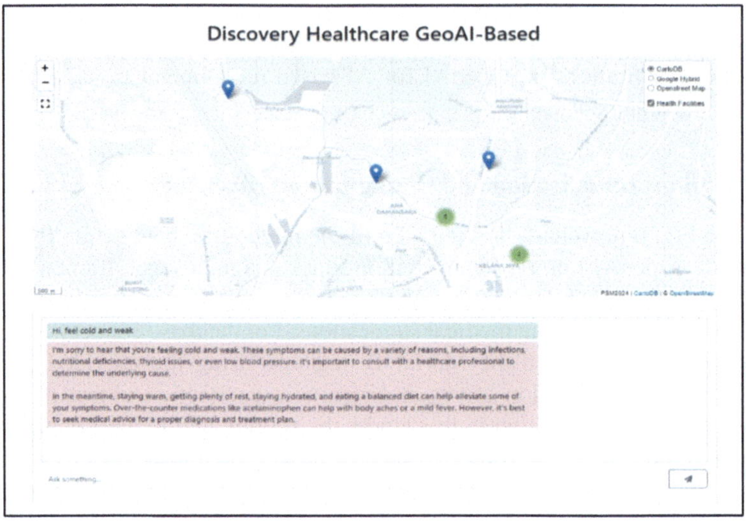

Fig. 8. Shows the interface of this study.

Figure 9 shows the recommendations of relevant healthcare locations based on the user's location, listing nearby healthcare facilities such as clinics and pharmacies that match the symptoms provided.

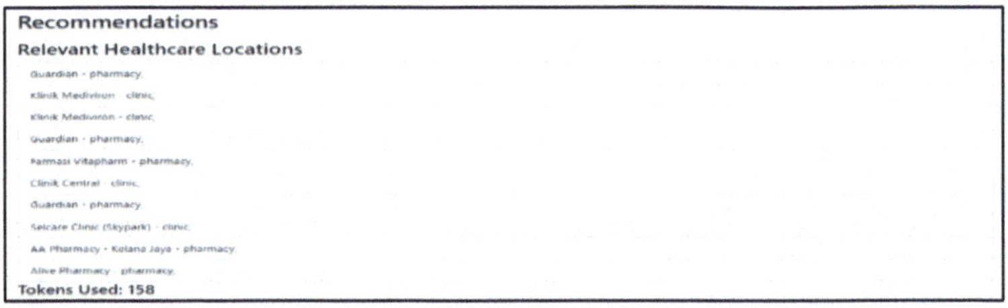

Fig. 9. Shows recommendations for relevant healthcare locations based on users' location.

4.1 Fine-Tuning Model

Fine-tuning is critical in machine learning, particularly for enhancing pre-trained language models. Fine-tuning involves taking a model already trained on a broad dataset and training it on a more specific dataset tailored to a particular task or domain. This process allows the model to adapt its knowledge and improve its performance in specific applications. Pre-trained models such as OpenAI's GPT-3.5-turbo and GPT-4 are initially trained on diverse datasets encompassing various topics. This extensive training enables the models to understand and generate human-like text across multiple contexts. However, these general-purpose models may only sometimes perform optimally for specialised tasks. Fine-tuning addresses this limitation by refining the model's abilities using a focused dataset relevant to a specific application. During the fine-tuning process, the pre-trained model is exposed to additional data, including examples and instructions related to the target task. This targeted exposure helps the model better understand the nuances and requirements of the specific task, enhancing its ability to generate more accurate and contextually appropriate responses. Upload the data in a JSON file (fine_tune_chat_data.json) to apply fine-tuning to the the OpenAI API. OpenAI's machine learning system will then perform the fine-tuning. Once the fine-tuning process is completed, the status will be marked as 'succeeded,' and the model's name will be provided. For example, the model's name for my fine-tuned model is ft.5- turbo-0125: personal:psm2023:9Z9bu0Da.

Fig. 10. It shows the process of uploading data (fine_tune_chat_data_jsnol) and fine-tuning using Open API.

To implement the fine-tuned model coding at the backend (app.py). First, define the parameter model (Fig. 10).

```
use_fine_tuned =data. Get ("use_fine_tuned", False)
```

If the `use_fine_tuned` condition is False, the GPT-3.5-turbo model will be activated. If the `use_fine_tuned` condition is True, the ft:gpt-3.5-turbo-0125:personal:psm2023:9Z9bu0Da fine-tuning model will be activated.

4.2 LLM Comparison: GPT3.5-turbo vs GPT 4

Recent advancements in natural language processing have significantly enhanced the capabilities of LLMs in understanding and generating human-like text across diverse domains, including healthcare. In analysing medical symptoms, this study investigates and compares the performance of two leading LLMs, GPT-3.5 Turbo and GPT-4. The evaluation focuses on their ability to provide accurate and contextually relevant responses to specific clinical prompts such as toothache, high fever with a headache, stomach- ache and nausea, cold and cough symptoms, back pain, red and itchy rash on limbs, eye pain and irritation, chest pain, deep breathing, difficulty breathing, and joint pain and elbows. By assessing these models, this study aims to elucidate their effectiveness in suggesting diagnoses, treatment options, and medical advice, thus contributing to advancing AI-driven healthcare applications and informing future developments in medical decision support systems. Below here are the example user prompts that will be used as the symptom analysis to compare using GPT-3.5-turbo and GPT 4:

> I'm experiencing a high fever and have had a throbbing headache for the past 24 h. (fever and headache)

The token usage for each generated response is also indicated, as the maximum token limit is 150. This limit is established to control the response length, manage costs, optimise performance, maintain relevance and focus, and efficiently manage resources (Table 1).

5 Findings Between GPT-3.5 and GPT-4

The analysis of AI responses to medical symptom descriptions reveals that GPT-4 provides more detailed, comprehensive, accurate, and context-specific responses compared to GPT-3.5-turbo, albeit with higher token usage and computational demands, indicating that while GPT-4 is better suited for tasks requiring higher detail and specificity, practical deployment must consider application-specific requirements and resource limitations.

6 Conclusion

This study successfully developed a GeoAI-augmented web platform to recommend healthcare facilities based on user-reported minor illnesses. By integrating a fine-tuned GPT-3.5 model and utilising (OSM) data, the platform addresses the limitations of existing healthcare facility locators, which often prioritise geographic proximity over specialised medical services. The system enhances user interaction and provides tailored facility suggestions based on specific conditions, significantly improving the user experience and the accuracy of healthcare recommendations. The comparative analysis of the GPT-3.5-turbo and GPT-4 models highlighted several critical differences in their performance. GPT-4 consistently produced more detailed and comprehensive responses than GPT-3.5-turbo, enhancing the quality and relevance of healthcare recommendations. For instance, GPT-3.5-turbo suggested primary dental care and over-the-counter pain

Table 1. The comparison of the Assistance prompt, also known as AI response symptoms, uses GPT-3.5-Turbo and GPT-4 for "fever and headache" symptoms.

Symptoms and token usage	GPT-3.5-turbo	GPT-4
Fever and headache	Your symptoms of a high fever and throbbing headache could indicate that you have an infection, such as the flu or a viral illness. I recommend you take some over-the-counter pain relievers such as acetaminophen or ibuprofen to help reduce the fever and alleviate the headache. Stay hydrated, get plenty of rest, and consider contacting a healthcare provider for further evaluation and treatment if your symptoms persist or worsen Tokens used: 141	I'm sorry to hear that you're not feeling well. Your symptoms may suggest a variety of conditions, including flu, sinusitis, or possibly something more serious like meningitis. You must get medical help immediately, as these conditions require proper medical treatment. Over-the-counter fever reducers and pain relievers might help you feel slightly better in the immediate term, but they will not treat the underlying issue. Hydration is also essential. Remember, this advice does not replace a doctor's consultation Token used: 142

relief for toothache with 194 tokens. In comparison, GPT-4 included a broader range of remedies and emphasised the need for professional care with 196 tokens. As an impact, this pattern was observed across various symptoms, indicating that GPT-4's enhanced capabilities can lead to better user satisfaction and more accurate recommendations. However, the increased token usage of GPT-4 requires careful management to adhere to application constraints, such as token limits in the system.

6.1 Implications and Future Work

The developed platform's ability to provide contextually relevant healthcare recommendations significantly advances health informatics. By leveraging fine-tuned AI models and integrating detailed geographic data, the tool enhances the accuracy and relevance of healthcare facility locators to match users' specific conditions and locations. Future research should optimise GPT-4's token efficiency, explore additional fine-tuning techniques, and expand the platform to include more complex medical conditions and user feedback mechanisms. In conclusion, our GeoAI-augmented web platform represents a significant step forward in personalised healthcare recommendations, improving precision and relevance by combining advanced AI models with precise geo-location data. The comparative analysis of GPT-3.5 and GPT-4 highlights the importance of ongoing AI advancements for better digital health services.

References

1. Bohr, A., Memarzadeh, K.: The rise of artificial intelligence in healthcare applications. In Artificial Intelligence in Healthcare, pp. 25–60 (2020)
2. Fansi Tchango, A., et al.: DDXPlus: A new dataset for automatic medical diagnosis. Adv. Neural Inf. Process. Syst. **35**, 31306–31318. Available at: https://figshare.com/articles/dataset/DDXPlus_Dataset/20043374 (2022). Accessed 31 Dec 2023
3. HT Mushonga F Banda A Mulolwa 2017 Developing a web-based GIS for health facilities mapping, monitoring and reporting: a Zambian Ministry of health case study South Afr. J. Geomats. 6 3 321
4. Zhou B., Yang, G., Shi, Z., Ma, S.: Natural Language Processing for Smart Healthcare. Available at: http://arxiv.org/abs/2110.15803 (2021). Accessed 1 Jan 2024
5. Luo, H., Li, S.W., Glass, J.: Knowledge of conversational Symptom Detection with Graph Memory Networks. Available at: http://arxiv.org/abs/2101.09773 (2021). Accessed 5 Jan 2024
6. Devlin, J., et al.: BERT: Pre-training of Deep Bidirectional Transformers for Language Understanding, pp. 4171–4186. Available at: https://github.com/tensorflow/tensor2tensor. Accessed 1 Jan 2024
7. T Nguyen-Mau AC Le DH Pham VN Huynh 2024 An information fusion-based approach to context-based fine-tuning of GPT models Inf. Fusion 104 102202
8. DD Miller EW Brown 2018 Artificial intelligence in medical practice: the question to the answer? Am. J. Med. 131 2 129 133
9. Taloni, A., et al.: Comparative performance of humans versus GPT-4.0 and GPT-3.5 in the self-assessment program of American Academy of Ophthalmology. Sci. Rep. **13**(1) (2023).
10. Wu, C., Varghese, A.J., Oommen, V., Karniadakis, G.E.: GPT vs Human for Scientific Reviews: A Dual Source Review on Applications of ChatGPT in Science. Available at: https://arxiv.org/abs/2312.03769v1 (2023). Accessed 6 Feb 2024

Machine Learning for Semi-Automated Land Parcel Plotting

Muhammad Ghaly Kurniawan[1,2], Malumbo Chipofya[1](✉), and Dimo Todorovski[1]

[1] Faculty ITC, University of Twente, Hallenweg 8, 7522 NH Enschede, The Netherlands
m.c.chipofya@utwente.nl

[2] Ministry of Agrarian Affairs and Spatial Planning/National Land Agency, Jakarta, Indonesia

Abstract. The Indonesian cadastral quality improvement process, established in 2018, faces huge challenges due to its huge land area, diverse natural and social conditions, and historical mapping practices. With 15 million land parcels yet to be plotted accurately, in this work we propose a solution based on machine learning to speed up the process of plotting parcels documented on old land certificates and maps. Our approach is a heuristic-driven geospatial data matching procedure that semi-automatically searches for possible locations for plotting land parcels. The heuristic aspect of the approach is its basis in the manual plotting method in use presently in Indonesia. We identified eight causes of unplotted parcels and for two of these causes we have implemented an optimization algorithm model defined over five geometric parcel attributes. The model is then optimized using the RCGA algorithm. The model performed with recall rates of 88.3% and 57% on test data and real data respectively, and precision rates of 98.8% and 91% also respectively for test data and real data. After the incorporation of text information in old cadastral maps, the recall value for real data improved to 91%.

Keywords: Land administration · Cadastral quality improvement · Unplotted land parcels · RCGA optimization algorithm · Machine learning · Spatial data matching

1 Introduction

Indonesia is one of the countries that is working on cadastral quality improvements. Cadastral quality improvement in Indonesia started in 2018 in the city of Surakarta and continues to other major cities such as Jakarta, Batam, Pontianak, Bali, and Surabaya. The large land area, the diversity of its nature and social structure, and historical mapping practices and legal precedents bring challenges in the replotting process of old cadastral maps in Indonesia. There are approximately 126 million land parcels in Indonesia, and 100 million of them are already mapped in the GIS database. From that number, there are approximately 21 million parcels that are not yet plotted in the correct location [1]. To plot that huge number of parcels, using a manual searching method is not feasible in the short- to near-term. Meanwhile, delaying the time of plotting creates the potential for land conflicts. Such conflicts can escalate into violent conflict if not well managed and anticipated [2].

A. Abdul-Rahman et al. (Eds.): GeoWeek 2024, LNGC, pp. 95–111, 2025.
https://doi.org/10.1007/978-3-031-86654-8_8

In this work, we propose a solution based on machine learning to speed up the process of plotting parcels documented on old land certificates and cadastral maps. The challenges of this task can be categorized into technical challenges and legal ones. The legal challenges appear as a result of previous regulations that allowed land certificates to be issued without the maps on them. Government Regulation 10/1961 of the Republic of Indonesia [3] stated that if the measurement letter (Surat Ukur) cannot be produced for some reason, a replacement certificate can be issued that has the same function as the certificate. It means that those types of certificates have the same legal force as the normal ones and also need to be plotted in the current cadastral database. The big question then arises of how to plot parcels for which the corresponding land certificates have no maps or spatial data attached to them.

The technical challenges arise in cases where geospatial information of some sort is attached to the land certificate but for which either the reference system cannot be identified or a local coordinate system was used or, in the worst case, only spatial references of a qualitative nature are included (as in metes and bounds) [4]. In the common case, this occurs due to the limited availability of national control points (Titik Dasar Teknik) and limitations in surveying technology at the time when these maps were produced covering the period from around 1960 and in some areas up until now [5].

2 Literature Review

The work presented in this paper relies on prior research on the cadastral quality improvement process, the use of machine learning in the cadastral improvement process, and optimization algorithms in machine learning. Due to space limitations, we will limit our review of the literature to those approaches most relevant to our results.

2.1 Cadastral Quality Improvement Process

The various data acquisition techniques used in cadastral mapping make the combination of old datasets with new datasets challenging. The process of digitizing and plotting old cadastre data into modern cadastral databases has been done in Spain [6], Turkey [7] Malaysia [8], and Israel [9] using various methods and techniques. From those activities, [10] defines the importance of having accurate documentation of the past product in boosting the success of the process.

The cadastral quality improvement process was also done in Indonesia by first converting the old analog data into digital format [11]. A mobile application was later used in the process to accommodate public participation which resulted in the validation of 5970 certificates in three months [12]. The process continues in some major cities' pilot-project stages of cadastral quality improvement. In the cadastral quality improvement process in Indonesia, land parcels are classified based on document availability and quality [13]. The process focuses on certificates produced in the past but not plotted yet in the current database due to limited spatial or textual data.

2.2 Machine Learning on the Cadastral Quality Improvement Process

Technological innovation has proven to create an efficient process in land administration, including the cadastral quality improvement process [14]. One of the innovations is the automatic parcel boundary extraction from satellite images using deep learning-based models such as Convolutional Neural Networks and Fully Convolutional Networks [15–17]. In addition, the extraction of textual information automatically from analog data using character recognition has been used for parcel identification [18, 19]. A recent work used several optimization algorithms including Random Forest and Genetic Algorithms to combine the cadastral maps with other maps as a way to improve the quality of cadastral data in Iran [20].

2.3 Optimization Algorithm on Machine Learning

The optimization algorithm is one of the common processes in machine learning to obtain the optimal solution for a certain problem [21]. Various functions such as model-free, gradient-based, and metaheuristic algorithms are available for model optimization. A proper analysis of the problem is necessary to correctly select the appropriate optimization function [22].

Our approach uses a real-coded genetic algorithm (RCGA) [23] to optimize the contribution of spatial attributes to the determination of the optimal matches of unmapped parcels to ground locations as determined by the current cadastral dataset. Our formulation of the problem as detailed in the methodology below lends itself to a solution using RCGA optimization: 5 weight variables give 5 dimensional search space but the specification of the optimization remains simple using RCGA as compared to e.g. Random Forest or Simulated Annealing.

Genetic algorithms are metaheuristic optimization algorithms inspired by a natural selection process, including the mutation and selection of the fittest traits of a solution method [23]. The strength of this algorithm in finding a solution from a complex search space makes it broadly applicable [24]. In particular, genetic algorithms have shown reliable performance in weighting problem domain attributes in a variety of applications [25, 26].

3 Methodology

The research used two methods in general: qualitative data analysis on data obtained via interviews informed the choice of heuristic cues used in the design of the machine learning model and quantitative data analysis was used to evaluate the performance of the machine learning model.

3.1 Semi-Structured Interview

Semi-structured expert interviews were conducted during the first phase of the data collection to gather information related to the cause of the land parcel being unplotted and the current process of plotting the land parcel. The ability to set an open-ended question with the opportunity to add some questions during the interview enables a deeper analysis of information related to certain problems [27]. The interview process was held in four Indonesian cities—Jakarta, Yogyakarta, Surabaya, and Serang—and involved 11 respondents with various roles within the cadastral mapping domain in Indonesia. The set of roles covered by the respondents group covered research, survey technicians, surveyors, and managers to ensure that comprehensive insight could be gained into the current cadastral quality improvement process in Indonesia.

3.2 Geospatial Data Matching Using Geometric Variables

Geospatial data matching was done using five geometric variables: polygon's area (A), number of vertices (n), perimeter (P), minimum bounding rectangle (MBR), and the Arkin Graph Angle (AGA). The formulas to calculate each of these variables, which due to space limitations will not be detailed in this paper, can be found in [28–30]. The geometric matching for each variable was calculated using Eqs. (1), (2), (3), (4), and (5) below respectively, before the weighting was optimized.

$$A_{match} = 1 - \frac{|A_1 - A_2|}{A_1 + A_2} \tag{1}$$

$$n_{match} = 1 - \frac{|n_1 - n_2|}{n_1 + n_2} \tag{2}$$

$$P_{match} = 1 - \frac{|P_1 - P_2|}{P_1 + P_2} \tag{3}$$

$$MBR_{match} = \frac{MBR_1 \cap MBR_2}{MBR_1 \cup MBR_2} \tag{4}$$

$$AGA_{match} = \frac{min(AGA_1, AGA_2)}{max(AGA_1, AGA_2)} \tag{5}$$

The matching values range from 0 to 1, with 0 for no match and 1 for a perfect match. The geometric matching is defined as the overall matching calculated by multiplying each individual matching value with the corresponding weight as shown in Eq. (6).

$$OM = \omega_1.A_{match} + \omega_2.n_{match} + \omega_3.P_{match} + \omega_4.MBR_{match} + \omega_5.AGA_{match} \tag{6}$$

where ω_i for $i \in [1, 5]$ is the weight for the ith matching variable.

3.3 Weight Optimization Based on RCGA Algorithm

The role of the weights in Eq. (6) is to balance out the influence of the different matching values so as to maximize the number of correctly matched parcel to ground location pairs. Below we describe the steps of the RCGA algorithm to calculate the geometric matching between the unplotted land parcels and the location in the cadastre database as used in our approach.

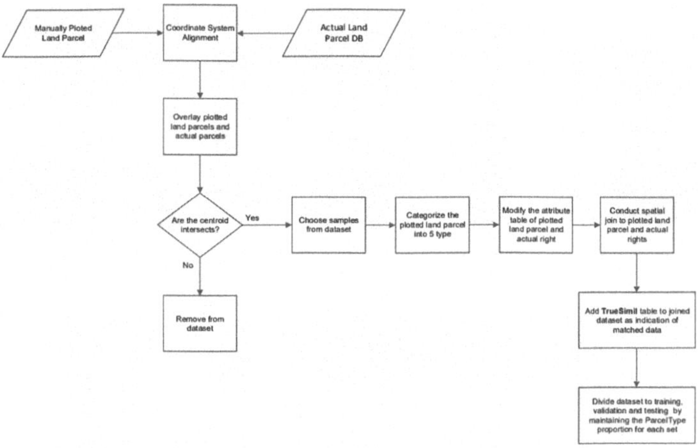

Fig. 1. Dataset Preparation Workflow

3.3.1 Dataset Preparation

The sample dataset for training was prepared with several processes of manual pairing. As illustrated in Fig. 1, the pairing process was done between the unplotted land parcel and the actual right on the database based on the spatial join and the parcel's identification number similarity. The label of TrueSimil was then added to each pair indicating a match to be used in the training phase.

The data was categorized into 5 classes in which the proportion will be maintained in the training, validation, and testing set to ensure a representative contribution of each condition to the model. The 5 categories are described in Fig. 2. This strategy is meant to prevent the model from learning only from good examples with very similarly shaped pairs by ensuring that the other conditions are included in the model learning phase as well. There were 1199 parcels used in the Sample dataset with Category 2 being the category with the largest proportion with a percentage of 44.9%. This was followed by category 1 (38.1%), category 3 (11.3%), category 4 (5.3%), and category 5 with 0.4%.

3.3.2 Define the Objective Function

The objective function used in our approach is given by Eq. (7). It computes the mean square error (MSE) between the true similarities of parcels (set to 1, denoted TrueSimil) and the predicted matches (denoted PredMatch) for each pair calculated using the weight vector $\omega = (\omega_1, \omega_2, \omega_3, \omega_4, \omega_5)$.

$$ObjectiveFunction = \frac{1}{n}\Sigma_{i=0}^{n-1}(TrueSimil_i - PredMatch_i)^2 \tag{7}$$

where n is the number of parcels in the training/validation/testing set;

$TrueSimil_i$ is the true similarity for parcel pair ɩ, initially set to 1;

$PredMatch_i(\omega)$ is the predicted similarity for parcel pair i, calculated using the weights ω.

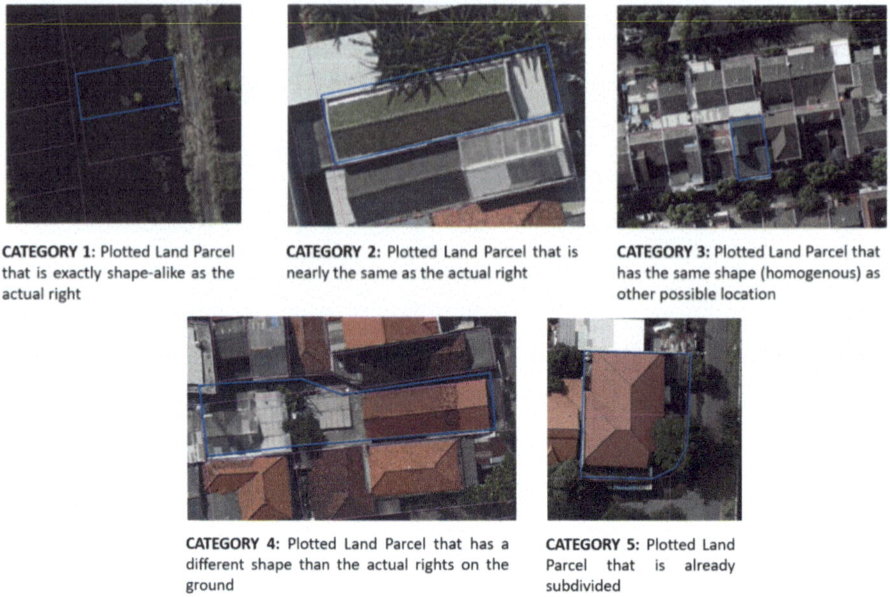

Fig. 2. Categorization of the sample data

3.3.3 RCGA Model for Weight Optimization

Weight optimization works by searching for a weight vector that minimizes the objective function. The RCGA algorithm involves the mutation and crossover operators from a selected population to find the generations of weights with better solutions [31]. The process iterates until the desired fitness score (calculated using the inverse of the objective function) is reached, resulting in a fine-tuned weight for each geometric variable component. Figure 3. Explains the process of RCGA calculation in finding the value of fine-tuned weight.

3.4 Geospatial Data Matching Using Textual Attribute Matching

Textual matching was added to the calculation to enhance the model performance in finding the location to plot the land parcels. This technique was applied to the top three candidate locations with the highest geometric match score, preventing a longer calculation time. The attribute used in this research was the village name information, surrounding parcels NIB (parcel identification number), and the registered area information. In our experiments, to protect the privacy of land ownership information, the NIB numbers were mapped to randomized numbers and ownership data was completely removed from the records.

3.5 Model Performance Evaluation

Model performance was evaluated using standard Precision and Recall measures. Precision is calculated by dividing the correct matched location (denoted as matched) with the

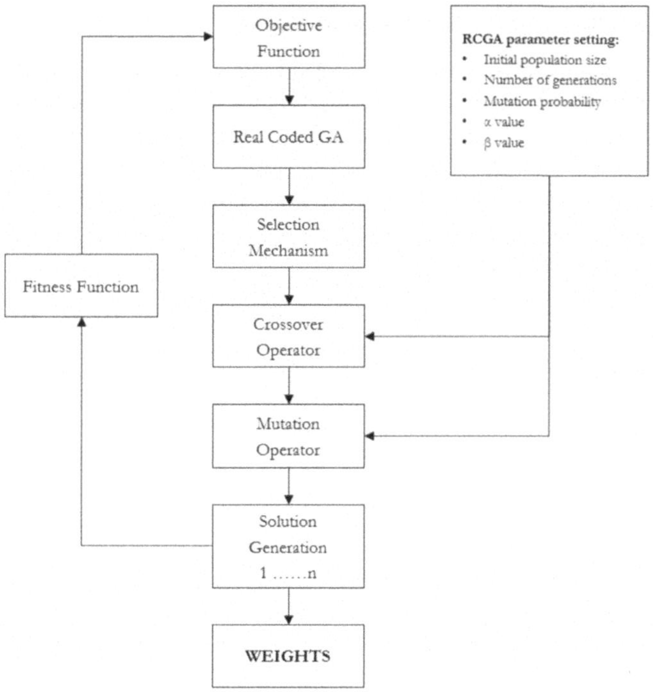

Fig. 3. RCGA Optimization Workflow (modified from [30])

addition of the matched location and the wrong location prediction (denoted as wrong). It represents the accuracy of the matches identified by the model. Recall value is calculated with an almost similar formula, but changing the wrong location components with the unmatched prediction. Unmatched prediction is a situation where the model cannot differentiate the matches between several candidates, representing the model's ability to identify all relevant matches [32]. An F1 score was added to understand the correlation between those two values, representing the balance between precision and recall.

$$Precision = \frac{correct}{correct + wrong} \tag{8}$$

$$Recall = \frac{corect}{correct + unmatched} \tag{9}$$

$$F1 = \frac{2(precision * recall)}{precision + recall} \tag{10}$$

4 Results

The results section will be divided into the results from the open-ended interviews that creates a model design and the quantitative analysis of the model calculation in finding the best-fitting location for the land parcels.

4.1 Current Method of Plotting the Land Parcels

Two methods are gathered from the open-ended interviews that were used in the current plotting process: the Studio Process and the Field Verification. Showed in Fig. 4, the studio process involves a combination of attribute filtering and manual filtering based on the parcel's shape and area. This process inspires the development of the machine learning model by combining the geometric aspect of parcels and the attribute of the land parcels.

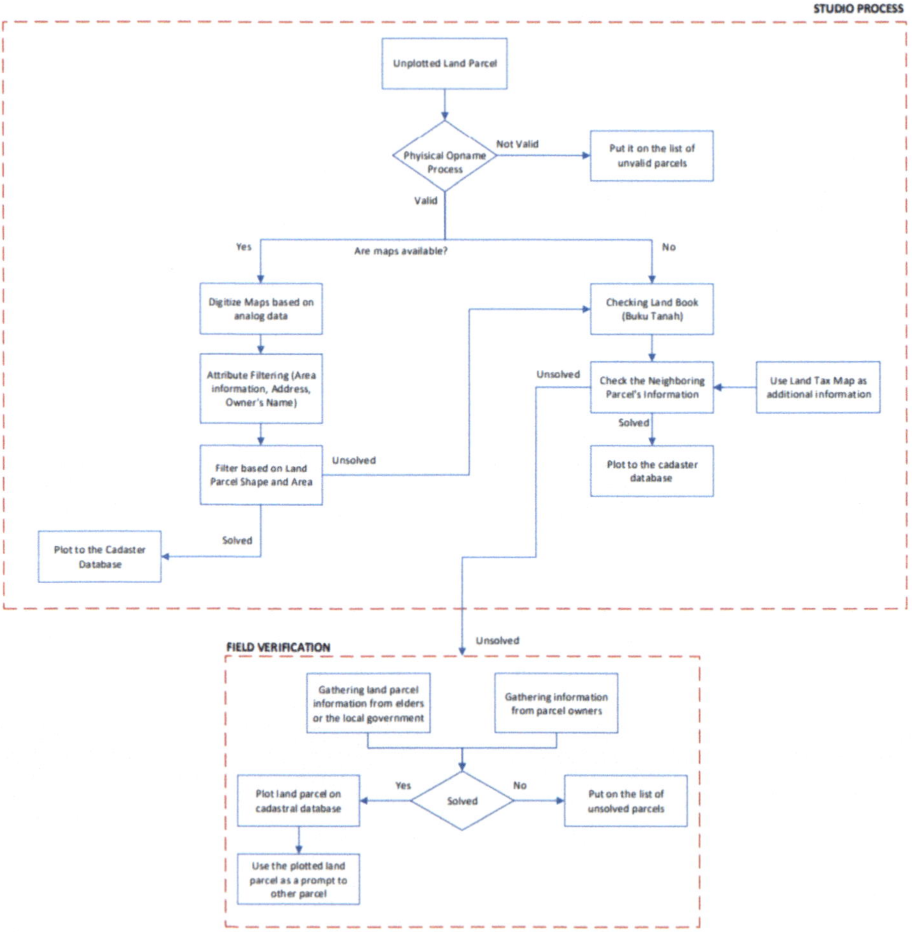

Fig. 4. Current Methods of Parcel's Plotting

4.2 Design of the Machine Learning Model to Plot the Land Parcel

The machine learning model design was inspired by the studio process of the plotting process that was done in the cadastral quality improvement process. It involves shape matching combined with attribute matching using the information available in the existing documents (Fig. 5). One respondent highlighted "*We are doing a one-by-one checking based on the visual shape of the land parcel.*" The other respondent noted the use of attribute matching in the current process, "*We are looking into the legal documents such as the land book to find the information about the registered area, the owners, and the parcels's identification number (NIB).*"

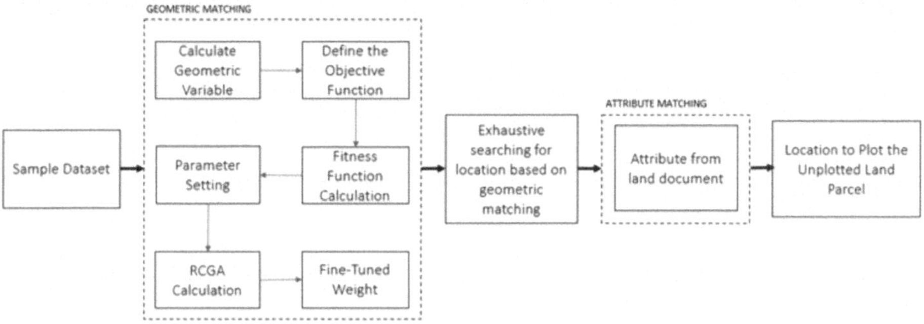

Fig. 5. Design of The Machine Learning Model

4.3 RCGA Based Weight Optimization

The output of RCGA calculations was the fine-tuned weight for each geometric variable. Before getting into the result, the parameter related to the RCGA needs to be optimized to find a better solution. The first parameter was the chromosome length which was set into a value of 5 in line with the number of variables used. The next parameter was the α value which was set to 0.5 to ensure the same probability of exploration and exploitation, preventing early convergence while keeping the possibility of searching for a better solution. The next parameter is the number of generations that are evaluated based on the fitness score over the generations.

From the experiment depicted in Fig. 6, it can be obtained that 240 generations were optimal to generate the highest fitness value. Based on this value, the initial population size was determined based on the ratio of 1/5 as suggested by previous works [30, 33]. The last was the mutation-related parameter (β value and mutation probability) which was determined using the same strategies by testing all possible combinations that generate the highest fitness value. The optimum value of the RCGA Parameter is listed in Table 1.

Based on the optimum parameter and the RCGA calculation, the fine-tuned weight could be obtained as listed in Table 2. The number of vertices (n) became the parameter with the highest weight indicating its importance in determining the geometric match, followed by the MBR, perimeter, polygon area, and the AGA.

Table 1. Optimum Value of The RCGA Parameter

RCGA Parameter	Optimum Value
Initial Population Size	48
Number of Generations	240
Chromosome Length	5
Mutation Probability	0.01
α value	0.5
β value	0.3

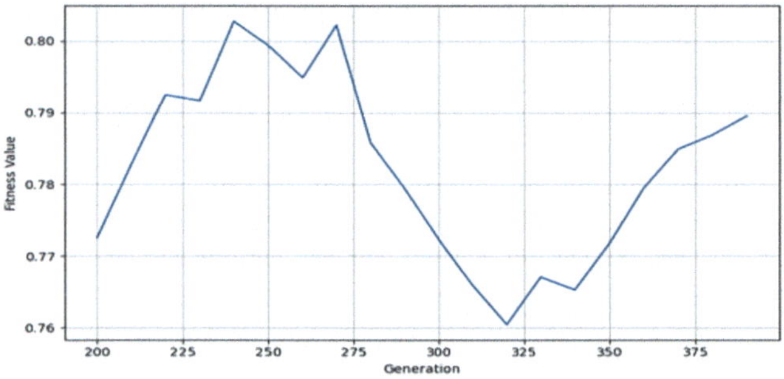

Fig. 6. Fitness Values Over Generations

Table 2. Fine-Tuned Geometric Variable Weight

Geometric Variable	Weight
Polygon's Area (A)	0.139
Number of Vertices (n)	0.304
Perimeter (P)	0.197
Minimum Bounding Rectangle (MBR)	0.222
Arkin Graph Angle (AGA)	0.138

4.4 Performance of Geometric Matching

Using the weight, the initial matching based on geometric variables was done. It was tested in two different databases: The Sample Data and The Indonesian Cadastral Database (The KKP Database), obtaining a wider analysis of the model performance. We also tested two different variable combinations: 4 variables (A, n, P, MBR) and 5 variables (A, n, P, MBR, AGA). For the first test, the model had a satisfying performance with a precision value of 98.75% and a recall value of 88.27%. Using 4 variables did not result in a significant difference in the sample dataset. The second test was done to search parcel locations in the KKP Database, allowing us to test the model performance against a real cadastral dataset. There was no significant difference in the precision value with a satisfying result of above 90% precision value on both combinations. However, there was a significant drop in the recall value due to overlapping rights and the homogenous shape of candidate locations. There was also a notable difference between the use of 4 variables and the 5 variables with an 8% gap (Table 3).

Table 3. The result of Geometric Matching

Tested on	Correct	Wrong	Unmatched	Precision	Recall	F1-Score
Sample Data (4 variables)	155	3	23	98.10%	87.08%	0.92
Sample Data (5 variables)	158	2	21	98.75%	88.27%	0.93
The KKP Database (4 variables)	27	2	28	93%	49%	0.64
The KKP Database (5 variables)	31	3	23	91%	57%	0.70

4.5 Performance of Attribute Matching

To enhance the model's performance from the previous matching strategy, textual attribute matching was applied with the parcels tested in the KKP Database. From three matching techniques, we found that the use of a unique identifier such as the parcel's registered area (i.e. which is similar to neighbourhood name or the British concept of borough) generates the highest performance. Compared with other attributes that contain a redundant record, the last strategy is proven to give the best recall value of 91%. The neighbouring NIB had the second best performance with a recall value of 87%, enabling the elimination of candidate locations with a homogenous shape. The last strategy of using the village name gives the lowest result but still significantly improves the recall value compared with the result before the attribute matching (Table 4).

5 Discussion

5.1 Learning from the Present Data

The satisfying performance of the model still has some aspects to be improved. First, it still needs to differentiate candidate locations with similar shapes. Using only geometric matching proved that the matching between those locations could not be detected resulting in 21 land parcels with unmatched results from the Sample Dataset.

As illustrated in Fig. 7. The unplotted with similar-shape candidate location gave more than one result with the highest matching score. The other matching strategy - attribute matching based on unique attributes- is needed to filter those locations into

Table 4. The result of Attribute Matching

Matching Strategy	Correct	Wrong	Unmatched	Precision	Recall	F1-Score
Before Attribute Matching	31	3	23	91%	57%	0.70
"Village Name" Matching	43	3	11	93%	80%	0.86
"The Neighboring NIB" Matching	47	3	7	94%	87%	0.90
"Registered Area" Matching	48	4	5	92%	91%	0.91

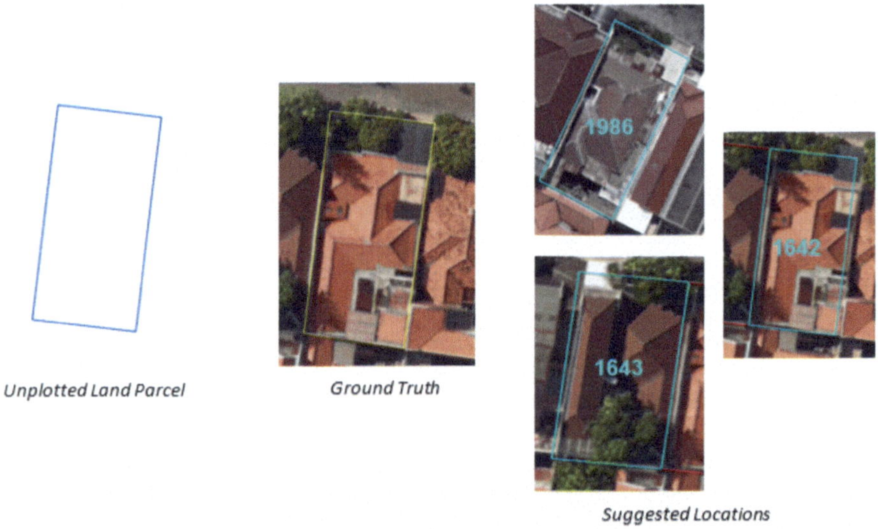

Fig. 7. Unmatched Results on the Similar-Shape Location

the selected locations. The other condition in the cadastral database that generated the unmatched result was the overlapping right on the ground, which also occurred due to the candidate locations having identical shapes.

The other example that affected the model performance was predicting the land parcels that have been subdivided in the present time as presented in Fig. 8. Due to the unavailability of locations that geometrically matched, the model will suggest another location, which is incorrect, resulting in a wrong-match result.

5.2 Towards the Future

To implement the model in the current cadastre system, several factors need to be considered including the justification for selecting geometric variables based on the cadastre data condition.

Unplotted Land Parcel *Ground Truth* *Suggested Location*

Fig. 8. Wrong-Matched Results on the Subdivided Parcel

5.2.1 Selected Geometric Matching Variable

To effectively run the phase of geometric matching, which takes the longest time, the matching variables need to be chosen wisely based on the problem's complexity. Table 5. Shows the comparison between a 4 variables and 5 variables combination based on F1-Score, its ability to minimize unmatched results, and the processing time.

Table 5. The result of Attribute Matching

Considered Factor	4 Variables Combination (n, MBR, P, A)	5 Variables Combination (n, MBR, P, A, AGA)
F1-Score on the Sample Data	0.92	0.93 (+1%)
F1-Score on the KKP Database	0.64	0.70 (+9%)
Unmatched Result on the Sample Data (parcels)	23	21 (−8%)
Unmatched Result on the KKP Database (parcels)	28	23 (−17%)
Average Processing Time (s)	1.62 (−18%)	1.98

Despite having higher average calculation time, the 5 variables combination was chosen to accurately search for the location in the cadaster database with complex conditions. The inclusion of the AGA variable enables the detection of matching on two geospatial data at different scales. Based on this, using the 5 variables combination in a complex database with the possibility of having different scales between unplotted parcels and the candidate location is recommended, while the 4 variables combination can be used for minimizing the searching time when the search space is particularly large.

5.2.2 Factors Related to the Model Implementation

Based on the experiments using two different datasets, several factors need to be considered: Firstly, the quality of digitized land parcels and the attributes of land parcels which are gathered from a manual input process prone to human error, have strong influence on the reliability of the matching result because the accuracy of the model relies on the accuracy of the data entered into the calculations.

Secondly, The characteristics of the candidate's location, including the accuracy of the location within the current cadastral database or the unavailability of the digitized possible location (i.e. the entire region is not mapped), also affects the model's accuracy. Because the model is searching for a match among the available polygons, it is important to have the accurate location in the database. Figure 9. Shows the effect of not having a good candidate location on the model's prediction.

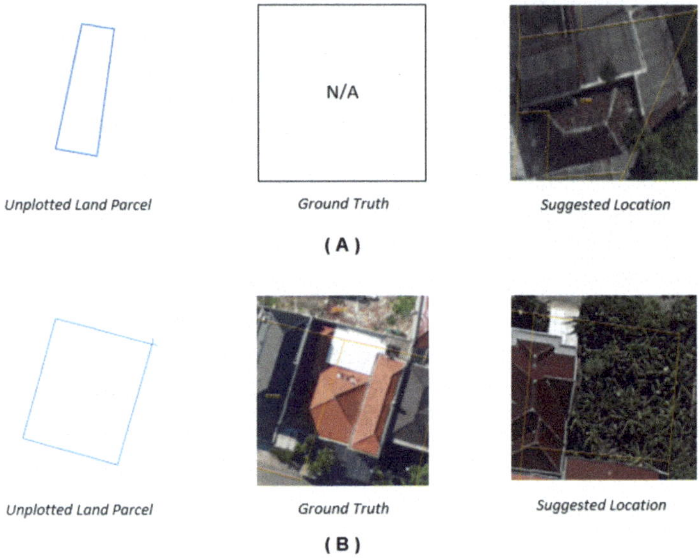

Fig. 9. Suggested location on (A) not available candidate location and (B) the candidate location that doesn't reflect the real conditions.

The last factor is the coverage of the cadastral database: the model was trained using the data from Surabaya, which is categorized as a big city with various land uses such as housing, industrial area, and commercial. The performance of the model in other types of cities with different land use conditions is still unknown. The assumption is that if the model works well in a city with complex land use, it will work in a smaller city with a simpler land use configuration. Further research is required in this area.

Despite the factors that affects the model's accuracy, it is important to have adequate computational resources, especially when dealing with large datasets.

6 Conclusion

The process of plotting old land parcels done in the Cadastral Quality Improvement Process could be accelerated using machine learning-based geospatial data matching. Using two matching strategies: geometric and attribute, the best-fitting location for plotting a land parcel will be determined based on the similarity between the unplotted land parcel and the available location on the parcels database.

In the geometric matching process, an RCGA optimization algorithm was used to find the weight for each geometric matching variable. The result of geometric matching was tested in two databases: Sample Dataset and the KKP Database. The model can accurately predict the location in both databases with precision values of 98% and 91% respectively. There are improvements needed especially in the testing phase using the KKP Database to differentiate matching in the homogenous shape candidate location and overlapping rights. The attribute matching was used based on three parcel attributes with the "registered area" giving the best result. It improves the recall value from 57% to 91%, showing the ability of the model to detect matches in the real condition of cadastre data.

Because the initial search was based on the geometric aspect, the model has some limitations on determining the parcels that have already been subdivided in the present time. The model also struggles to find the location in which the polygon is not available on the database since the matching calculation was based on the vector operation. The future improvement of the model should involve testing the model with data from different cities to understand its performance under various data conditions.

References

1. Ministry of ATR/BPN.: Rapat Kerja Nasional. Jakarta (2023)
2. LJ Alston GD Libecap B Mueller 2000 Land reform policies, the sources of violent conflict, and implications for deforestation in the Brazilian Amazon J. Environ. Econ. Manage 39 162 188 https://doi.org/10.1006/jeem.1999.1103
3. The Government of Republic Indonesia.: Government Regulation 10/1961. https://peraturan.bpk.go.id/Details/72692/pp-no-10-tahun-1961 (1961). Accessed 13 Mar 2024 [Online]. Available: https://peraturan.bpk.go.id/Details/72692/pp-no-10-tahun-1961
4. Ary Sucaya, I.K.G.: Application and validation the Land Administration Domain Model in a real life situation (A case study in Indonesia). University of Twente, Enschede (2009)
5. Handono, A.B., Suhattanto, M.A., Nugroho, A.: Strategi Percepatan Peningkatan Kualitas Data Pertanahan di Kantor Pertanahan Kabupaten Karanganyar. J. Tunas Agrar. **3**(3) (2020)
6. Femenia-Ribera, C., Mora-Navarro, G., Santos Pérez, L.J.: Evaluating the use of old cadastral maps. Land Use Policy **114**, 105984. https://doi.org/10.1016/j.landusepol.2022.105984(2022)
7. O Yildiz Ç Erden 2020 Cadastral updating: The case of Turkey Surv. Rev. 53 379 335 348 https://doi.org/10.1080/00396265.2020.1759982
8. Hashim, N.M., Omar, A.H., Ramli, S.N.M., Omar, K.M., Din, N.: Cadastral database positional accuracy improvement. In: International Archives of the Photogrammetry, Remote Sensing and Spatial Information Sciences—ISPRS Archives. International Society for Photogrammetry and Remote Sensing, pp. 91–96. https://doi.org/10.5194/isprs-archives-XLII-4-W5-91-2017(2013)
9. Klebanov, M., Doytsher, Y.: Cadastral Triangulation: A Block Adjustment Approach for Joining Numerous Cadastral Blocks (2009)
10. Grant, D.B., Mccamley, G., Mitchell, D., Enemark, S., Zevenbergen, J.: Upgrading Spatial Cadastres in Australia and New Zealand: Functions, Benefits & Optimal Spatial Uncertainty (2018)

11. Sabekti, W.S.: A Conversion Strategy to Improve the Quality of Cadastral Map and to Support the Registration Process: Indonesian Case (2010)
12. Aditya, T., Sucaya, I.K.G.A., Nugroho Adi, F.: LADM-compliant field data collector for cadastral surveyors. Land Use Policy **104**, 105356. https://doi.org/10.1016/J.LANDUSEPOL.2021.105356(2021)
13. T Aditya PB Santosa Y Yulaikhah N Widjajanti D Atunggal M Sulistyawati 2021 Title validation and collaborative mapping to accelerate quality assurance of land registration Land Use Policy 109 105689https://doi.org/10.1016/j.landusepol.2021.105689
14. FAO, UNECE, FIG.: Digital transformation and land administration—Sustainable practices from the UNECE region and beyond. FAO; UNECE (United Nations Economic Commission for Europe). https://doi.org/10.4060/cc1908en(2022)
15. Gafurov, A.: Automated Mapping of cropland boundaries using deep neural networks. AgriEngineering **5**(3), 1568–1580. https://doi.org/10.3390/AGRIENGINEERING5030097(2023)
16. Wudye Tareke, B.: Visible Cadastral Boundary Extraction Using VHR Remote Sensing Images: A Deep Learning Approach. University of Twente, Enschede (2022)
17. H Zhang 2021 Automated delineation of agricultural field boundaries from Sentinel-2 images using recurrent residual U-Net Int. J. Appl. Earth Obs. Geoinf. 105 102557https://doi.org/10.1016/J.JAG.2021.102557
18. Franken, J., Florijn, W.: Rebuilding the Cadastral Map of the Netherlands: The Artificial Intelligence Solution. FIG (2021)
19. Wouters, R., Meijerink, G., Vaandrager, R., Zavrel, J.: Extracting information from deeds by optical character recognition (OCR) and text interpretation. In FIG Congress, Sydney Australia (2010)
20. Hajiheidari, A., Delavar, M.R., Rajabifard, A.: Smart urban cadastral map enrichment—A machine learning method. ISPRS Int. J. Geo-Inf. **13**(3), 80. https://doi.org/10.3390/IJGI13030080(2024)
21. Sun, S., Cao, Z., Zhu, H., Zhao, J.: A Survey of Optimization Methods from a Machine Learning Perspective (2019)
22. Claesen, M., De Moor, B.: Hyperparameter search in machine learning. In MIC 2015: The XI Metaheuristics International Conference [Online]. Available: https://www.codalab.org/competitions/2321 (2015). Accessed 28 Mar 2024
23. JH Holland 1992 Adaptation in natural and artificial systems: an introductory analysis with applications to biology, control, and artificial intelligence Adaptation in Natural and Artificial Systems https://doi.org/10.7551/MITPRESS/1090.001.0001
24. F Herrera M Lozano JL Verdegay 1998 Tackling real-coded genetic algorithms: operators and tools for behavioural analysis Artif. Intell. Rev. 12 265
25. D Li L Luo W Zhang F Liu F Luo 2016 A genetic algorithm-based weighted ensemble method for predicting transposon-derived piRNAs BMC Bioinformatics 17 1 1 11 https://doi.org/10.1186/S12859-016-1206-3/TABLES/9
26. K Varpa K Iltanen M Juhola 2014 Genetic algorithm based approach in attribute weighting for a medical data set J. Comput. Med. 2014 1 11 https://doi.org/10.1155/2014/526801
27. B DiCicco-Bloom BF Crabtree 2006 The qualitative research interview Med. Educ. 40 4 314 321 https://doi.org/10.1111/J.1365-2929.2006.02418.X
28. Caldwell, D.R.: Unlocking the Mysteries of the Bounding Box. Seattle: Ala Map and Geography Round Table, Jan. 2005. Accessed: Apr. 03, 2024. [Online]. Available: https://archive.ph/20120721210709/http://purl.oclc.org/coordinates/a2.htm
29. E Arkin L Chew D Huttenlocher K Kedem J Mitchell 1991 An efficiently computable metric for comparing polygonal shapes IEEE Trans. Pattern Anal. Mach. Intell. 13 209 216 https://doi.org/10.1145/320176.320190

30. Ruiz-Lendínez, J.J., Ureña-Cámara, M.A., Ariza-López, F.J.: Geo-Information a Polygon and Point-Based Approach to Matching Geospatial Features. https://doi.org/10.3390/ijgi6120399(2017)
31. S Katoch SS Chauhan V Kumar 2021 A review on genetic algorithm: past, present, and future Multimed. Tools Appl. 80 5 8091 https://doi.org/10.1007/S11042-020-10139-6
32. Xavier, E.M.A., Ariza-L, F.J., ´ Opez, Ure˜na, M.A., Ariza-López, F.J., Ureña, M.A.: A survey of measures and methods for matching geospatial vector datasets. ACM Comput. Surv. (CSUR) **49**(2). https://doi.org/10.1145/2963147(2016)
33. Correa Espinal, A.A., Rodriguez Velasquez, E., Jaramillo, J.S.C.: Diseño de Experimentos para la Evaluación de un Algoritmo Genético de Programación de Producción. *SISTEMAS, CIBERNÉTICA E INFORMÁTICA* **7**(1) (2010)

Assessing Height Determination Accuracy in Digital Terrain Models (DTM) Derived from UAV Data Using Sarawak Geocentric Datum

Ramzi Abdillah[1], Hajah Siti Romiza Romali[1], Madeline Anak Ambit[1], Mohd Faizury Abol Hassan[1], Stephen Ling Jin Huat[1], and Wan Anom Wan Aris[2(✉)]

[1] Lands and Surveys Department Sarawak, Kuching, Malaysia
ramzia@sarawak.gov.my
[2] Geomatic Innovation Research Group, Faculty of Built Environment and Surveying, Universiti Teknologi Malaysia, Iskandar Puteri, Malaysia
ramzia@sarawak.gov.my

Abstract. In the realm of geospatial sciences, the utilisation of Unmanned Aerial Vehicles (UAVs) has emerged as a transformative tool for acquiring high-resolution spatial data. UAVs, equipped with advanced sensors, offer an unprecedented capability to capture detailed terrain information, essential for applications ranging from topographic mapping to environmental monitoring. This study focuses on evaluating the accuracy of height determination in Digital Terrain Models (DTM) derived from UAV data, with a specific emphasis on utilising the Sarawak Geocentric Datum. The objective is to ensure reliable and precise elevation information for diverse applications in the Sarawak region. The methodology involves the strategic deployment of Ground Control Points (GCPs) for georeferencing, adhering to the Sarawak Geocentric Datum. Coordinate system consistency is maintained throughout data collection and processing, guaranteeing accurate height determination. Vertical accuracy assessment is conducted by comparing UAV-derived heights against ground truth points, employing statistical measures such as root mean square error (RMSE) for quantitative evaluation. To enhance data reliability, rigorous data quality control procedures are implemented, addressing errors during acquisition and processing. Spatial resolution and sampling density are carefully considered to meet the specific requirements of the study area, capturing fine-scale terrain features accurately. The outcomes of this research aim to contribute valuable insights to geospatial practitioners, land-use planners, and infrastructure developers relying on precise elevation information in the Sarawak region. By evaluating the accuracy of UAV-derived DTMs within the Sarawak Geocentric Datum, this study addresses a critical aspect of geospatial data quality and supports informed decision-making for applications such as Native Customary Right (NCR) survey planning, environmental monitoring, land management, and infrastructure planning.

Keywords: Digital Terrain Models · Unmanned Aerial Vehicle · Height Determination · Sarawak Geocentric Datum · Vertical Datum

A. Abdul-Rahman et al. (Eds.): GeoWeek 2024, LNGC, pp. 112–128, 2025.
https://doi.org/10.1007/978-3-031-86654-8_9

1 Introduction

An effort to establish a unified vertical network in Sabah and Sarawak has undergone very tedious survey operations, lack of access routes to connect the levelling network due to geographical setting constraints and is too costly to be fully implemented [14]. The vertical network establishments in Peninsular Malaysia, Sabah, and Sarawak were constructed separately. The unification of vertical datum was completed in 1998 with the adoption of Peninsular Malaysia Geodetic Vertical Datum [15]. A consistent and accurate adjusted height of the benchmark has been achieved in the adjustment of the precise levelling network of Peninsular Malaysia on a datum defined by the MSL height at Port Klang [16]. Meanwhile, levelling networks in Sabah and Sarawak are still not unified and always refer to various vertical datums [17].

The alternative approach to spirit levelling for the creation of vertical datum is through geoid modelling [21]. Geoid is a fundamental surface to which orthometric height is referred [22]. Nowadays, the Global Navigation Satellite System (GNSS) has been accepted by people for practical height determination. GNSS measurements give ellipsoidal height, which is geometrical height, and in practice, people need physical height or orthometric height. Therefore, in order to transform ellipsoidal height from Global Positioning System (GPS) measurement into orthometric height, precise geoids need to be used to obtain orthometric height easily, faster, and at a lower cost [23].

This study discussed and analysed the unmanned aerial vehicle (UAV) data for determining the height using the Sarawak geocentric datum independently of conventional vertical networks. UAV-based mapping has proven comparatively labour-intensive and costly compared to conventional methods. However, the accuracy of height determination in Digital Terrain Models (DTM) derived from UAV data remains a critical consideration, particularly when employing the Sarawak Geocentric Datum. Therefore, it is questionable whether to define the utilisation of height derived from UAV with respect to Sarawak Geocentric Datum, which has better accuracy than the existing vertical datums in Sarawak, using a new definition of vertical datum.

2 Objective

The objective is to explore the extent to which the accuracy and reliability of DTM derived from UAVs data can be utilised for various applications in the Sarawak region under the Sarawak Geocentric Datum by experimenting with the comparison of various techniques of height data collection.

i) Comparison values between the levelling survey (Mean Sea Level) and GPS Network Heighting (Orthometric Height)
ii) Comparison values between DTM (Orthometric) and values between levelling surveys (Mean Sea Level)
iii) Comparison values between DTM (Orthometric) and DTM (Mean Sea Level)
iv) Comparison values between DTM MSL height from UAV and MSL height from levelling

3 Methodology

3.1 Study Area

This study takes place in the Kampung Santubong area, Kuching, Sarawak, approximate latitudes 1o 42′ 56.80″ N—1o 43′ 25.36″ N and longitudes 110o 18′ 58.30″ E—110o 19′ 32.41″ E. It covers an area of approximately 0.35km^2 and exhibits topographical features akin to the surroundings of Mount Santubong and the coast. It is important to note that the size of the area under investigation is valuable information, as it provides context for understanding the scope and scale of the study (Fig. 1).

3.2 Project Workflow

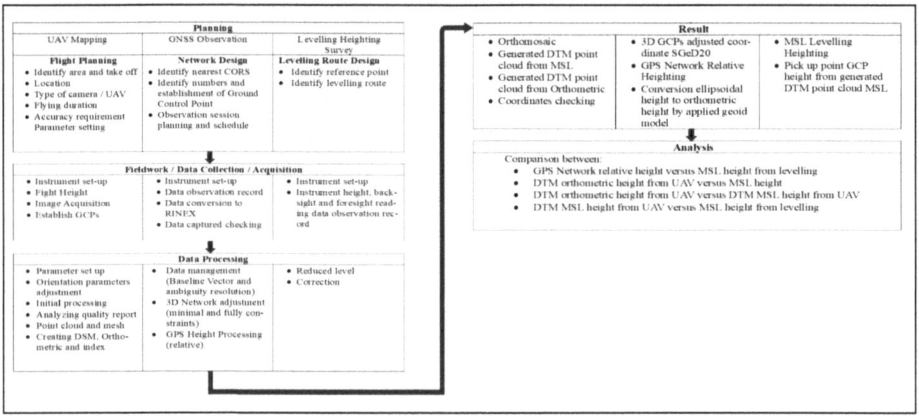

Fig. 1 Project workflow

3.3 Project Planning

Designing a GPS (Global Positioning System) network involves planning and configuring a system of GPS receivers and associated infrastructure to provide accurate positioning and navigation information. The study area covers approximately 0.35 km^2 in size, with nearby CORS stations selected, namely BAUU, SANT, and SMHN, which are in an ideal state such that shorter baseline distances are preferable.

In the context of a GPS (Global Positioning System) network, a baseline design often refers to the development of a fundamental architecture or configuration for the network. This design involves determining the locations, attributes, and connections of GPS stations to ensure accurate and reliable positioning data. Four (4) hours of concurrent observation by three (3) CORS stations with a static approach, while the three (3) base receivers must operate concurrently 20 min with two base stations in a fast static approach; however, they do not need to be in a continuous session.

3.4 Establishing the Ground Control Points

Ground Control Points (GCPs) play an essential role in geo-referencing and evaluating the accuracy of DTMs. The team conducted reconnaissance to assess the suitability of each GCP's location to avoid multipath factors during GPS observation. In this study, a total of eleven (11) GCPs (2 base stations and 9 GCPs) were established. All points were identified, placed at various heights, and uniformly distributed to cover the entire area. The base stations and GCPs were observed accordingly, following the GPS observation schedule. The GCPs are marked on the ground with 1 × 1 m targets which were designed in blue and orange canvas with a highly reflective material. The coordinates of GCPs were measured in Sarawak Geocentric Datum (SGeD20) using GNSS device technology.

3.5 Survey Flight

To collect aerial data, an EbeeX Sensefly RTK type Fix wing that allows high-resolution aerial photography is employed. In this study, we used eMotion software for planning the flight and to control EbeeX's mission. In the automatic mode, several vital parameters are set for the drone, including a mapping area of 1.288 km^2, a flight height of 265m, 75% forward overlap, and 75% side-lap. The position of each camera was measured at the time of shooting by a GNSS/INS receiver mounted on the drone, with an average accuracy of 2.5 m along the axes. This information is stored in each image's file and used for processing imagery to obtain photogrammetric products. The standalone approach is applied in the flying research. GCPs are directly referenced in the processing of the flight data findings. The "output" that the S.O.D.A. camera produces at a height of 265 m with the Ground Sampling Distance (GSD) is 6 cm.

3.6 Levelling Heighting

To enable the comparison of height data, the team conducted a levelling height survey using a closure loop at 23 stations, consisting of 9 GCPs, 2 base stations and 16 levelling stations. Vertical height levelling processing involves determining accurate elevations or heights of points on the earth's surface using levelling techniques. In this study, the team utilised a total station and prism as the traditional method to measure height differences between points, with Base Station G118 serving as the known reference for orthometric height derived from the vertical datum.

3.7 Data Processing and Results

In this data processing phase, the team have collected data and processes of:

a) RINEX GNSS data for coordinate determination of base stations and GCPs usingMATLAB software,
b) Aerial photo data processed using Pix4D mapper software.
c) GPS heighting and levelling in manual processing using Microsoft excel / MATLAB.

Table 1 CORS data value

CORS STATION	X	Y	Z
BAUU	−2195753.461	5986279.703	155933.6241
SANT	−2217571.922	5977271.923	188420.5974
SAMHN	−2232071.672	5972701.638	160781.0513

3.8 GNSS Data Processing

The reference CORS data value for the GNSS as in Table 1:

In a network adjustment with a minimum constraint, the adjustment process is carried out with the least possible number of constraints. Constraints, in this context, are known values or fixed points used to control and refine the adjustment. A minimum constraint adjustment allows for greater flexibility in the network solution, with fewer restrictions imposed on the adjustment process. Meanwhile, Network Adjustment Fully Constraint is performed with all possible constraints applied to the network. Constraints in this context refer to known values, fixed points, or other conditions used to control and improve the adjustment.

To verify the accuracy of adjustments between known and estimated values at the stations BAU CORS Station (Table 2) and SANT and SMHN (Table 3) as shown.

Table 2 3D difference at station BAUU

	BAUU		
	X	Y	Z
SGeD20	−2195753.461	5986279.703	155933.6241
Estimated	−2195753.47	5986279.717	155933.6203
Difference,	−0.0093	−0.0142	−0.0038
3D Difference (m)	0.017394539		

Table 3 3D difference at Station SANT and SMHN

	SANT			SMHN		
	X	Y	Z	X	Y	Z
SGeD20	−2217571.922	5977271.923	188420.5974	−2232071.672	5972701.638	160781.0513
Estimated	−2217571.91	5977271.889	188420.6033	−2232071.68	5972701.657	160781.0536
Difference,	0.0123	−0.0344	0.0059	−0.0084	0.0195	0.0023
3D Difference (m)	0.037006216			0.021456498		

Thus, we obtained the final coordinates for the two (2) base station (Table 4) and all the 9 GCPs (Table 5) as in GPS Radial Processing.

Table 4 Final XYZ value for Base Station

Base Station	X	Y	Z
G118	−2214447.153	5978387.345	190050.0614
G121	−2213589.719	5978698.609	189816.5002

3.9 Network GPS Height Processing

Processing GPS height, also known as GPS elevation data, involves in determining accurate vertical positions for specific points on the Earth's surface. This process typically utilises signals from Global Navigation Satellite Systems (GNSS), such as GPS, to measure elevations.

MyGEOID is the short front of the Malaysian geoid models consist of the following:

i .Peninsular Malaysia—WMGEOID04 and
ii .Sarawak and Sabah—EMGEOID05

The geoid models are hybrid ones, combining the gravimetric geoids with datum transformations and GPS ellipsoid heights on levelled benchmarks. The known orthometric height (Benchmark K312) of the base station G118 obtained from Malaysia Mapping and Survey Department (JUPEM) serves as a reference, enabling us to calculate the GPS relative heighting for GCPs, as shown in (Tables 6, and 7).

3.9.1 MSL Levelling Height Processing

As explained in 3.5, the team carried out a levelling survey along the 2.7 km closure loop. The MSL height levelling data were processed based on the Mean Sea Level equation. Result MSL Height for close loop levelling at reference base station G118 is—0.018cm (Table 8).

3.9.2 UAV Data Processing

In this study, Pix4D mapper was used to process UAV images acquired in the field. The data processing procedure of Pix4D mapper includes 4 steps which are (i) initial processing, (ii) analyzing quality report, (iii) point cloud and mesh and (iv) creating DSM, Orthometric and index.

Figure 2 and 3 displays the results of accuracy per GCP generated with both orthometric and MSL heights. The RMS error for both results are within centimeter accuracy.

The accuracy XY of the GCP/Check Point that has been given the accuracy tolerance of 0.02m and Z of the GCP/Check Point also set the accuracy tolerance of 0.02m. Each computed 3D point has initially been detected on the images (2D key point). On each image, the detected 2D key point has a specific position. When the computed 3D point is projected back to the images, it has a re-projected position. The distance between the initial position and the re-projected one gives the re-projection error (Figs 4 and 5).

Table 5. Final XYZ value for GCPs

Start	End	DX(m)	DY(m)	DZ(m)	X (m)	Y (m)	Z (m)	3D Difference (mm)	Mean X	**Mean Y**	**Mean Z**
G118	A102	220.5338	73.5911	149.3988	−2214226.61950	5978460.93590	190199.46020				
G121	A102	−636.9016	−237.6629	382.9598	−2214226.62040	5978460.94610	190199.46000	10.24158226	−2214226.61995	5978460.94100	190199.46010
8G118	A112	878.7323	305.2848	214.4689	−2213568.42100	5978692.62960	190264.53030				
G121	A112	21.2934	−5.973	448.0295	−2213568.42540	5978692.63600	190264.52970	7.789737074	−2213568.42320	5978692.63280	190264.53000
G118	A113	398.3656	132.4427	194.2281	−2214048.78770	5978519.78750	190244.28950				
G121	A113	−459.0769	−178.8123	427.7823	−2214048.79570	5978519.79670	190244.28250	14.05844977	−2214048.79170	5978519.79210	190244.28600
G118	A115	658.8094	231.8981	−52.5263	−2213788.34390	5978619.24290	189997.53510				
G121	A115	−198.6295	−79.3752	181.0304	−2213788.34830	5978619.23380	189997.53060	11.06435672	−2213788.34610	5978619.23835	189997.53285
G118	A116	417.2062	144.2153	−111.927	−2214029.94710	5978531.56010	189938.13410				
G121	A116	−440.2255	−167.0385	121.6226	−2214029.94430	5978531.57050	189938.12280	15.61057333	−2214029.94570	5978531.56530	189938.12845
G118	A117	143.8715	45.4976	−14.249	−2214303.28180	5978432.84240	190035.81240				
G121	A117	−713.5622	−265.7704	219.3063	−2214303.28100	5978432.83860	190035.80650	7.063285384	−2214303.28140	5978432.84050	190035.80945
G118	A122	619.3389	224.5591	−225.836	−2213827.81440	5978611.90390	189824.22580				
G121	A122	−238.0886	−86.7086	7.7277	−2213827.80740	5978611.90040	189824.22790	8.10308586	−2213827.81090	5978611.90215	189824.22685
G118	A123	395.4686	143.5264	−281.932	−2214051.68470	5978530.87120	189768.12910				
G121	A123	−461.9661	−167.7383	−48.377	−2214051.68490	5978530.87070	189768.12320	5.924525311	−2214051.68480	5978530.87095	189768.12615
G118	A124	267.0234	90.1007	−119.623	−2214180.12990	5978477.44550	189930.43800				
G121	A124	−590.4163	−221.1544	113.9437	−2214180.13510	5978477.45460	189930.44390	12.02746816	-2214180.13250	5978477.45005	189930.44095

Table 6 Reference Station Height

Reference GCP	Ellipsoidal height (m)	Geoid Height (m)	Orthometric Height (m)
A118	48.7834	31.786	15.511 (From JUPEM)

Table 7 Result Network GPS Relative Height

GCP	Ellipsoidal height (m)	Geoid Height (m)	dh	dN	dH	Orthometric Height (m)
A102	45.687	31.776	3.097	−0.010	3.107	12.404
A112	36.353	31.750	12.430	−0.036	12.466	3.045
A113	40.464	31.768	8.319	−0.018	8.337	7.174
A115	35.872	31.762	12.911	−0.024	12.935	2.576
A116	35.772	31.772	13.012	−0.014	13.026	2.485
A117	41.051	31.781	7.733	−0.005	7.738	7.773
A122	37.499	31.766	11.285	−0.020	11.305	4.206
A123	37.573	31.775	11.211	−0.011	11.222	4.289
A124	36.953	31.778	11.831	−0.008	11.839	3.672
G121	35.907	31.757	12.877	−0.029	12.906	2.605

This process aims to evaluate the variance between observation values converted to BT68 and the points selected from the orthomosaic after processing, as depicted in Table 9.

4 Analysis

The holistic analysis outlined above aims to provide a thorough understanding of the accuracy of height determination in DTMs derived from UAV data, specifically considering the nuances introduced by the Sarawak Geocentric Datum. By dissecting georeferencing precision, coordinate system consistency, vertical accuracy, data quality, spatial resolution, and terrain complexity, this research ensures a comprehensive evaluation, offering insights crucial for the improvement and application of UAV-derived DTMs in the Sarawak region.

The study endeavour entailed a meticulous examination of various elevation datasets to ascertain their accuracy and reliability. This investigation involved comparing the relative height with the MSL height determined through levelling techniques, a fundamental aspect of assessing the vertical positioning accuracy (Table 10). In this comparison, it is acceptable the differences between the two sets of data, which are measured in centimeter accuracy.

Table 8 Result MSL Levelling Height

Station	IH	Height		HEIGHT	Corr	Final Heght
		Target	Delta H			
K312	1.489			15.511	0.000	15.511
B1	1.469	1.500	−4.220	11.280	0.001	11.281
A117	1.596	1.597	−3.459	7.693	0.002	7.695
B2	1.535	1.535	4.683	12.437	0.002	12.439
A102	1.591	1.606	0.030	12.396	0.003	12.399
B3	1.595	1.598	−4.996	7.393	0.004	7.397
A113	1.469	1.45	−0.453	7.085	0.005	7.090
B5	1.557	1.529	−2.687	4.338	0.005	4.343
B6	1.538	1.518	−0.873	3.504	0.006	3.510
A112	1.568	1.530	−0.393	3.119	0.007	3.126
B7	1.497	1.480	0.474	3.681	0.008	3.689
B8	1.622	1.620	0.340	3.898	0.009	3.907
B9	1.656	1.638	−1.122	2.760	0.009	2.769
G121	1.582	1.565	−0.216	2.635	0.010	2.645
B10	1.560	1.538	1.472	4.151	0.011	4.162
B11	1.541	1.515	0.104	4.300	0.012	4.312
A123	1.520	1.745	0.113	4.209	0.013	4.222
B12	1.595	1.665	−1.562	2.502	0.013	2.515
B13	1.565	1.430	−0.096	2.571	0.014	2.585
B14	1.580	1.440	1.453	4.149	0.015	4.164
B15	1.541	1.568	−0.569	3.592	0.016	3.608
B16	1.514	1.620	1.874	5.387	0.016	5.403
B17	1.530	1.452	4.082	9.531	0.017	9.548
K312		1.498	5.930	15.493	0.018	15.511
STN		**FS**	**IH**	**TH**	**DH**	**RL**
B10		A122	1.56	1.58	0.102	4.244
B14		A124	1.58	0.1	−2.018	3.626

The comparison between the DTM orthometric height extracted from UAV data and the MSL height. The comparison involves the processed orthometric height values against the height values obtained using conventional surveying methods. Checks are conducted between different datasets, and validation is performed against known ground control points. The results are displayed in Table 11.

GCP Name	Accuracy XY/Z [m]	Error X [m]	Error Y [m]	Error Z [m]	Projection Error [pixel]	Verified/Marked
A112 (3D)	0.020/ 0.020	0.017	0.014	0.031	0.694	17 / 17
A113 (3D)	0.020/ 0.020	-0.032	0.001	0.023	0.655	13 / 13
G118 (3D)	0.020/ 0.020	0.021	0.005	-0.028	0.607	10 / 10
A122 (3D)	0.020/ 0.020	0.032	-0.004	-0.016	0.701	16 / 16
A123 (3D)	0.020/ 0.020	0.003	-0.003	-0.035	0.780	13 / 13
A102 (3D)	0.020/ 0.020	0.025	-0.020	0.057	0.507	8 / 8
A117 (3D)	0.020/ 0.020	-0.002	0.005	-0.049	0.744	15 / 15
A124 (3D)	0.020/ 0.020	-0.043	-0.015	0.106	0.941	19 / 19
A116 (3D)	0.020/ 0.020	-0.009	-0.002	-0.054	0.561	13 / 13
Mean [m]		0.001414	-0.002033	0.003715		
Sigma [m]		0.024186	0.009705	0.051083		
RMS Error [m]		0.024228	0.009915	0.051218		

Fig. 2 Result accuracy per GCP with Orthometric Height

GCP Name	Accuracy XY/Z [m]	Error X [m]	Error Y [m]	Error Z [m]	Projection Error [pixel]	Verified/Marked
A112 (3D)	0.020/ 0.020	0.014	0.013	0.070	0.610	17 / 17
A113 (3D)	0.020/ 0.020	-0.032	-0.017	-0.088	0.566	12 / 12
G118 (3D)	0.020/ 0.020	0.014	0.009	-0.029	0.590	10 / 10
A122 (3D)	0.020/ 0.020	0.027	-0.006	0.003	0.601	17 / 17
A123 (3D)	0.020/ 0.020	0.001	-0.004	-0.054	0.750	13 / 13
A102 (3D)	0.020/ 0.020	0.057	-0.036	0.163	0.534	8 / 8
A117 (3D)	0.020/ 0.020	-0.016	0.008	-0.122	0.582	15 / 15
A124 (3D)	0.020/ 0.020	-0.040	-0.007	0.082	0.952	19 / 19
Mean [m]		0.003214	-0.005018	0.003189		
Sigma [m]		0.030091	0.014941	0.089830		
RMS Error [m]		0.030263	0.015761	0.089886		

Fig. 3 Result accuracy per GCP with MSL Height

Additionally, the analysis extended to contrasting DTM orthometric height with DTM MSL height derived from UAV Table 12. Lastly, the study scrutinised the comparison between DTM MSL height extracted from UAV data and MSL height obtained through traditional levelling methods shown in Table 13.

The finding is that DTMs are often generated through interpolation techniques, which may introduce errors, particularly in areas with sparse data or abrupt changes in terrain elevation. For areas with significant differences in elevation, generating DTMs may have limitations in accurately representing vertical changes in terrain, especially in areas with steep slopes or abrupt changes in elevation.

5 Expectation Result

The expectations outlined above collectively contribute to the advancement of reliable and accurate geospatial datasets, fostering confidence in the utility of UAV-derived DTMs for applications related to environmental monitoring, infrastructure development, and sustainable land management.

Expect heightened vertical accuracy in the UAV-derived DTMs, validated through accuracy assessments against ground truth points, which is crucial for reliable elevation information in geospatial applications. The assessment revealed a high level of accuracy

Fig. 4 Orthomosaic Kampung Santubong, Kuching

GCP Name	Accuracy XY/Z [m]	Error X [m]	Error Y [m]	Error Z [m]	Projection Error [pixel]	Verified/Marked
A112 (3D)	0.020/ 0.020	0.015	0.012	0.034	0.695	17 / 17
A113 (3D)	0.020/ 0.020	-0.034	-0.007	0.032	0.653	13 / 13
G118 (3D)	0.020/ 0.020	0.011	-0.005	-0.025	0.605	10 / 10
A122 (3D)	0.020/ 0.020	0.028	-0.007	-0.008	0.701	16 / 16
A116 (3D)	0.020/ 0.020	-0.010	0.009	-0.072	0.509	11 / 11
Mean [m]		0.001765	0.000429	-0.007731		
Sigma [m]		0.021883	0.008206	0.039267		
RMS Error [m]		0.021954	0.008218	0.040021		

0 out of 4 check points have been labeled as inaccurate.

Check Point Name	Accuracy XY/Z [m]	Error X [m]	Error Y [m]	Error Z [m]	Projection Error [pixel]	Verified/Marked
A123		-0.0056	-0.0061	-0.0296	0.7775	13 / 13
A102		0.0187	-0.0343	0.0799	0.5050	8 / 8
A117		-0.0141	-0.0057	-0.0435	0.7396	15 / 15
A124		-0.0576	-0.0217	0.1343	0.9464	19 / 19
Mean [m]		-0.014649	-0.016933	0.035275		
Sigma [m]		0.027559	0.011914	0.074509		
RMS Error [m]		0.031210	0.020704	0.082437		

Fig. 5 Result accuracy per GCP with utilizing 5 control points and 4 check points

in height determination within the Sarawak Geocentric Datum, with vertical discrepancies, as measured by RMSE, within acceptable limits, demonstrating the reliability of the UAV-derived DTM for diverse applications.

Table 9 Determine the difference between the known coordinates and the coordinates selected from the orthophoto as a check point.

	OBSERVATION COORDINATES		COORDINATES FROM ORTHOPHOTO		Difference		Point assign as
Station	Easting	Northing	Easting	Northing	E	N	
A102	2070537	5190274	2070537	5190274	−0.037	−0.044	CHECK POINT
A112	2069839	5190339	2069839	5190339	0.022	−0.001	GCP
A113	2070350	5190319	2070350	5190319	0.012	−0.065	GCP
A115	2070071	5190072	2070071	5190072	−0.020	0.030	CHECKPOINT
A116	2070328	5190013	2070328	5190013	0.009	−0.046	GCP
A117	2070619	5190110	2070619	5190111	0.056	−0.033	CHECKPOINT
A122	2070111	5189899	2070111	5189899	0.026	−0.032	GCP
A123	2070349	5189843	2070349	5189843	0.015	−0.028	CHECKPOINT
A124	2070488	5190005	2070488	5190005	−0.034	−0.025	CHECKPOINT
G118	2070769	5190124	2070769	5190125	0.048	−0.014	GCP

Table 10 Result of Comparison: Orthometric Height versus MSL Height

Station	GPS Relative Height	Levelling	Difference
	Orthometric (m)	MSL (m)	
A102	12.404	12.399	0.005
A112	3.045	3.126	−0.081
A113	7.174	7.09	0.084
A117	7.773	7.695	0.078
A122	4.206	4.244	−0.038
A123	4.289	4.222	0.067
A124	3.672	3.626	0.046
G118	15.511	15.511	Reference
G121	2.605	2.645	−0.04

Anticipate validation results against existing high-quality elevation datasets, establishing the reliability and accuracy of the UAV-derived DTMs by benchmarking against established standards, especially for engineering projects.

Table 11. Result of Comparing DTM orthometric height from UAV derivation versus MSL height from levelling.

Station	DTM Orthometric	DSM Orthometric	MSL	Diff DTM Ortho vs MSL	Diff DSM Ortho vs MSL
A102	12.401	12.43	12.399	0.002	0.031
A112	3.073	3.046	3.126	−.053	−0.08
A113	7.11	7.103	7.09	0.02	0.013
A115	2.669	2.596			
A116	2.576	2.57			
A117	7.771	7.726	7.695	0.076	0.031
A122	2.567	4.228	4.244	−1.677	−0.016
A123	1.888	4.319	4.222	−2.334	0.097
A124	1.584	3.649	3.626	−2.042	0.023
G118	15.402	15.499	15.511	−0.109	−0.012
G121	2.643	2.639	2.645	−0.002	−0.006
B2	12.435	12.453	12.439	−0.004	0.014
B3	7.275	7.385	7.397	−0.122	−0.012
B5	4.34	4.349	4.343	−0.003	0.006
B6	3.478	3.429	3.51	−0.032	−0.081
B7	3.499	3.631	3.689	−0.19	−0.058
B8	3.857	3.851	3.907	−0.05	−0.056
B9	2.805	2.739	2.769	0.036	−0.03
B10	2.625	4.099	4.162	−1.537	−0.063
B11	1.77	4.306	4.312	−2.542	−0.006
B12	2.627	2.63	2.515	0.112	0.115
B13	2.8	2.844	2.585	0.215	0.259
B14	1.25	4.189	4.164	−2.914	0.025
B15	3.567	3.573	3.608	−0.041	−0.035

6 Recommendations

a .Continue to use accurate GCPs and refine their distributions to further enhance georeferencing precision.

b .Applying SGeD20 as a datum for surveying and mapping will happen soon, and with this usage, we may make survey work more efficient. Coordinated transformation and projection will be less complicated and definitely more accurate.

c .Regularly update and calibrate UAV sensors to maintain data accuracy and consistency over time.

Table 12 Result of Comparison DTM&DSM orthometric height from UAV versus DTM&DSM MSL height from UAV

Station	DTM Orthometric	DSM Orthometric	DTM MSL	DSM MSL	Diff DTM Ortho vs DTM MSL	Diff DSM Ortho vs DSM MSL
A102	12.401	12.43	11.612	11.66	0.789	0.77
A112	3.073	3.046	2.474	2.576	0.599	0.47
A113	7.11	7.103	6.404	6.397	0.706	0.706
A115	2.669	2.596	1.725	1.661	0.944	0.935
A116	2.576	2.57	1.503	1.491	1.073	1.079
A117	7.771	7.726	6.979	6.93	0.792	0.796
A122	2.567	4.228	1.323	2.992	1.244	1.236
A123	1.888	4.319	0.354	2.774	1.534	1.545
A124	1.584	3.649	0.228	2.638	1.356	1.011
G118	15.402	15.499	14.826	14.904	0.576	0.595
G121	2.643	2.639	1.687	1.702	0.956	0.937
B2	12.435	12.453	11.94	11.836	0.495	0.617
B3	7.275	7.385	6.52	6.594	0.755	0.791
B5	4.34	4.349	3.779	3.71	0.561	0.639
B6	3.478	3.429	2.936	2.913	0.542	0.516
B7	3.499	3.631	2.876	2.858	0.623	0.773
B8	3.857	3.851	3.083	3.02	0.774	0.831
B9	2.805	2.739	1.916	1.853	0.889	0.886
B10	2.625	4.099	1.521	2.974	1.104	1.125
B11	1.77	4.306	0.356	2.96	1.414	1.346
B12	2.627	2.63	1.395	1.348	1.232	1.282
B13	2.8	2.844	1.607	1.575	1.193	1.269
B14	1.25	4.189	−0.158	3.283	1.092	0.906
B15	3.567	3.573	2.712	2.677	0.855	0.896

d .Establishing a Homogeneous Height Datum for Sarawak involves selecting the most compatible datum from the three (3) datums used in Sarawak. Later, we may suggest using all 28 CORS for benchmarking heights and fully utilising our CORS for height measurement.

e .For other applications to support decision-making processes in various fields, including environmental monitoring, land management, and infrastructure planning.

Table 13 Result of Comparing DTM & DSM MSL height from UAV versus MSL height from levelling

Station	DTM MSL	DSM MSL	MSL	Diff DTM MSL vs Levelling	Diff DSM MSL vs Leveling
A102	11.612	11.66	12.399	−0.787	−0.739
A112	2.474	2.576	3.126	−0.652	−0.55
A113	6.404	6.397	7.09	−0.686	−0.693
A115	1.725	1.661		1.725	
A116	1.503	1.491		1.503	
A117	6.979	6.93	7.695	−0.716	−0.765
A122	1.323	2.992	4.244	−2.921	−1.252
A123	0.354	2.774	4.222	−3.868	−1.448
A124	0.228	2.638	3.626	−3.398	−0.988
G118	14.826	14.904	15.511	−0.685	−0.607
G121	1.687	1.702	2.645	−0.958	−0.943
B2	11.94	11.836	12.439	−0.499	−0.603
B3	6.52	6.594	7.397	−0.877	−0.803
B5	3.779	3.71	4.343	−0.564	−0.633
B6	2.936	2.913	3.51	−0.574	−0.597
B7	2.876	2.858	3.689	−0.813	−0.831
B8	3.083	3.02	3.907	−0.824	−0.887
B9	1.916	1.853	2.769	−0.853	−0.916
B10	1.521	2.974	4.162	−2.641	−1.188
B11	0.356	2.96	4.312	−3.956	−1.352
B12	1.395	1.348	2.515	−1.12	−1.167
B13	1.607	1.575	2.585	−0.978	−1.01
B14	−0.158	3.283	4.164	−4.322	−0.881
B15	2.712	2.677	3.608	−0.896	−0.931

7 Conclusion

The study concludes that DTMs derived from UAV data using the Sarawak Geocentric Datum are accurate and reliable for height determination. The comprehensive assessment methodology employed ensures the validity of elevation information, providing a robust foundation for various geospatial applications in the Sarawak region.

References

1. J. Syst. Eng. Electron. 21(4) (2010)—Jianjun Zhang et al.: Analysis of unmanned aerial vehicle navigation and height control system based on GPS
2. George, C. Zuffada.: Theoretical description of a bistatic system for ocean altimetry using the GPS signal. Radio science. **38**(5): 10–19 (2003)
3. Jiménez-Jiménez, S.I., Ojeda-Bustamante, W., de Jesús Marcial-Pablo, M., Enciso, J.: Digital terrain models generated with low-cost UAV photogrammetry: Methodology and accuracy. ISPRS Int. J. Geo-Inf. **10**, 285. https:// doi.org /10.3390/ ijgi1 0050285 (2021)
4. Serifoglu Yilmaz, C., Gungor, O.: Comparison of the performances of ground filtering algorithms and DTM generation from a UAV-based point cloud. Geocarto Int. **33**, 522–537 (2021)
5. Rangel, J.M.G., Gonçalves, G.R., Pérez, J.A.: The impact of number and spatial distribution of GCPs on the positional accuracy of geospatial products derived from low-cost UASs. Int. J. Remote Sens. **39**, 7154–7171 (2018)
6. Ewertowski, M.W., Tomczyk, A.M., Evans, D.J.A., Roberts, D.H., Ewertowski, W.: Operational framework for rapid, very-high resolution mapping of glacial geomorphology using low-cost unmanned aerial vehicles and structure-from-motion approach. Remote Sens. **11**, 65 (2021)
7. Rosnell, T., Honkavaara, E.: Point cloud generation from aerial image data acquired by a quadrocopter type micro unmanned aerial vehicle and a digital still camera. Sensors **12**, 453–480 (2012)
8. Agüera-Vega, F., Carvajal-Ramírez, F., Martínez-Carricondo, P., Sánchez-Hermosilla López, J., Mesas-Carrascosa, F.J., GarcíaFerrer, A., Pérez-Porras, F.J.: Reconstruction of extreme topography from UAV structure from motion photogrammetry. Meas. J. Int. Meas. Confed. **121**, 127–128 (2018)
9. ISPRS Annals of the Photogrammetry, Remote Sensing and Spatial Information Sciences, Volume II-2/W2, 2015 Joint International Geoinformation Conference 2015, 28–30 October 2015, Kuala Lumpur, Malaysia—accuracy assessment of lidar-derived digital terrain model (DTM) with different slope and canopy cover in tropical forest region
10. Pavlis, N.K., Holmes, S.A., Kenyon, S.C., Factor, J.K.: An earth gravitational model to degree 2160: EGM2008, presented at the 2008 general assembly of the European Geosciences Union, Vienna, Austria, 13–18 (2008)
11. Unification of vertical Datum in Sabah and Sarawak, Othman A.H., Omar, K., Othman. R., Mat Som, Z.A Opaluwa, Y.D
12. An investigation of the vertical control network of peninsular Malaysia using a combination of levelling, gravity, gps and tidal data. Universiti Teknologi Malaysia, phd thesis. Shahrum Ses and Azhari Mohamed. (2009)
13. Height Modernization Using Fitted Geoid Model and MyRTKNet. Universiti Teknologi Malasysia. Msc Thesis
14. Mehrabi, H., Voosoghi, B.: Optimal observational planning of local GPS networks: assessing an analytical method. Journal of Geodetic Science **4**(1), 87–97 (2014)
15. Martínez-Carricondo, Patricio, et al. Assessment of UAV-photogrammetric mapping accuracy based on variation of ground control points. International j. Appl. Earth Obs. Geoinformation **72**, 1–10 (2018)
16. Yazid, N. M., et al.: The implementation and advancement of a regional geodetic vertical datum. IOP conference series: Earth and environmental science. **880**, 1. IOP Publishing. (2021)
17. Ismail, M.K., et al.: Establishment of new fitted geoid model in Universiti Teknologi Malaysia. Int. Arch. Photogramm. Remote. Sens. Spat. Inf. Sci. **42**, 27–33 (2018)

18. Kim, Su-Kyung, et al.: On determining orthometric heights from a corrector surface model based on leveling observations, GNSS, and a geoid model. J. Appl. Geod. **12**(4), 323–333 (2018)
19. Fusami, A. A, Edan J. D., Takana A.: Local orthometric height based on a combination of GPS-derived ellipsoidal height and geoid model: A review paper. J. Geod. Sci. **13**(1), 20220158 (2023)
20. Beretta, Filipe, et al.: Topographic modelling using UAVs compared with traditional survey methods in mining. REM-Int. Eng. J. **71**, 463–470 (2018)
21. Guidelines for the use of the Malaysian Geoid Model (Mygeoid)—Director general of survey and mapping circular No. 10 OF 2005 Our Reference: JUPEM 18/7/2.148 (87) 6 Sept 2005
22. ASPRS accuracy standards for digital geospatial data https://www.asprs.org/a/society/divisions/pad/Accuracy/Draft_ASPRS_Accuracy_Standrds_for_Digital_Geospatial_Data_PE&RS.pdf
23. https://www.unavco.org/software/geodetic-utilities/geoid-height-calculator/geoid-height-calculator.html
24. https://earth-info.nga.mil/index.php?dir=wgs84&action=wgs84#tab_egm2008
25. https://geographiclib.sourceforge.io/cgi-bin/GeoidEval

Assessing the Impact and Efficiency of Cadastral Modernisation on Coordinate Systems for Orthophoto Production: A Comprehensive Evaluation for Town Planning

Mohd Zunica Mohd Sapi'ee, Mohd Iskandar Yusof(✉), Awan Amal, Wong Miliw, and Abg Ahmad Azlan Abg Ariffin

Lands and Surveys Department Sarawak, Kuching, Malaysia
iskandaryusoof1979@gmail.com

Abstract. Assessing the impact and efficiency of cadastral modernisation on coordinate systems for orthophoto production is a complex yet vital task. Cadastral modernisation involves the updating and digitalising of land records, which is crucial for accurate orthophoto production. The coordinate system used in cadastral modernisation greatly influences the accuracy and efficiency of orthophoto production. Therefore, a comprehensive evaluation of this relationship is of great significance. In this context, using accurate coordinate systems is paramount for ensuring that orthophotos are mapped correctly onto the digital cadastral data. This study analysed the output of the coordinate system from orthophoto production that was mapped to the cadastral plan to become the 3D cadastral. Apart from that, the significant use of the integration of cadastral plan and orthophoto output for urban planning analysis. Currently, the preparation of site analysis and the development plan proposal by the urban planners in the department only utilise the output from the 2D cadastral plan and orthophoto plan without a clear methodology which is unpractical, unrealistic, and unresponsive to the environment. From these challenges, the study analyses the significant use of the orthophoto output for the urban planners to do the slope and contour analysis and volume analysis for the new development project and analyse how effectively the existing development complies with the town planning guidelines. The study is based on the results obtained from the UAV surveys conducted at Kampung Santubong, Kuching, Sarawak, by improving the survey and mapping practices through the SGeD20 coordinate system. The findings of this study have explored how the cadastral plan can be overlayed with 3D spatial data to facilitate the use of urban planning analysis.

Keywords: Development Control Standard · SGeD20 coordinate system · 3D Cadastre

A. Abdul-Rahman et al. (Eds.): GeoWeek 2024, LNGC, pp. 129–142, 2025.
https://doi.org/10.1007/978-3-031-86654-8_10

1 Introduction

The geocentric datum is based on the Earth-centred Earth-fixed (ECEF) coordinate system, with its origin and orientation defined by a best-fit ellipsoid. It follows a specific configuration, aligning the X-axis with the mean Greenwich meridian, the Z-axis with the Earth's rotation, and the Y-axis with the right-hand system. The International Terrestrial Reference Frame (ITRF) serves as the foundation for the global, regional, and national realisations of geocentric datum, ensuring consistency with the International Terrestrial Reference System (ITRS). Sarawak has recently established Sarawak Geocentric Datum 2020 (SGeD20) in line with ITRF2014, replacing the previous Borneo Triangulation 1968 (BT68) datum. Sarawak's Land and Survey Department is currently studying SGeD20, and it has yet to be utilised for surveying and mapping activities.The ITRF plays a vital role in maintaining the integrity and interoperability of the Global Navigation Satellite Systems (GNSS). Concerning town and country planning, the geocentric datum is very significant because it is based on 3D spatial data. However, the department's planners are still using the 2D Cadastral map in the planning analysis, especially site selection and subdivision of the building plot. This limited information will lead to the failure of planning and decision-making, incur operational costs, and delay project implementation. To make optimal use of this, the UAV is used in this study to produce a 3D orthophoto. The output from the 3D orthophoto overlapped on the cadastral layer and became a 3D Cadastre. This 3D Cadastre can facilitate the planning application. The study is focused on the regions of Santubong, Kuching, and Sarawak. Within this area, the coordinate reference location area is 0.5 km × 0.7 km, approximately Latitude 1o42′56.80″ N, 1o43′25.36″ E, approximately Longitude 110o18′58.30 E, 110o19′32.41″ E, with 10 GCPs. This study holds considerable significance as it aims to analyse building setback, slope, and volume by utilising 3D Orthophoto. Furthermore, the study highlights the contribution to the potential advancement of cadastral modernisation on coordinate systems into the production of 3D orthophoto cadastral for town and country planning analysis. The outcomes of this study can potentially enhance the overall efficiency and effectiveness of land information and administration, supporting various decision-making processes and providing up-to-date and precise land data for a wide range of applications.

2 Coordinate System in Sarawak Used for Land Development

Some reviews on general information about the geodetic datum, Borneo Triangulation 1968, the Sarawak Geocentric datum, GNSS surveying, 3D Orthophoto, 3D Cadastre Slope Analysis, Volume and Cut and Fill Analysis, and Sarawak Development Control Standard (Building Setback).

2.1 Geodetic Datum

A geodetic datum is a fundamental reference surface for various coordinate systems used in Earth mapping, allowing for an accurate representation of size and shape. Throughout history, numerous distinct datums have been utilised, evolving in response to changing perspectives on the world [6]. Geodetic datums are established based on a geocentric terrestrial frame and often use a best-fit ellipsoid aligned with the ECEF coordinate system with revolution at specific size and shape. Additionally, the geodetic datum encompasses parameters that define a coordinate system, including origin, scale, and orientation to the 3D Cartesian system [7]. The adoption of geocentric data occurs globally, regionally, and nationally, following a hierarchical structure. Global geocentric datums are adopted first, followed by regional and national datums, owing to their compatibility with the International Terrestrial Reference System (ITRS) [8]. Continuous enhancements in data analysis strategies are undertaken to determine the optimal approach for determining the ITRF [9]. Furthermore, the ITRF is a critical resource in ensuring the reliability and compatibility of the GNSS and other applications that require precise positioning data [10].

2.2 Borneo Triangulation 1968 (BT68)

The classical geodetic method used to build survey control networks by taking angle measurements and some spare distances is known as triangulation. In this manner, triangular points or benchmarks are spread out all over the country. This document seeks to review the geodetic survey work that was done in East Malaysia and Brunei from 1918 to 1968, which covers approximately 204,119 sq km. However, classical geodetic datum is not directly compatible with present surveying trends based on Global Navigation Satellite System technologies. The Borneo Triangulation, 1968 (BT68), is a geodetic network that covers Sabah and Sarawak in East Malaysia. It was established with the origin located in Bukit Timbalai on the island of Labuan. The Directorate of Overseas Surveys (DOS) from the United Kingdom readjusted the major control of East Malaysia, including Sabah, Sarawak, and Brunei, resulting in the creation of the BT68 network. Some other developments have been made to enhance this network such as the Borneo West Coast Triangulation (1930–1942) of Brunei and Sabah, Borneo East Coast Triangulation (1955–1960) of Sarawak, and a prolongation of West Coast Triangulation in Sabah. However, between 1961 and 1968 some new points were surveyed for further enhancement.

2.3 Sarawak Geocentric Datum 2020 (SGeD20)

The SGeD20 is a geodetic reference system specific to the region of Sarawak, established in 2020 by the Land and Survey Department. SGeD20 is aligned with the Global Reference Frame, specifically to the ITRF2014. To support this datum, the Sarawak Geocentric Datum Network (SGeDnet) was developed, which comprises of GNSS Continuously Operating Reference Station (CORS) infrastructure. SGeDnet serves as a reliable source for accurate positioning and surveying activities within Sarawak, ensuring compatibility with the global reference framework.

2.4 GNSS Surveying

GNSS surveying uses GNSS signals, along with a GNSS receiver and antenna, to pinpoint the shape, perimeter, and location of objects or points in space in relation to other shapes, perimeters, or points. This replaces the old-fashioned optical and mechanical surveying techniques that rely on theodolites and distance measuring tools. GNSS surveying, with the help of CORS, enables the modelling of distance-dependent errors, such as ionosphere and troposphere delays, thus improving the accuracy and reliability of GNSS surveying. Because of its dependability and ongoing improvements, GNSS has become one of the most widely utilised technologies for everything from navigation to surveying and mapping of large-scale research projects like geodynamics and geophysics. This technology has also been used in cadastral surveys, which establish the borders of land, in addition to engineering surveys such as mapping and topographical surveys. The measurements obtained using GNSS surveying, specifically in Malaysia, are currently not subjected to any form of validation, primarily involving the Network Real-Time Kinematic (N-RTK) technique, in contrast to traditional instrumentation.

2.5 3D Orthophoto

The Orthophoto map contains no linework, and symbols are now replaced with an Ortho photographic image. The emergence of new technology has introduced Unmanned Aerial Vehicles (UAVs) as an alternative to traditional photogrammetry in creating orthophoto maps. This research paper will demonstrate how UAV can be used for orthophoto map creation. In UAV photogrammetry, a photographic measurement platform refers to any photogrammetric vehicle that is either remotely operated or semi-autonomous or even fully autonomous but without any human pilot on-board. Such platforms substitute the large format air cameras employed in conventional photogrammetry with small format digital camera lenses. For the general area, the amount of forward overlap and side overlap is 75% and 60%, respectively. For forest area, the amount of forward overlap and side overlap is 85% and 70%, respectively, as the UAV flight altitude will be higher. The whole workflow is introduced in the research paper which involved capturing 461 images with a DJI Phantom 4 Pro equipped with a camera model FC6310. PIX4D Capture was used to fulfil this flight mission. Using Agisoft PhotoScan, an orthophoto map over an area of 250, 458 m^2 was generated. This output provides a high-quality orthophoto map for UTHM Pagoh Campus that gives an overview of the place with updated information for its users [5].

Within the civil engineering and architectural construction, the calculation of earthwork volumes stands as a crucial element during both the design and construction phases; hence, precise calculation is imperative. Furthermore, given the significance of earthwork management, this investigation implemented a three-dimensional (3D) modelling approach for earthwork calculation and administration by employing an unmanned aerial vehicle (UAV). The model considered multiple image acquisition angles (45°, 60°, and 75°) at altitudes of 50 m and 100 m to ensure precision in earthwork calculations and 3D modeling. The images were segregated into eight distinct cases: cases 1–4 comprised images captured at 50 m, while cases 5–8 involved images from a 100-m altitude (case 1: 90°, case 2: 90° + 45°, case 3: 90° + 60°, case 4: 90° + 75°, case 5: 90°, case 6: 90° + 45°, case 7: 90° + 60°, case 8: 90° + 75°). Three separate evaluations were conducted on the collected data.

First, checkpoints were set up to evaluate the accuracy of the orthophoto along with comparing earthwork volumes calculated through GPS and UAS as well as evaluating the 3D model. Case 2 (Root Mean Square Error)—Lower the RMSE, to get the better Orthophoto accuracy. Case 2 was the most optimal in earthwork volume prediction and as-build model two level comparison. The results showed that the 3D model for earthwork volume calculation and management was best generated using a vertical image with high oblique images of 40 to 50°. Also, it is better to shoot from 50 m or even closer if the UAV will not be affected by obstacles.

2.6 3D Cadaster

Initial categorisation of 3D Parcels and the initial basis of exploration in subsequent work into appropriate matching formats for database representations exchange and data capture encodings. The categories listed from the least amount of complexity to the greatest are:

a. 2D spatial unit (prism of 3D space): A prism in the context of three-dimensional configurations, with its bases defined by a shape that occupies two dimensions.
b. Building form type: A unit of area established by dimensions (which can be absolute or relative) in an existing or proposed.
c. Form (e.g., apartment).
d. 2D shape with upper or lower surface: Semi-open spatial unit.
e. Polygonal slice spatial unit (a 2D shape with ad upper and lower surface).
f. Single-valued stepped spatial unit: With horizontal and vertical boundaries only.
g. Multi-valued stepped spatial unit: As above but now multi valued.
h. General 3D spatial unit: Defined also by boundaries other than horizontal and vertical.

General 3D space category partially: open/closed volume planar/curved boundaries, multi-valued single/multi-volume, etc. Combining dimension per unit area descriptions with the limited number of 3D land admins in a very company DB and transfer format encodings may be another perceptible challenge to beat [1]. According to the Dutch cadastral registration, every real-estate object (cadastral object) is considered 2D (a parcel). This has resulted into various challenges, such as multi-level construction, more infrastructures above or underground, numerous wires, apartments atop other apartments/offices/shops. However, these physical 3D entities do not fall under the category of cadastral objects because they appear as 2D parcels on the cadastral map. The need for an advanced land information system that incorporates 3D data to ensure sustainable, uniform, and efficient land registration in the Netherlands. The research, a collaboration between the Department of Geodesy and the Netherlands' Kadaster, focuses on merging cadastral requirements with technological advancements. The proposed solution involves enhancing the current 2D cadastral system with 3D capabilities to better register and manage complex 3D situations [2].

2.7 Heighting (DTM)

The advancement of technology has significantly broadened the use of digital photogrammetry in various mapping-related fields. Digital orthophotos, a key photogrammetric product, have gained widespread adoption in the private sector due to their ease of interpretation. This increased utilisation has driven the demand for accurate maps, prompting research into the factors influencing their accuracy. The resolution and scale of aerial photos or satellite images, the quality and scale of the Digital Elevation Model (DEM), and the characteristics of land topography are crucial elements affecting orthophoto accuracy. This study specifically investigates the impact of DEM and land topography on the accuracy of orthophotos.

The article discusses both global and local quality parameters. Global quality parameters, determined using check points, evaluate a specific measurement technique such as airborne laser scanning (ALS) or digital stereo photogrammetry across an entire area of interest using a few parameters. In contrast, local quality parameters assess the quality of a DTM at a detailed level, describing the accuracy of each grid point within the model. The use of Unmanned Aerial Vehicles (UAVs) and the Global Positioning System (GPS) has revolutionized the creation of accurate, reliable, and high-resolution maps over the past decade. With these technologies, digital elevation maps (DEMs) and orthophotos of an area can be generated quickly and efficiently. UAVs, equipped with high-resolution cameras and GPS, capture detailed aerial images and positional data, which are then processed to produce precise geospatial information. This advancement has significantly enhanced the capability to map and analyse various terrains in a short amount of time. Resolution of maps were compared when maps were produced. More detailed and high-resolution maps were created with 30 m when compared with 50 m elevation flight height. As a result of the study, it was understood that flight height should be determined depending on the structure of the land, accuracy and precision expected from the work [11].

2.8 Sarawak Development Control Standard (DCS)

Chapter 8 of Sarawak Development Control Standard—Setback of Building from Lots Boundaries. The minimum setback of a building from its lot boundary is defined as the shortest distance between the external wall of the main building and the relevant lot boundary. Setback is measured from relevant lot boundary to the external walls of the building, either at the ground level or at a higher level, whichever is the nearest. The siting of a building in relation to its lot boundary needs to be relegated for various purposes. This is usually done through the control of setbacks of the building from its boundary. In the residential scheme, the open space created provided private grounds for landscaping or gardening, drying of clothes, car parking and outdoor activities. The setback requirement also ensures adequate separation between adjacent buildings thereby creating a firebreak as well as providing a certain degree privacy and enough natural lighting and ventilation.

2.9 Slope Analysis, Volume and Cut and Fill Analysis

In many engineering projects, large parcels of land are needed for the site, and substantial amounts of material must be moved to create the necessary embankments, cuttings, foundations, basements, lakes, and other specified features. Finding suitable land and materials can be costly. Therefore, to ensure project's profitability for the construction company, it is crucial that engineers make precise measurements of the areas and volumes involved. Accurate measurements are essential to include appropriate cost estimates for earthwork quantities in the tender documents. This precision helps in budgeting and planning, ensuring that the project remains financially viable.

3 Methodology

This chapter explains the methodology of this study. Phase 1 is the administrative work for acquiring current data set, Phase 2 shows the flow of the establishment of GNSS control network, Phase 3 describes the heighting calculation at the CORS station and control point while Phase 4 shows how to produce UAV orthophoto image. Finally, Phase 5 shows the process of producing the result and analysis.

3.1 Phase 1: Administrative Work: Acquiring Current Data Set (Cadastral Layer) and RINEX Data.

In this phase, the acquisition of preliminary data such as Cadastral Lot Layer was obtained from the Land and Survey Department Sarawak. The data are required to be overlaid on the result of the orthophoto. RINEX data was obtained through CORS station and for this study, the group had chosen the Santubong CORS' station.

3.2 Phase 2: Establishment of GNSS Control Network

At this stage, the field observation process involves two techniques, namely static observation, and fast static observation. The static observation is a process to get the coordinate value of the premier control point while the fast static observation is used to obtain the coordinate of the GCP.

3.3 GNSS Static and Fast Static Observation

Static and fast static observation uses 5 receivers and 3 CORS stations. Three receivers are installed at the premier control points, which are stations A108, A118, and A121 for four hours observation. Two receiver devices are used to create observation into GCP (A124, A117, A123, A122, A116, A115, A114, A112, A113 & A102) for twenty minutes. Theselection of the control point for the premier and GCP are very important because it affects the final accuracy of the coordinate of the control point. A sky view that covers an area of at least 60% is highly emphasised, while multipath impact factors should also be taken into consideration to minimise the error of the observation data. The control station should be away from the building or any structure. Then, the equipment should be set up at all control stations according to the observation schedule that has been made. When preparing equipment for the control station, the data and the height of the receiver should be recorded. The elevation mask angle should not be less than 15° (Table 1).

Table 1 GPS Observation Session

Group	Point (Receiver)						
Expected Time	1030	1115	1145	1230	1315	1400	1430
	–	–	–	–	–	–	–
	1115	1145	1230	1315	1400	1430	1600
G1	A108	A108	A113	A102	A108	A108	A108
G2	A121	A121	A121	A121	A116	A115	A121
G3	A118	A118	A123	A122	A121	A121	A121
G4	A124	A117	A118	A118	A118	A118	A118
G5	A114	A112	A108	A108	A108	A108	A108

3.4 GNSS Data Management

After the data collection is completed, the observed data is downloaded from the instrument. The download procedure is referred to as the instrument user manual. The observation data is downloaded using a USB flash drive or directly transferred to the PC by using a USB cable. The raw data is then converted into RINEX for further analysis. From the RINEX file, the observation data and navigation message can be obtained which was used to determine the position of the control points. The antenna height was computed based on the slant measurement and antenna offset.

3.5 GNSS Data Processing

Data processing is performed using the Satlab Geobiz Solution (SGS) software to get the coordinates of the GNSS control points. GNSS data processing is divided into four (4) steps, baseline processing, network adjustment for minimum constraint, network adjustment for fully constraint and accuracy checking. Baseline processing uses observation from two (2) receivers to calculate the relative distance between the receivers. Errors are then minimised using double differencing with proper error modelling. The output of baseline processing are baseline vectors and variance-covariance of the baseline. A minimum constraint is performed to check the measurement. One (1) CORS is fixed as the known point and another two (2) CORS are used to compare the adjusted coordinate with the true coordinate provided by the Land and Survey Department. The 3D differences of the checked CORS are calculated and compared to detect any result that have abnormal difference with limit of 5 cm. Full constraint is performed to get the final coordinate of the control points. After the minimum constraint was performed, two best CORS are chosen to be held fixed as known points and the third CORS are used to check the accuracy of the network adjustment. Again, accuracy checking was performed by comparing the 3D difference of the adjusted coordinate and the true coordinate of the CORS. The difference must be less than 3 cm to be considered as a good network adjustment.

3.6 Phase 4: Production UAV Orthophoto

The production of images is done by drone using Pix4DMapper to produce orthophoto and digital surface models. The processing method is using the post Processing Kinematic (PPK) method. Basically, the processing is in the SGeD20. However, a transformation process was carried out at the coordinate stage of 3D to convert the coordinates to the BT68 using Sarawak Coordinate Transformation Software (SCT20) programme. This transformation must be implemented due to the absence of projections in the Pix4DMapper software. In addition to ensure the spatial data results on the orthophoto will be in line with the cadastral layer of the lot which is used as one of the layers in this study later.

3.7 Phase 5: Analysis and Result

Analysis and Result using Picx4DMapper and ArcGIS software to study the setbacks, slope analysis and volume calculation using orthophoto UAV Image together with the cadastral layer lot. The planning for flying path for UAV within the subject area with altitude 265 m, resolution 6.00 cm / px, latitude and longitude overlap is about 75% and generated 262 photos.

4 Analysis Difference Coordinate Parcel Lot with Orthophoto

POINT	PARCEL COORDINATES		ORTHOPHOTO COORDINATES		DIFF	
	X	Y	X	Y	X	Y
POINT1	2,069,703.97	5,190,342.80	2,069,704.01	5,190,342.74	-0.04	0.06
POINT2	2,069,847.51	5,190,341.24	2,069,847.62	5,190,340.70	-0.1	0.54
POINT3	2,069,923.63	5,190,325.70	2,069,923.24	5,190,325.29	0.39	0.4
POINT4	2,069,697.82	5,190,189.50	2,069,697.77	5,190,188.99	0.05	0.51
POINT5	2,069,828.02	5,190,232.34	2,069,828.48	5,190,232.79	-0.47	-0.45
POINT6	2,069,902.60	5,190,225.57	2,069,902.05	5,190,226.27	0.56	-0.7
POINT7	2,070,300.53	5,189,892.53	2,070,301.02	5,189,892.57	-0.49	-0.04
				total	-0.1	0.32
				average	-0.01429	0.045714

Fig. 1 Finding Analysis Different Coodinate between Parcel Lot and Orthophoto

The difference coordinate analysis between parcel lot coordinates and the orthophoto is done based on the selected points as shown in Table 5.1. The results of the comparison between the parcel lot coordinates are measured using Timbalai BT68 Datum and the orthophoto coordinates that uses GCP (Ground Control Point) which are measured using Datum SGeD20. It can be viewed as shown in Table 5.1. In total there was a displacement difference of −0.1 m between easting of the parcel coordinates compared to the orthophoto coordinates, while there was displacement of 0.32 m difference on the northing coordinates. However, on average, the coordinates at the point selected for the purpose of this study are at −0.01429 at the easting coordinates and 0.045714 m split at the northing coordinates (Fig. 1).

5 Analysis the Urban Planning Development Guidelines

5.1 Building Setback Analysis

Building Setback Analysis is based on requirement from Sarawak Development Control Standard. The requirements of the building setback for affordable residential lot are minimum 6-m of frontage, side setback is 4.5 m (roadside will be 6 m) and rear setback is minimum 9 m. Based on the visualisation of the orthophoto analysis, the building setback dimensions does not comply with the building setback standard with its frontage setback of 5.908-m, left and right-side setback of 4.504 m and 7.273 m respectively and rear setback of 12.736 m. The analysis is very important for urban planners to analyse how much the contractors or developers complies to the building setback standards. It is also important for the enforcement officer as well as local authority to always do building construction inspections. The output of orthophoto coordinate system and cadastral plan (overlay 2D and Orthophoto) is very useful for the urban planners to facilitate the developer on the ground and the preventative initiative will takes place once the significant data from the orthophoto are fully utilised.

5.2 Plot Coverage

The total plot area for the subject lot 1159 is 504.023 sqm while the built-up area for the house is 53.536 sqm. According to Sarawak Development Control Standard, the plot coverage maximum must not exceed 25% from the total lot area. The UAV captured and measured the built-up area 53.536 sqm (10.6%) from total of 504.023 (100%). Based on the visualisation of the orthophoto analysis, the plot coverage as mentioned in the Figure 11 clearly showed that the plot coverage complies to the building setback standard with only 10.6% built up area compared to the plot area. The analysis is very crucial and significant for the urban planners to analyse how far the contractor or developer comply to the plot coverage standards. This is also important to the enforcement officer as well as local authority to enforce the Building Ordinance 1994's standards and the Sarawak Development Control Standard. The output of orthophoto coordinate system and cadastral plan (overlay 2D and Orthophoto) is also useful for the urban planners to guide the developer on the ground.

5.3 Slope Data Analysis

The slope analysis is very significant for the urban planners to analyse how far the development project is responsive to the environment such as terrain, slope and topography. This study tries to explore the proposal for the road connecting the Santubong Village Extension Scheme to the existing village with total length for the proposed road is 2.95 km and road width is 15 m. The urban planners aligned the road as shown in Fig. 2.

Based on the slope analysis standard, the urban planners must consider the parameter that slope with more than 25 degrees of the area cannot be developed as it covered the huge area because it will affect the environment in term of structural and characteristic terrain. Not to mention, slopes of even 20 degrees would also not be safe to develop.

Fig. 2 Proposed Road linking from the Santubong Village Extension Scheme connecting to the existing village

Again, if planners are still keen to proceed development within slope 20 degree, it will incur an expensive treatment engineering cost which will not be profitable in the long run. However, the slope that is undulating, gentle and flat is allowed for development and the best way to construct is to follow the contour or topography line. Based on the analysis, most of the coverage area will be linked by the road, with slope that is within below 15 degrees. The northern side has the highest point of about 14 m which is slightly covered by small portion of slope 20 degree to 26°. From the data of the orthophoto image analysis, it is significant clear proof for the urban planners to understand especially during the site selection process. The study also tries to overlay the contour line with the cadastral plan to give the urban planners an indicator on how the proposed roading system will follow the terrain significantly without cutting off the contour line. The output from the combination of the coordinate system from cadastral plan and contour polygon line practically give the result to the urban planners to consider the best practice while doing the site analysis. A side from the data, it can strategically be able to generate the 3D Cadastral plan with combinations of layer which is parcel lot and 3D Orthophoto.

5.4 Volume Analysis

The study also analyses the volume of the soil involved once the road project will be implemented. The study tries to understand from the DTM and software analysis, how the cubic metric or volume for the cutting of the land and the filling involved, will take into consideration the cost of the development project. Additionally, this factor is also essential for the valuers to calculate the project cost profile and the land price of the subject area. Not only that, but this indicator is very important for the urban planners to design the project implementation according to schedule and construction period. Data from Fig. 3 below show the overall slope height at the study area. Variants from 12 m and up to 14 m difference in heights were analysed for the need to cut and fill the land to the platform level required to build the road.

From the analysis, the volumetric calculations from the respective software give the result that the cut volume is about 9815.40 cubic metres and fill volume is about 6.149 cubic metres. This data proved that to achieve 12 m platform level of the proposed road, more volume needs to be cut. The volume of the excess soil from the cutting of the land

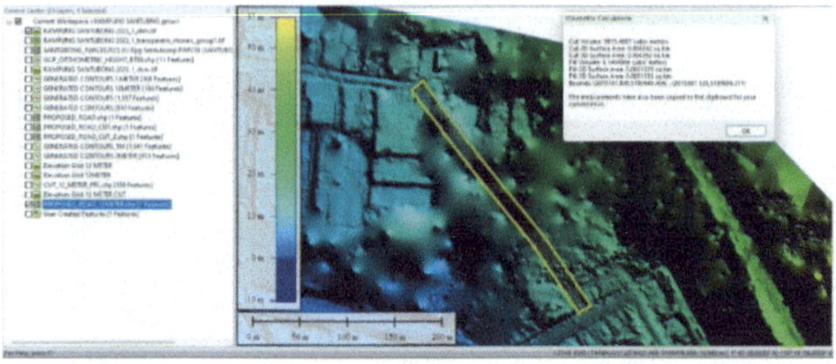

Fig. 3 Cut and Fill at Platform 12 m

soil can generate income for the states by selling to the licensee/buyer by charging the royalty and soil price at average of RM19.00 per cubic metre. This information would also facilitate the real estate management to analyse that the profit for the land royalty is about RM 186,375.76.

5.5 Analysis for Proposed 13 Metres Road Platform Level

Data from Image Fig. 5 below show the overall slope height at the study area. Variants from 12 m and up to 14 m difference in heights analysed and the need to cut and to fill the land to the platform level required to build the road. The data shows from image Fig. 22 that the soil can be cut and produce up to 5797.86 cubic metres of soil and only needs 463.09 cubic metres of soil to fill in the lowest part of the proposed road platform level of 13 m (Fig. 4).

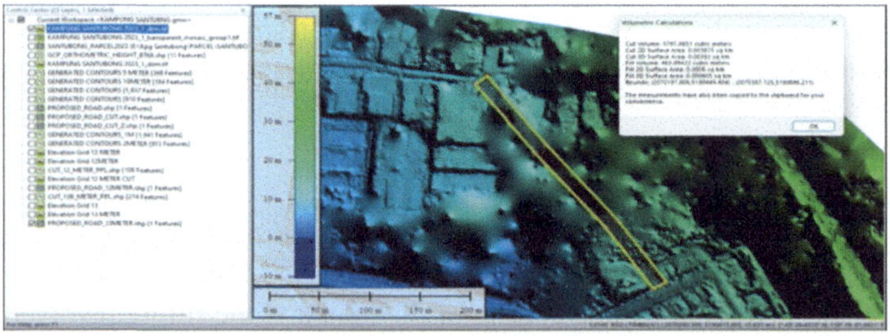

Fig. 4 Cut and Fill at Platform 13 m

The volume of the excess soil from the cutting the land soil can generate income for the states by selling it to the licensee/buyer by charging the royalty and soil price at average of RM19.00 per cubic meter for 5334.77 cubic metres. This information would also facilitate the real estate management to analyse the profit for the land royalty is about RM101360.63. However. The profit is less than the cutting for levelling 12 m as mentioned above with difference of RM85,015.13.

5.6 Analysis for Proposed 14 Metres of Road Platform Level

Data from Fig. 6 show the overall slope height at the study area. Variants from 12 m and up to 14 m of difference in heights were analysed to produce the need to cut and to fill the land to the platform level required to build the road.

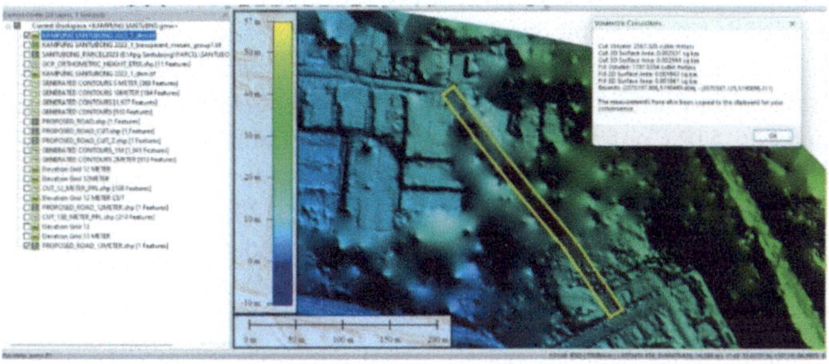

Fig. 5 Cut and Fill at Platform 14 m

The data from Fig. 6 shows that the soil can be cut and produce up to 2567.32 cubic metres of soil but only needs 1707.03 cubic metres of soil to fill in the lowest part of the proposed road platform level of 13 m.

6 Result

The technical analysis shows a minimal coordinate difference of −0.1 m between parcel lots and orthophotos, unaffected by reference system variations (BT68 and SGed20). Plot coverage analysis reveals compliance with building laws (10.6%), but actual setbacks often deviate from approved plans. Planners must ensure compliance with setback standards to maintain development quality. Slope analysis using DTM 3D data highlights the importance of elevation and contour lines in planning. Cut and fill analysis suggests setting platform levels at 12 or 13 m to minimize earthwork. Orthophotos are essential for planning, necessitating improved methodologies in cadastral mapping.

Acknowledgement. We would like to express our sincere appreciation to Prof. Dr. Tajul Arrifin Musa, Dr. Wan Anom Binti Wan Aris, Dr. Norhadija Binti Darwin from UTM, and ASS Stanley anak Nyawai from the Land and Survey Department Kuching Division for assisting in contributing data and survey operations for this study. Special credit is also given to UTM's undergraduate Geomatics Engineering students who have given support for data processing and analysis.

References

1. Van Oosterom, P.: Best Practices 3D Cadastres from fig report, No.17 Publication)
2. Stoter, J.: Towards a 3D Cadastre: Where do cadastral needs and technical possibilities meet?" (2021)

3. Ayhan, E., Erden, Ö., Atay, G., Tunç, E.: Digital Orthophoto generation with aerial photos and satellite images and analyzing of factors which affect accuracy, pp. 8–13. Shaping the ChangeXXIII FIG CongressMunich, Germany, October (2006)
4. (PDF) Digital orthophoto generation with aerial photos and satellite images and analyzing of factors which affect accuracy. Available from: https://www.researchgate.net/publication/237797550_Digital_Orthophoto_Generation_with_Aerial_Photos_and_Satellite_Images_and_Analyzing_of_Factors_which_Affect_Accuracy Accessed 10 Sep 2024. Borneo triangulation 1968: a review and the way forward, Joannes Jawan (2023)
5. Ngadiman, N., Kaamin, M., Sahat, S., Mokhtar, M., Ahmad, N. F. A., Kadir, A. A., Razali, S.N.M.: Production of orthophoto map using UAV photogrammetry: A case study in UTHM Pagoh campus. In AIP Conference Proceedings **1** AIP Publishing (2016)
6. Janssen, V., McElroy, S.: The Australian height datum turns 50: Past, present and future. Proceedings of APAS Webinar Series 2021 (AWS2021). **24**, 3–27
7. Zhiping, L., Yunying, Q., Shubo, Q.: Introduction to geodetic datum and geodetic systems. Geodesy. (2014). https://doi.org/10.1007/978-3-642-41245-5
8. Leow, H.X., Din, A.H.M., Zulkifli, N.A., Yong, C.Z., Adzmi, N.H.M.: Assessing global geodetic datum between international terrestrial reference frame 2014 and world geodetic system 1984 using GPS Observation. J. Adv. Geospatial Sci. Technology. **2**(2), 90–116 (2002)
9. Altamimi, Z., Boucher, C., Sillard, P.: New Trends for the realization of the international terrestrial reference system. Adv. Space Res. **30**(2), 175–184 (2002)
10. Kaplan, ED., Hegarty, C.: Understanding GPS/GNSS: Principles and application, Third Edition (2017)
11. Celik, M.O., Bunyanunel, F., Kusak, L., Kanun, E., Alptekin, A.: The effect of different flight heights on generated digital products: Dsm and Orthophoto. Mersin Photogrammetry J. e-ISSN 2687–654X. (2019)

Marine Information System

Projection of Sea Surface Salinity Using Ensemble of Selected CMIP6 GCMs of Bay-of-Bengal

Mahiuddin Alamgir[1,1](✉), Mazlan Hashim[1,2], and Shamsuddin Shahid[3]

[1] Geoscience and Digital Earth Centre (INSTeG), Research Institute of Sustainable Environment (RISE), Universiti Teknologi Malaysia (UTM), 81310 Johor Bahru, Malaysia
mmalamgir@utm.my

[2] Faculty of Built Environment and Surveying, Universiti Teknologi Malaysia (UTM), 81310 Johor Bahru, Malaysia

[3] Department of Water and Environmental Engineering, Faculty of Civil Engineering, Universiti Teknologi Malaysia (UTM), Skudai, 81310 Johor, Malaysia

Abstract. Changing ocean properties threaten coastal communities and marine ecosystems global. Sea surface salinity (SSS) is a vital indicator for monitoring the hydrological cycle within the ocean system. This study projected potential future changes in SSS over the Bay of Bengal under shared socioeconomic pathways (SSPs) using Global Climate Model (GCM) data from the Coupled Model Intercomparison Project Phase 6 (CMIP6). We applied Quantile Mapping (QM)and Taylor Diagrams (TD) to downscale model outputs at the resolution of 0.5 considering the Ocean Reanalysis System 5 (ORAS5) Data as the reference. However, most of the regions exhibit an SSS increase ranging from 33 psu to 34 psu in the historical period from 1970 to 2014. The SSP370 results indicate that near-future (2020–2059) changes range from −6.22 to 26.60 PSU at the upper confidence level and −0.82 to 26.66 PSU at the lower confidence level. In contrast, far-future (2066–2099) changes range from −1.06 to 23.53 PSU at the upper confidence level and −8.45 to 23.55 PSU at the lower confidence level. Significantly, decrease in south eastern part of Bangladesh and western part of Myanmar, but low salinity (PSU) is observed east to south of Sri Lanka and Indian coastal areas compared to central part of BOB in the near future (2020–2059) and far future (2066–2099). These findings focus on robust adaptation planning to protect coastal communities and marine ecosystems from the complex challenges of climate change.

Keywords: Sea surface salinity · Bay of Bengal · Quantile mapping · Shared socioeconomic pathways and CMIP6

1 Introduction

Climate change, primarily driven by anthropogenic activities, contributes to changes in global sea surface characteristics like sea surface salinity [5, 21, 37]. Sea surface salinity (SSS) is commonly used as a metric for monitoring the hydroclimatic cycle within the

A. Abdul-Rahman et al. (Eds.): GeoWeek 2024, LNGC, pp. 145–161, 2025.
https://doi.org/10.1007/978-3-031-86654-8_11

global climate system [10, 17, 27], reflecting climate variability [48] and oceanographic processes such as ocean circulation [9, 20] sea-level changes and height [7, 15], and planetary waves [30]. The physical features of global sea levels have changed in the past and are projected to continue at a significant rate by the end of the 21st century due to climate change impacts [23, 44]. These changes in sea surface attributes are a major concern for millions living in low-lying coastal zones and the ecosystems these areas support [24, 44]. Mathematical models of the marine environment, and earth systems are vital tools for understanding and forecasting SSS. The simulation accurateness of SSS is one of the most important indicators for model examination [12, 13, 25]. According to the latest Intergovernmental Panel on Climate Change (IPCC) report (AR6 2021), global mean sea level has exhibited a consistent rise by 1.3 mm/year from 1901 to 1971, followed by a faster pace of 1.9 mm/year from 1971 to 2006, and a notable acceleration to 3.7 mm/year between 2006 and 2018 [28]. Sea level rise due to climate change is expected to worsen extreme weather events such as storm surges, cyclones, high waves, and accelerated erosion [35]. It's important to note that regional sea levels can deviate significantly from the global average due to the influence of ocean dynamics, varying heating patterns, and gravitational and solid Earth processes [35]. In addition to human influence, natural internal variability caused by phenomena like El Niño Southern Oscillation (ENSO), the Indian Ocean Dipole (IOD), the Pacific Decadal Oscillation (PDO), and the Southern Annular Mode (SAM) can contribute to inhomogeneous sea level changes across the globe [30, 35]. Oceanic circulation and currents also manifest external forces, such as latitudinal changes in monsoon winds and solar radiation. These can be considered atmospheric and oceanic interactions despite regional differences in wind systems [35]. SSS is the most critical oceanographic parameters for analyzing ocean gradients and vertical motion [43]. They are also considered key factors controlling marine environments and gas flux [19]. The SSS-stratified layer acts as a barrier, hindering vertical mixing between the surface layer and deeper, colder water, thereby affecting SST [1]. Changes in SSS serve as a key indicator of climate change and can influence ocean circulation patterns [32]. A few of studies have used CMIP6 models to estimate sea surface salinity in the Bay of Bengal and sorrunding seas [4, 22, 39] explored future change of SSS along tropical Indian Ocean, BOB using an earlier generation of GCMs named CMIP3. Asok et al. [2] studied the Indian Ocean Dipol and BOB. Additionally, the study suggests that the interaction between the North and South BOB can affect the Indian Ocean Dipole (IOD) events, which more influences in temperature variation and, ocean heat content (OHC). However, this study is partial and is based only on Indian Ocean Dipole and a short historical simulation time. This study aims to future change of sea surface salinity using seventeen GCM CMIP6 models. We found that four out of the seventeen GCMs were best suited to replicate the Ocean Reanalysis System 5 (ORAS5) data. We employed the QM method to refine the coarse resolution GCM bias correction. The SSS for two future periods, 2020–2059 and 2060–2099, are estimated for four SSPs using the selected GCMs in comparison to the reference period of 1970–2014.

2 Study Area and Data

2.1 Study Area

The Bay of Bengal (BOB) primarily extend in the south of Bangladesh-to-southeast India-south west Myanmar across between 10° and 23°N latitude and 80° to 97°E longitude (Fig. 1). Shroyer et al. [40] study found Sea surface salinity (SSS) in the BOB generally ranges from 28 to 34 PSU (Practical Salinity Unit). However, it can vary significantly due to factors such as river discharge, seasonal rainfall and ocean currents. In the northern part of the bay, near the mouth of the river, the salinity can be as low as 28–30 PSU due to abundant inflow of freshwater from rivers such as the Ganges and the Brahmaputra [38]. Moreover, Madhupratap et al. [26] obtained in the central and southern parts of the BOB, salinity levels are generally higher, ranging from 32–34 PSU. The BOB is home to numerous coastal and island communities and a variety of marine biota, particularly surrounded by mangrove forests along the vast coastal areas of India and Bangladesh [3, 8, 16]. Nevertheless, any change in sea surface characteristics in the BOB could potentially cause economic disruption to coastal and island inhabitants, loss of property and life, and eventually the extinction of marine species.

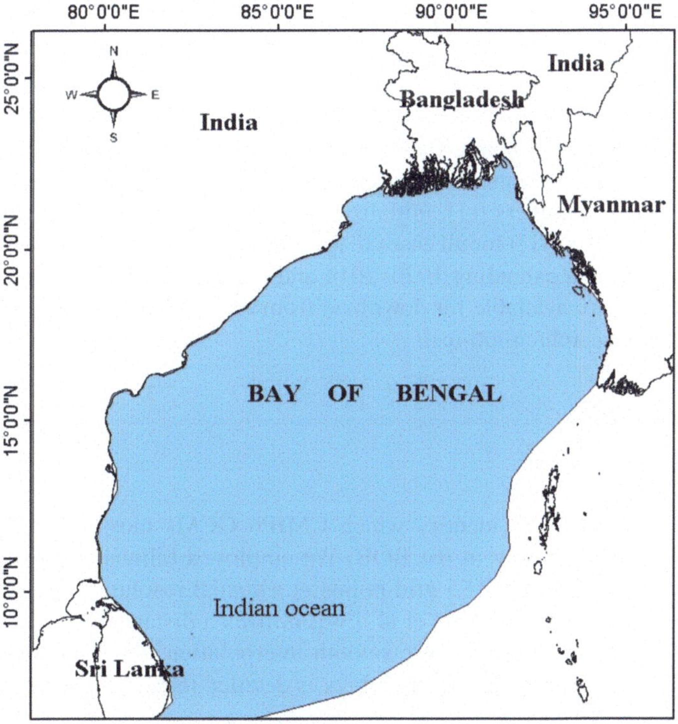

Fig. 1. Location map of the study area

2.2 Datasets

2.2.1 The Ocean Reanalysis System 5 (ORAS5) Data

Monthly sea surface salinity data from ORAS5, with a spatial resolution of 0.25°, were used in this study. The European Centre for Medium-Range Weather Forecasts (ECMWF) OCEAN5 system produces an extensive monthly resource called the ORAS5 Global Ocean Reanalysis Dataset, which extend over 1958 to the present [31, 45]. The ORAS5, with a spatial resolution of 0.25° and a monthly temporal resolution, is considered more accurate than numerical model results. It integrates extensive observed data into the simulation, providing a more accurate demonstration of the historical ocean state in the Arctic Ocean [18]. The original data was stored by Network Common Data Form (NetCDF) format. The data can be available from the following website: https://cds.climate.copernicus.eu/cdsapp#!/dataset/reanalysis-oras5?tab=form.

2.2.2 CMIP6 Global Circulation Model (GCM)

To gain a deeper understanding of climate change and its effects, the CMIP model was proposed 20 years back via the Working Group on Coupled Modeling (WGCM) under the World Climate Research Program (WCRP) [11, 29, 42]. We exploited simulations from seventeen CIMP6 GCMs on a monthly scale for sea surface salinity (SSS) to identify the most suitable models for replicating these variables in the Bay of Bengal. These GCMs were established by collaborating with various international research organizations of different countries around the world, which is shown in Table 1. The GCMs were chosen because they offered monthly historical simulations and future projections for four Shared Socio-Economic Pathway (SSPs) such as SSP126, SSP245, SSP370, and SSP585. Further, we considered historical single-run data labeled as r1i1p1f1 (representing ensemble members [r1], initialization states [i1], physical parameterizations [p1], and forcing index [f1]) for all cases to assure fairness in the comparison of various GCMs. Historical data extending 1970–2014 and projection data covering 2015–2100 from these GCMs are available for download from the following website: https://esgf-node.ipsl.upmc.fr/search/cmip6-ipsl/.

3 Methodology

3.1 Procedure

In this study, we aimed to identify which CMIP6 GCMs most accurately simulate observed sea surface salinity in the BOB. We employed bilinear interpolation to re-grid all GCM datasets to ORAS5 grid points at a spatial resolution of $0.25° \times 0.25°$. This method, as described by Sein et al. [39], utilizes a distance-weighted average of the four surrounding points to ensure smooth interpolation. We then applied the QM approach to adjust for bias. The methodology is detailed in the following steps:

- Every GCMs were interpolated to align with ORAS5 grid points using bilinear interpolation at a spatial resolution of $0.25° \times 0.25°$.
- Bias in the re-gridded GCMs' SSS data for the period 1970 to 2014 was corrected using the Quantile Mapping (QM) method.

Table 1. The list of the general circulation models (GCMs) evaluated in this study.

No	Model name	Institution	Country	Resolution
1	ACCESS-CM2	Australian Community Climate and Earth System Simulator coupled model	Australia	1.2° × 1.8°
2	ACCESS-ESM1-5	Australian Community Climate and Earth System Simulator (ACCESS) and Earth System Model (ESM)	Australia	1.2° × 1.8°
3	CanESM5	Canadian Earth System Model version 5	Canada	2.8° × 2.8°
4	CanESM5-1	Canadian Earth System Model version 5	Canada	2.8° × 2.8°
5	CESM2-WACCM	Community Earth System Model version 2 with the Whole Atmosphere Community Climate Model	USA	1.9° × 2.5°
6	CMCC-CM2-SR5	Euro-Mediterranean Centre on Climate Change coupled climate model (standard configuration)	Europe	1° × 1°
7	CMCC-ESM2	Fondazione Euro-Mediterranean Center on Climate Change	Italy	0.9° × 0.9°
8	EC-Earth3	EC-Earth consortium	Europe	0.7° × 0.7°
9	EC-Earth3-Veg	EC-Earth consortium	Europe	0.7° × 0.7°
10	EC-Earth3-Veg-LR	EC-Earth consortium	Europe	0.7° × 0.7°
11	INM-CM4-8	Institute for Numerical Mathematics	Russia	1.5° × 2.0°
12	INM-CM5-0	Institute for Numerical Mathematics	Russia	1.5° × 2.0°
13	MIROC6	Model for Interdisciplinary Research on Climate Version 6	Japan	1.4° × 1.4°
14	MPI-ESM1-2-LR	Max Planck Institute Earth System Model	Germany	1.875° × 1.86°
15	MRI-ESM2-0	Meteorological Research Institute Earth System Model Version 2.0	Japan	1.125° × 1.125°
16	NorESM2-LM	The Norwegian Earth System Model	Norway	2.5° × 1.89°
17	NorESM2-MM	The Norwegian Earth System Model	Norway	2.5° × 1.89°

- A Multimodel Ensemble (MME) was constructed using the top four GCMs identified during the ranking procedure.
- The MME of the selected GCMs was used to project changes in sea surface salinity under four Shared Socioeconomic Pathways (SSPs) for the near future (2020–2059) and the far future (2060–2099).
- Upper and lower band uncertainties were established by calculating the 95th percentile, mean, and 5th percentile of all GCMs across all SSPs.
- The procedure was repeated for SSS projections using the top-ranked four GCMs' models.

3.2 Quantile Mapping

QM is a bias correction algorithm developed by Panofsky and Brier [33] based on cumulative distributive function. It is commonly used to adjust the distribution of model outputs with the observation. QM formulated by [14] as follows:

$$P_0 = h(P_m) \tag{1}$$

where P_0 is height or salinity, h is a transfer function and is downscaled height or salinity. The distribution of data set and parameters of the distribution are determined and then the following expression is used for bias correction:

$$P_o = F_o^{-1} F_m(P_m) \tag{2}$$

where F_o^{-1} is the inverse CDF (or quantile function) corresponding to observed (P_o) height or salinity, F_m is the cumulative distribution function (CDF) of downscaled height or salinity. P_o and P_m shows the observed and simulated data, respectively.

3.3 Statistical Analysis Using Taylor Diagram (TD)

The TD, which reveals how closely a pattern fitted observations, offers a brief statistical summary for assessing the simulation of sea surface salinity [41]. Three statistical parameters are included in the diagram i.e. the correlation of coefficient (r) measures the relationship between the model and observed data, the spatial standard deviation (s), and the centered root-mean-square error (RMSE) (E') procedures the difference between the model and observed data. Therefore, the TD was employed to provide a brief statistical summary for appraising the simulation of Sea Surface Salinity (SSS).

$$r = \frac{\frac{1}{N}\sum_{n=1}^{N}(m_n - \overline{m})(o_n - \overline{o})}{\delta_m \delta_o} \tag{3}$$

$$\delta_m = \sqrt{\frac{1}{N}\sum_{n=1}^{N}(m_n - \overline{m})^2} \tag{4}$$

$$\delta_o = \sqrt{\frac{1}{N}\sum_{n=1}^{N}(o_n - \overline{o})^2} \tag{5}$$

$$E' = \sqrt{\frac{1}{N}\sum_{n=1}^{N}[(\mathrm{o_n} - \overline{\mathrm{o}}) - (\mathrm{m_n} - \overline{\mathrm{m}})]^2} \quad (6)$$

where $\mathrm{m_n}$ and $\mathrm{o_n}$ are the values of "model" and "reference" fields at each grid point, respectively; N is the total number of grid points of the field; and the overbar indicates the overall mean of a field.

Subsequently various fields utilize different units of measurement, their statistics must be non-dimensionalized to be displayed on the same graph. Therefore, the RMS difference and the two standard deviations were normalized using the observed standard deviation of each field prior to plotting. If a model shows low correlation (signifying low similarity) but also low RMS error (signifying close agreement with observations), it becomes challenging to determine whether the simulation result is good or bad. Therefore, it is essential to define a skill score to evaluate the capability of model simulation. In this study, we used the Taylor Diagram (TD) Score, which is defined as follows:

$$\mathrm{TD} = |1 - r| + \left|1 - \frac{\delta_m}{\delta_o}\right| + \frac{E'}{\delta_o} \quad (7)$$

where r, δ, and E' carry the similar meaning as in the previous formulas.

4 Results

4.1 Observed, Raw and Bias-Corrected SSS

Figure 2(a) visualizes the spatial distribution of SSS for observed ORAS5 data, the top-ranked raw GCM outputs, and the bias-corrected subset of GCMs using the QM algorithm. The bias-corrected values from the top four performing GCMs closely align with the observed ORAS5 data after subtracting the simulated SSS from the observation. Most regions exhibit an SSS increase ranging from 33 psu to 34 psu across the historical period from 1970 to 2014. Notably, to appears decrease in south-eastern part of Bangladesh and western part of Myanmar. But low salinity (psu) is observed east to south of Sri Lanka and Indian coastal areas compared to central part of BOB.

4.2 Valuation of the Simulation Skill of SSS by the CMIP6 Models

The TD in Fig. 3 displays the replicated SSS in the BOB along the CMIP6 models. The observed SSS is specified by the black color line, which is situated at the point with $\mathrm{r} = 1$, $\mathrm{s} = 1$, and $\mathrm{E_0} = 0$. A simulation point that is closer to the observation point demonstrates a higher level of contract with the observations. The CMIP6 models show good arrangement with the observation in this study (Fig. 3). However, the result found negative correlations between models and observed the except MIROC6, EC-Earth3-Veg, EC-Earth3-Veg-LR, NorESM2-LM, and NorESM2-MM models. Where with correlation coefficients of −0.2 to 0.2. The Bay of Bengal (BOB) exhibits the poorest simulation performance, characterized by widely varying correlation coefficients ($-0.6 < \mathrm{r} < 0.95$), spatial standard deviations ranging from 0 to 1.5, and the largest centered root mean square (RMS) differences, which fall between 2.9 and 5.2.

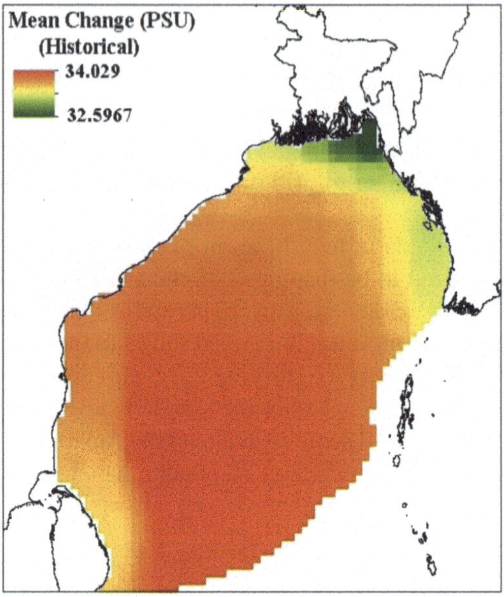

Fig. 2. Observed ORAS5 data were obtained through quantile mapping. The sea surface salinity (SSS) at each grid point was averaged over the period from 1970 to 2014 to prepare the figure.

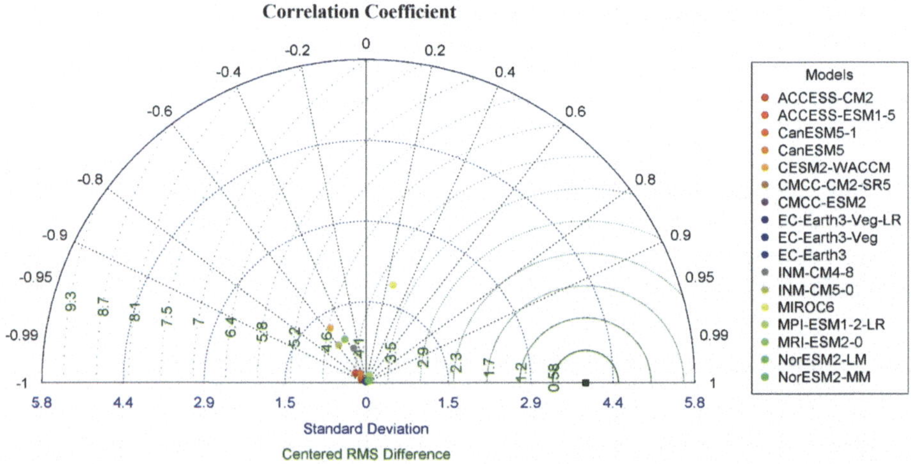

Fig. 3. Taylor Diagram of simulated annual mean sea surface salinity (SSS) in BOB in the CMIP6 models. The blue color indicates the normalized spatial standard deviation. The green color indicates the centered RMS difference, and black solid color correlation coefficient.

4.3 Percentages Changes of SSS Compared to Base Period Under SSP126 Scenario

The spatial distribution of sea surface salinity (SSS) for two periods, the near future (2020–2059) and the far future (2060–2099), across the BOB under SSP126 (Fig. 4). A similar pattern of SSS distribution is observed across all SSP126 within the study area. While, to the far future and worst-case scenarios, there is a larger spatial spread of increased SSS during the near future and for lower in this scenario. This predicted increase up to 15.501% and 23.025% PSU is most noticeable along south and south-east part of Bangladesh and west part of Myanmar. Under this scenario, a decline in SSS is projected for the far future throughout the BOB.

4.4 Percentages Changes of SSS Compared to Base Period Under SSP245 Scenario

Figure 5 shows the distribution of SSS the near future (2020–2059) and the far future (2060–2099), across the BOB under SSP245. In near future the projected changes (0.82% to 26.243% PSU) within the range of observed SSS values (2.089% to 28.803% PSU) under the SSP245 scenario, the result suggested the potential impacts on marine ecosystems, climate, and human activities dependent on sea surface salinity [34], similarly, the projected changes for far future shows (0.142–27.938% PSU) throughout the BOB, consequences of SSS created the surface density and stronger stratification [6, 27].

4.5 Percentages Changes of SSS Compared to Base Period Under SSP370 Scenario

Figure 6 illustrate the uncertainty in the projected SSS, calculated using SSP370 upper and lower bound values. In this scenario the change of lower bound uncertainty in SSS shows minimal variation between the near and distant future (2020–2099). Projected changes in sea surface salinity (SSS) for the near future in range from 0.6220% to 26.666% PSU, compared to historical SSS of 2.041% to 28.714% PSU. In contrast, for the far future, the projected change ranges from −1.006% to 23.552%, compared to observed values of 1.363% to 27.06%. However, south east region Bangladesh and some part of Myanmar. The far future shows a higher prevalence of the greatest values of upper uncertainty compared to the near future. The highest uncertainty is found, particularly in the far future.

4.6 Percentages Changes of SSS Compared to Base Period Under SSP585 Scenario

Figure 7 illustrates the changes in the areal distribution of sea surface salinity (SSS) under the SSP585 scenario. The upper boundary uncertainty ranges from 0.793% to 26.611% PSU for the far future and from −4.185% to 13.155% PSU for the near future. The mean change is projected to be between 2.029% and 28.655% PSU in the near future, and between 1.005% and 25.07% PSU in the far future. Additionally, the lower boundary uncertainty spans from 0.793% to 26.689% PSU for the near future and from

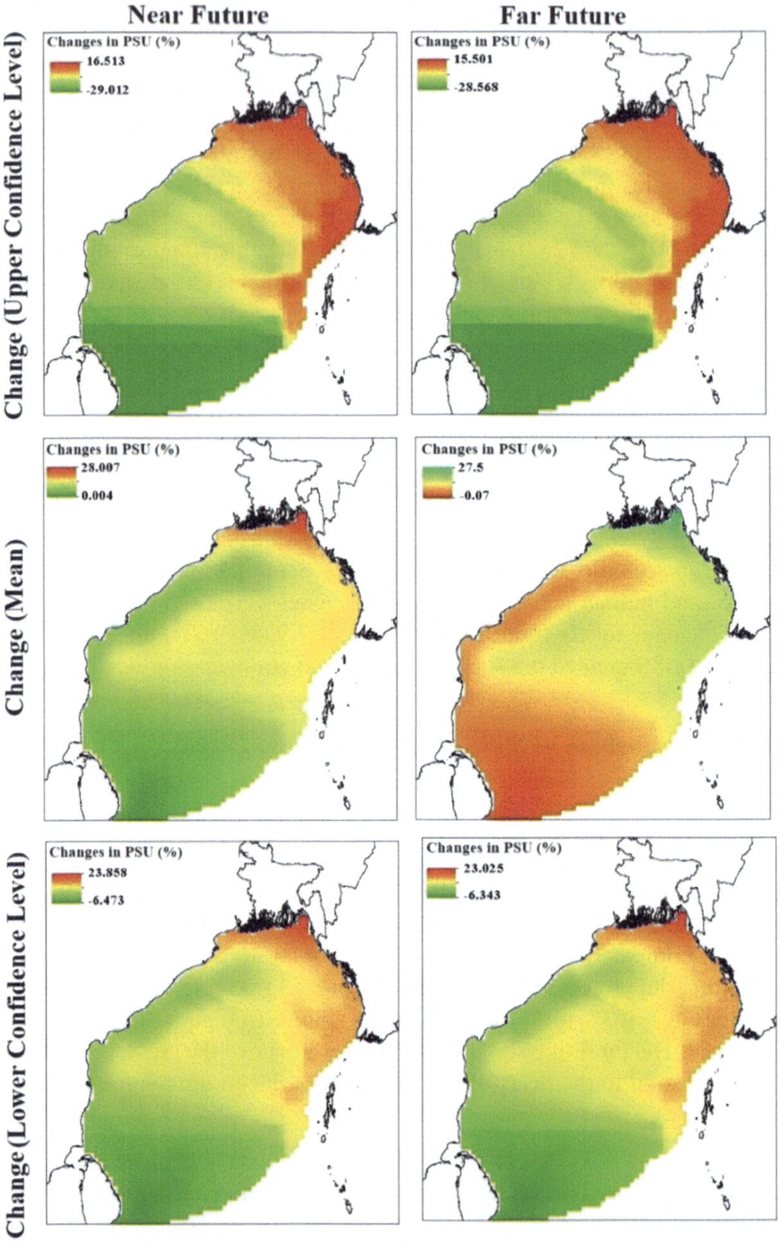

Fig. 4. Plot Projected changes in sea surface salinity (%) compared to the base period for SSP126 in the near (right column) and far (left column) futures with 95% confidence interval. The in the upper row shows the changes in lower bound of 95% confidence band, the lower row shows the changes in upper bound of the band and the middle row shows the projected mean changes.

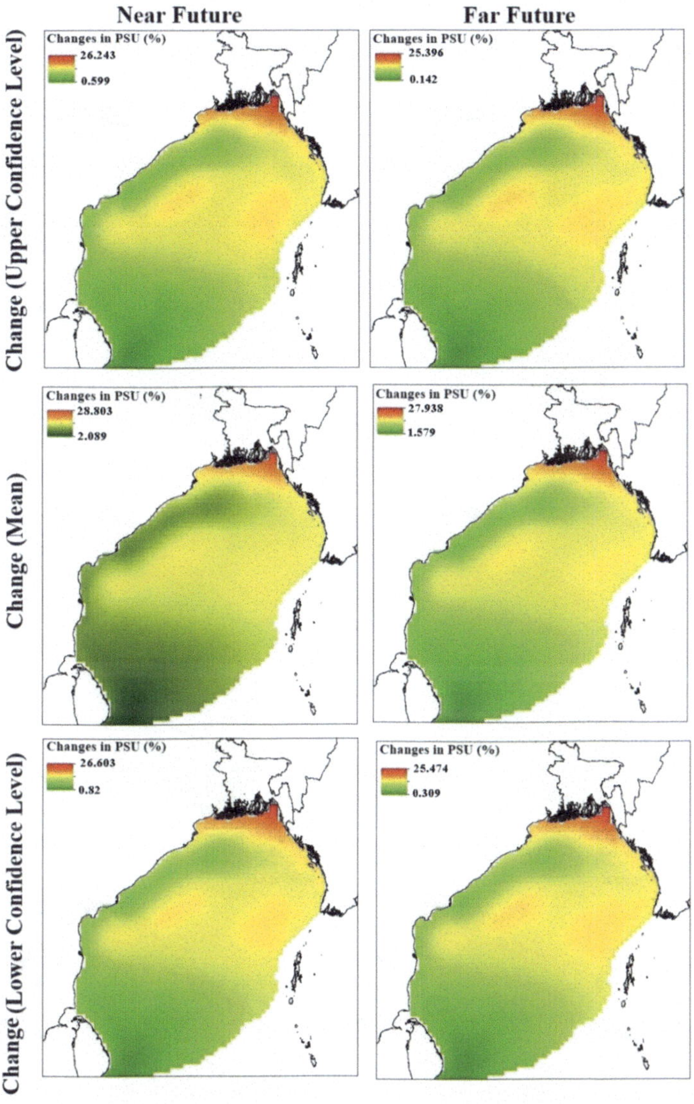

Fig. 5. Projected changes in sea surface salinity (%) compared to the base period for SSP245 in the near (right column) and far (left column) futures with 95% confidence interval. The plot in the upper row shows the changes in lower bound of 95% confidence band, the lower row shows the changes in upper bound of the band and the middle row shows the projected mean changes.

−1.358% to 21.079% PSU for the far future. This statement more clearly separates and explains the different projections and uncertainties for SSS under the SSP585 scenario, providing a clear distinction between near and far future changes in different places in the BOB. The impacts of projected changes in sea surface salinity (SSS) in the Bay of

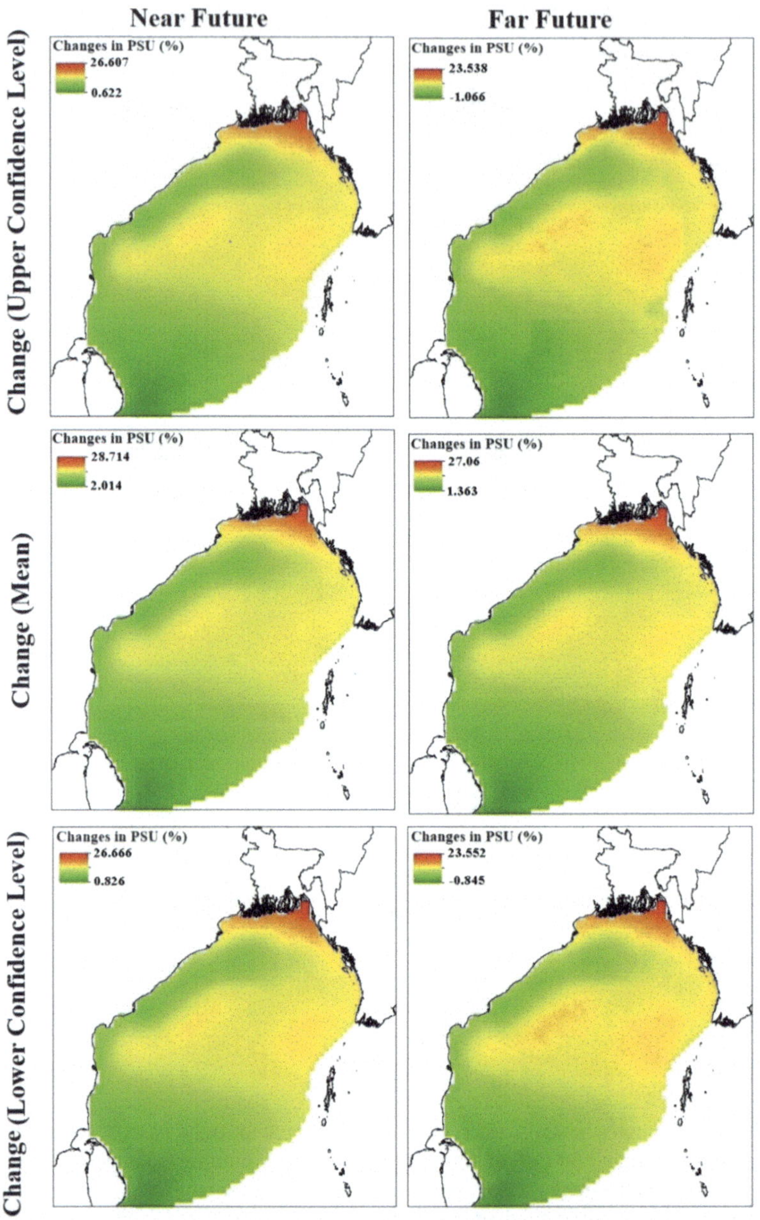

Fig. 6. Projected changes in sea surface salinity (%) compared to the base period for SSP370 in the near (right column) and far (left column) futures with 95% confidence interval. The plot in the upper row shows the changes in lower bound of 95% confidence band, the lower row shows the changes in upper bound of the band and the middle row shows the projected mean changes.

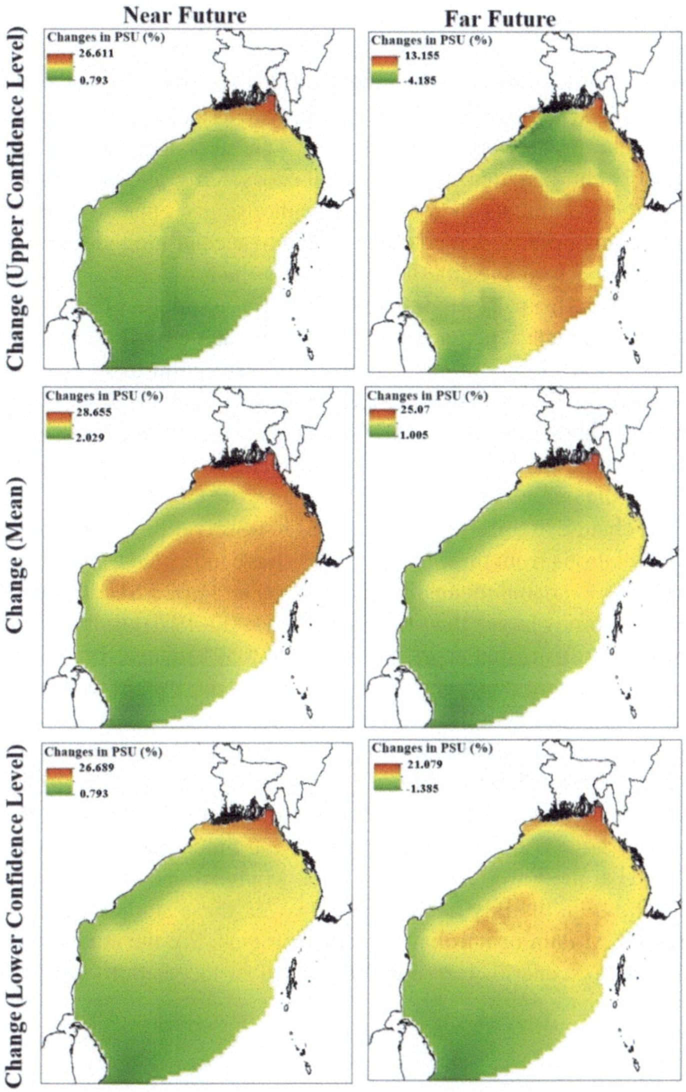

Fig. 7. Projected changes in sea surface salinity (%) compared to the base period for SSP585 in the near (right column) and far (left column) futures with 95% confidence interval. The plot in the upper row shows the changes in lower bound of 95% confidence band, the lower row shows the changes in upper bound of the band and the middle row shows the projected mean changes.

Bengal using SSP scenarios affect ecological dynamics, fisheries, climate systems, and socio-economic conditions (Cui et al. 2024).

5 Discussion

Climate change is significantly transforming the dynamics of our planet's oceans, resulting in a variety of physical changes. These changes are associated with sea level rise and changes in important parameters such as salinity. These changes have significant consequences that affect human, marine and coastal ecosystems, and global socioeconomic context. Due to the continuous increase in greenhouse gas emissions, the once stable oceans are undergoing unprecedented changes due to accelerating global warming. This study utilized a past performance method to identify the top-performing General Circulation Models (GCMs) from 17 models in CMIP6 for simulating observed sea surface salinity (SSS). We evaluated the agreement between each GCM's simulations of these variables and the ORAS5 data at each observation grid point using various statistical metrics, specifically the Root Mean Square Error (RMSE) and the Taylor Diagram (TD). The results indicate that while most CMIP6 models successfully capture the spatial patterns of sea surface salinity (SSS), they generally tend to overestimate SSS in the Bay of Bengal. During the historical period (1900–2014), the CMIP6 models do not show a clear interannual change in SSS trends compared to observations. However, since 1970, the historical simulations from the CMIP6 models have indicated a significant change in SSS. This result shows spatial mean change in 32.60 to 34.02 PSU during (1970–2014) which is around 45 times larger than the past-day (1900–1970) in the CMIP6 models. The SSP370 results indicate that near-future (2020–2059) changes range from −6.22 to 26.60 PSU at the upper confidence level and −0.82 to 26.66 PSU at the lower confidence level. In contrast, far-future (2066–2099) changes range from −1.06 to 23.53 PSU at the upper confidence level and −8.45 to 23.55 PSU at the lower confidence level. Additionally, SSP585 results show that near-future changes range from 0.79 to 26.61 PSU at the upper confidence level and 0.79 to 26.69 PSU at the lower confidence level. In contrast, far-future changes range from −4.18 to 13.15 PSU at the upper confidence level and −1.38 to 21.07 PSU at the lower confidence level. Many studies have projected changes in sea surface salinity in the Bay of Bengal. However, these studies are limited by their use of outdated GCM data or unreliable methodologies. So, the study fills this gap by applying four CMIP6 scenarios to project future sea surface salinity over the BOB.

6 Conclusions

This study projected sea surface salinity along BOB using the four best-selected GCM CMIP6 models. Our findings revealed that even under the best-case scenarios, such as in the SSP126 scenario, sea surface salinity (SSS) changes range from −29.01 to 16.51 PSU in the near future and from −28.56 to 15.50 PSU in the far future. In the SSP245 scenario, SSS changes range from 0.59 to 26.24 PSU in the near future and from 0.14 to 25.34 PSU in the far future. For the SSP370 scenario, SSS changes range from −6.22 to 26.60 PSU in the near future and from −1.06 to 23.53 PSU in the far future. Finally, in the SSP585 scenario, SSS changes range from 0.79 to 26.61 PSU in the near future and from −4.18 to 13.15 PSU in the far future, all with a 95% confidence interval. The projected to experience the highest changes south and south east of Bangladesh, south west of Myanmar, north and east coast of Srilanka and some part of India. The most concerning finding is the projected change in sea surface salinity (SSS) in the Bay of Bengal, which could significantly impact fisheries, wet land ecosystems, and marine biodiversity. Furthermore, the projections indicate diverse regional changes in SSS, with important implications for coastal communities and marine ecosystems. There is considerable uncertainty in these SSS projections. Future work could address this by utilizing a broader range of observational data, including tidal gauge measurements, to further validate the accuracy of GCMs. Integrating tidal gauge data together with existing satellite-based measurements and reanalysis data would provide a more comprehensive validation approach.

References

1. Akhil, V.P., Lengaigne, M., Krishnamohan, K.S., Keerthi, M.G., Vialard, J.: Southeastern Arabian sea salinity variability: mechanisms and influence on surface temperature. Clim. Dyn. **61**(7), 3737–3754 (2023)
2. Asok, A.B., Joshi, A.P., Warrior, H.V.: Ocean heat content variability in the Bay of Bengal: a CMIP6 model analysis with implications on the Indian Ocean Dipole. Curr. Sci. **126**(8), 00113891 (2024)
3. Banerjee, A., Kang, S., Meadows, M.E., Sajjad, W., Bahadur, A., Moazzam, M.F.U., Kirsten, K.L.: Evaluating the relative influence of climate and human activities on recent vegetation dynamics in West Bengal, India. Environ. Res. **250**, 118450 (2024)
4. Bhattacharya, B., Mohanty, S., Singh, C.: Assessment of the potential of CMIP6 models in simulating the sea surface temperature variability over the tropical Indian Ocean. Theoret. Appl. Climatol. **148**(1), 585–602 (2022)
5. Bindoff, Nathaniel, L., Willebrand, J., Artale, V., Cazenave, A., Gregory, J.M., Gulev, S., Hanawa, K., et al.: Observations: oceanic climate change and sea level 385–428 (2007)
6. Chaitanya, A.V.S., Durand, F., Mathew, S., Gopalakrishna, V.V., Papa, F., Lengaigne, M., Venkatesan, R., et al.: Observed year-to-year sea surface salinity variability in the Bay of Bengal during the 2009–2014 period. Ocean Dyn. **65**, 173–186 (2015)
7. Church, J.A., Gregory, J.M., Huybrechts, P., Kuhn, M., Lambeck, K., Nhuan, M.T., Woodworth, P.L., et al.: Changes in sea level. In: Houghton, J.T., Ding, Y., Griggs, D.J., Noguer, M., Van der Linden, M., Dai, X., Maskell, K., Johnson, C.A. (eds.) Climate Change 2001: The Scientific Basis: Contribution of Working Group I to the Third Assessment Report of the Intergovernmental Panel, pp. 639–694 (2001)
8. Dehairs, F., Rao, R.G., Chandra Mohan, P., Raman, A.V., Marguillier, S., Hellings, L.: Tracing mangrove carbon in suspended matter and aquatic fauna of the Gautami-Godavari Delta, Bay of Bengal (India). Hydrobiologia **431**, 225–241 (2000)
9. Delcroix, T., Henin, C., Porte, V., Arkin, P.: Precipitation and sea-surface salinity in the tropical Pacific Ocean. Deep Sea Res. Part I **43**(7), 1123–1141 (1996)
10. Durack, P.J.: Ocean salinity and the global water cycle. Oceanography **28**(1), 20–31 (2015)
11. Eyring, V., et al.: Overview of the coupled model intercomparison project phase 6 (CMIP6) experimental design and organization. Geosci. Mod. Dev. **9**(5), 1937–1958 (2016)
12. Fan, K., Wang, X., He, Z.: Control of salinity stratification on recent increase in tropical cyclone intensification rates over the postmonsoon Bay of Bengal. Environ. Res. Lett. **15**(9), 094028 (2020)
13. Fathrio, I., Manda, A., Iizuka, S., Kodama, Y. M., Ishida, S.: Evaluation of CMIP5 models on sea surface salinity in the Indian Ocean. In: IOP Conference Series: Earth and Environmental Science, vol. 54, no. 1, p. 012039. IOP Publishing (2017)

14. Gudmundsson, L., Bremnes, J.B., Haugen, J.E., Skaugen, T.E.: Downscaling RCM precipitation to the station scale using quantile mapping–a comparison of methods. Hydrol. Earth Syst. Sci. Discuss. **9**(5), 6185–6201 (2012)
15. Gulev, S. K., Grigorieva, V.: Last century changes in ocean wind wave height from global visual wave data. Geophys. Res. Lett. **31**(24) (2004)
16. Hasan, M., Uddin, M.M., Mahbub-E-Kibria, A.S.M., Das, M.K., Deb, D.: A GIS-based integrated study assessing the socio-physical vulnerability of the south-central and south-western coastal regions of Bangladesh. GeoJournal **89**(4), 163 (2024)
17. Helm, K.P., Bindoff, N.L., Church, J.A.: Changes in the global hydrological-cycle inferred from ocean salinity. Geophys. Res. Lett. **37**(18) (2010)
18. Jin, Y., Chen, M., Yan, H., Wang, T., Yang, J.: Sea level variation in the Arctic Ocean since 1979 based on ORAS5 data. Front. Mar. Sci. **10**, 1197456 (2023)
19. Kankaanpää, A., Tolvanen, A., Joensuu, L., Waller, K., Heikkinen, A., Kaprio, J., Sillanpää, E., et al.: The associations of long-term physical activity in adulthood with later biological ageing and all-cause mortality–a prospective twin study. *medRxiv* (2023)
20. Kim, H., Timmermann, A., Lee, S.S., Schloesser, F.: Rainfall and salinity effects on future Pacific climate change. Earth's Future **11**(8), e2022EF003457 (2023)
21. Kirezci, E., et al.: Projections of global-scale extreme sea levels and resulting episodic coastal flooding over the 21st Century. Sci. Rep. **10**(1), 1–12 (2020)
22. Kumar, V., Joshi, A.P., Warrior, H.V.: Study of sea surface salinity due to river fluxes using the CMIP6 models for the Bay of Bengal region. Curr. Sci. **124**(11), 00113891 (2023)
23. Li, Y., Ren, T., Kinney, P.L., Joyner, A., Zhang, W.: Projecting future climate change impacts on heat-related mortality in large urban areas in China. Environ. Res. **163**, 171–185 (2018)
24. Liu, M., Peng, J., Dong, J., Jiang, H., Xu, D., Meersmans, J.: Trade-offs of landscape connectivity between regional and interregional ecological security patterns in a junction area of Beijing-Tianjin-Hebei region. Appl. Geogr. **167**, 103272 (2024)
25. Liu, Y., Cheng, L., Pan, Y., Tan, Z., Abraham, J., Zhang, B., Song, J., et al.: How well do CMIP6 and CMIP5 models simulate the climatological seasonal variations in ocean salinity? Adv. Atmos. Sci. **39**(10), 1650–1672 (2022)
26. Madhupratap, M., Gauns, M., Ramaiah, N., Kumar, S.P., Muraleedharan, P.M., De Sousa, S.N., Muraleedharan, U., et al.: Biogeochemistry of the Bay of Bengal: physical, chemical and primary productivity characteristics of the central and western Bay of Bengal during summer monsoon 2001. Deep Sea Res. Part II: Topical Stud. Oceanogr. **50**(5), 881–896 (2003)
27. Marsh, R., Dey, D., Lenn, Y.D., Roberts, E.M.: Shifts from surface density compensation to projected warming, freshening and stronger stratification in the subpolar North Atlantic. Clim. Dyn. 1–27 (2024)
28. Masson-Delmotte, V., Zhai, P., Pirani, A.: Intergovernmental panel on climate change (IPCC), 2021: summary for policymakers. In: Connors, S.L., Péan, C., Berger, S., Caud, N., Chen, Y., Goldfarb, L., Zhou, B. (eds.) Climate Change 2021: The Physical Science Basis. Contribution of Working Group I to the Sixth Assessment Report of the Intergovernmental Panel on Climate Change, pp. 3–32. Cambridge University Press (2023)
29. Meehl, G.A., Covey, C., Delworth, T., Latif, M., McAvaney, B., Mitchell, J.F., Taylor, K.E., et al.: The WCRP CMIP3 multimodel dataset: a new era in climate change research. Bull. Am. Meteorol. Soc. **88**(9), 1383–1394 (2007)
30. Menezes, V.V., Vianna, M.L., Phillips, H.E.: Aquarius sea surface salinity in the South Indian Ocean: revealing annual-period planetary waves. J. Geophys. Res.: Oceans **119**(6), 3883–3908 (2014)
31. Meng, L., Zhuang, W., Zhang, W., Yan, C., Yan, X.H.: Variability of the shallow overturning circulation in the Indian Ocean. J. Geophys. Res. Oceans **125**(2), e2019JC015651 (2020)

32. Ouyang, P., et al.: Integrating biochar and bacteria for sustainable remediation of metal-contaminated soils. Biochar **5**(1), 63 (2023)
33. Panofsky, H.A., Brier, G.W.: Some applications of statistics to meteorology, 224 pp. Pa. State Univ., University Park, Pa (1968)
34. Röthig, T., Trevathan-Tackett, S.M., Voolstra, C.R., Ross, C., Chaffron, S., Durack, P.J., Sweet, M., et al.: Human-induced salinity changes impact marine organisms and ecosystems. Glob. Change Biol. **29**(17), 4731–4749 (2023)
35. Sajidh, C.K., Chatterjee, A.: Indian Ocean dynamic sea level, its variability and projections in CMIP6 models. Clim. Dyn. **61**(5), 2229–2252 (2023)
36. Samanta, S., Sahoo, J.R., Mishra, S.: Effect of dynamic strain ageing on flow stress and critical strain for jerky flow in Al-Mg alloys. Int. J. Plastic. **180**, 104053 (2024)
37. Sannino, G., et al.: Modelling present and future climate in the Mediterranean Sea: a focus on sea-level change. Clim. Dyn. **59**(1), 357–391 (2022)
38. Shasan, S., Sarkar, U.K., Ali, S., Kumari, S., Puthiyotti, M.: Status, ecological services and management of aquatic weeds of floodplain wetlands in India: an overview. Lakes Reserv. Res. Manag. **26**(1), 76–91 (2021)
39. Sein, D.V., Martyanov, S.D., Dvornikov, A.Y., Cabos, W., Ryabchenko, V.A., Mishra, A.K., Kumar, P., et al.: Future climate change in the Northern Indian Ocean as simulated with a high-resolution regional earth system model. Clim. Dyn. **62**(2), 911–932 (2024)
40. Shroyer, E.L., Gordon, A.L., Jaeger, G.S., Freilich, M., Waterhouse, A.F., Farrar, J.T., Mahadevan, A., et al.: Upper layer thermohaline structure of the Bay of Bengal during the 2013 northeast monsoon. Deep Sea Res. Part II: Top. Stud. Oceanogr. **172**, 104630 (2020)
41. Taylor, K.E.: Summarizing multiple aspects of model performance in a single diagram. J. Geophys. Res. Atmos. **106**(D7), 7183–7192 (2001)
42. Taylor, K.E., Stouffer, R.J., Meehl, G.A.: An overview of CMIP5 and the experiment design. Bull. Am. Meteor. Soc. **93**(4), 485–498 (2012)
43. Vazquez-Cuervo, J., García-Reyes, M., Gómez-Valdés, J.: Identification of sea surface temperature and sea surface salinity fronts along the California Coast: application using Saildrone and satellite derived products. Remot. Sens. **15**(2), 484 (2023)
44. Visvalingam, G., Krishnaraj, S., Andiyappan, R.K., Kamalapathy, R., Datchanamourthy, S.V., Lagudu, S.: Understanding the impact of climate-induced sea level rise on groundwater inundation in a low-lying coastal area: a numerical simulation in Southeast India. Reg. Stud. Marine Sci. **70**, 103401 (2024)
45. Winkelbauer, S., Mayer, M., Seitner, V., Zsoter, E., Zuo, H., Haimberger, L.: Diagnostic evaluation of river discharge into the Arctic Ocean and its impact on oceanic volume transports. Hydrol. Earth Syst. Sci. **26**(2), 279–304 (2022)
46. Yang, H., Gou, X., Xue, B., Ma, W., Kuang, W., Tu, Z., Zhang, J., et al.: Research on the change of alpine ecosystem service value and its sustainable development path. Ecolo. Ind. **146**, 109893 (2023)
47. Zhang, Q., Shen, Z., Pokhrel, Y., Farinotti, D., Singh, V.P., Xu, C.Y., Wang, G., et al.: Oceanic climate changes threaten the sustainability of Asia's water tower. Nature **615**(7950), 87–93 (2023)
48. Zhang, R.H., Zhou, G., Zhi, H., Gao, C., Wang, H., Feng, L.: Salinity interdecadal variability in the western equatorial Pacific and its effects during 1950–2018. Clim. Dyn. **60**(7), 1963–1985 (2023)

Seasonal Variability of Sea Surface Salinity in Malaysian Waters Using Space-Based Observation of SMOS

Syarif Abdul Asa'ad Jairin[1(✉)], Ami Hassan Md Din[1], Mat Nizam Uti[2], Mohammad Hanif Hamden[1], and Zamri Ismail[1]

[1] Geospatial Imaging and Information Research Group (Gi2RG), Faculty of Built Environment and Surveying, Universiti Teknologi Malaysia, 81310 Johor Bahru, Johor, Malaysia
saasaad2@graduate.utm.my, amihassan@utm.my

[2] Kolej Alam Bina, Universiti Teknologi Mara Cawangan Perak, Kampus Seri Iskandar, 32610 Seri Iskandar, Perak, Malaysia

Abstract. Sea surface salinity (SSS) is a crucial but often overlooked oceanic parameter that significantly influences the environment, mainly marine ecosystems. It plays a vital role in shaping the water cycle, predicting the global water balance, and enhancing our comprehension of ocean currents. Generally, Sea Surface Salinity (SSS) measurements obtained through in situ techniques, such as those conducted by ships, moored buoys, and profiling floats, are unable to provide extensive spatial and temporal resolution datasets. The advancement in space-based observation has introduced Sea Surface Salinity (SSS) measurements through the Soil Moisture and Ocean Salinity (SMOS) mission, enabling global SSS measurements. This information is crucial for climate studies, weather monitoring, and coastal surveillance. This paper focuses on evaluating the seasonal variability of sea surface salinity in Malaysian waters utilising data from the Soil Moisture and Ocean Salinity (SMOS) mission. The study extracts two years of corrected SSS data (from January 2022 to December 2023) from the European Space Agency (ESA) website. The extracted data is used to assess the SSS changes due to seasonal monsoon. Two major monsoons assessed include the Southwest monsoon (SW) and Northeast monsoon (NE). The analysis of seasonal variability reveals that the coastal area of the Malaysian Sea exhibits higher sea surface salinity during the SW monsoon compared to the NE monsoon. The study demonstrates that seasonal changes in Malaysia play a pivotal role in influencing the variations in sea surface salinity (SSS) over the Malaysian waters. Consequently, the findings of this paper could offer valuable insights into SSS conditions for local authorities, aiding in the mitigation and monitoring of the effects of climate change.

Keywords: Sea surface salinity · SMOS · Seasonal variability

1 Research Background

Ocean salinity exhibits variations across regions, seasons, and depths, driven by intricate interactions among the atmosphere, land, and ocean [1]. A key factor influencing sea surface salinity (SSS) variability is the equilibrium between the rate of precipitation,

A. Abdul-Rahman et al. (Eds.): GeoWeek 2024, LNGC, pp. 162–171, 2025.
https://doi.org/10.1007/978-3-031-86654-8_12

river discharge and evaporation. Precipitation, including rain and snow, introduces freshwater to the ocean, while evaporation extracts water, concentrating salt and elevating overall salt concentration [1, 2]. Understanding variations in SSS is crucial for unravelling climate patterns, as changes in salinity directly impact ocean circulation, freshwater distribution, and, consequently, global climate systems. Several studies have investigated Sea Surface Salinity (SSS) in Malaysian waters, employing diverse approaches, including salinity profiling and examining spatial variations [3] and space-based SSS validation [4]. Sea Surface Salinity (SSS) data is accessible through various sources, encompassing real-time and historical datasets, including in-situ measurements such as Conductivity, Temperature, and Depth (CTD) and ARGO floats. Additionally, ocean data platforms like Integrated Marine Observing System (IMOS) or the Global Ocean Observing System (GOOS), findings from oceanographic research cruises, and satellite sensors like NASA's Aquarius, ESA's Soil Moisture Ocean Salinity (SMOS), and NASA's Soil Moisture Active Passive (SMAP) missions contribute to the wealth of available information.

The Sea Surface Salinity (SSS) data for this study will be acquired from the European Space Agency's (ESA) online dissemination service. This service offers satellite-derived information on SSS, providing extensive coverage across various geographic areas with large spatial and temporal resolutions. Recent research in the realm of sea surface salinity has underscored the need for a nuanced understanding of this parameter, particularly in the context of Malaysia. A notable challenge identified in these studies is the scarcity of in-situ data, particularly from buoys, and the limited sharing of marine data by local authorities [5, 6]. This lack of data has hindered a comprehensive assessment of SSS in Malaysian waters, posing challenges to researchers aiming to decipher the seasonal variations in sea surface salinity. To bridge this gap, the study utilises sea surface salinity (SSS) data extracted from the SMOS mission via the European Space Agency's (ESA) online dissemination service. Employing satellite-derived information, the research conducts a comprehensive analysis of SSS variations in Malaysian waters. The objective is to contribute valuable insights into climate dynamics, marine ecosystems, and the regional implications of sea surface salinity changes.

1.1 Sea Surface Salinity (SSS)

Ocean salinity refers to the concentration of dissolved salts in the entire water column of the ocean, from the surface to the depths [7]. Ocean salinity is typically measured in parts per thousand (ppt) or practical salinity units (PSU) [8]. Sea Surface Salinity (SSS) is a crucial oceanographic parameter that measures the concentration of dissolved salts in the upper layer of the ocean, specifically at the sea surface. The basic principle of sea surface salinity is related to the water cycle and various processes that affect the salt content of the ocean [9]. Evaporation concentrates salts in seawater as only water molecules turn into vapour, leaving salts behind. Conversely, precipitation, river runoff, freezing and melting of ice, as well as vertical mixing and ocean circulation, play roles in either diluting or increasing the salinity of the ocean, creating a dynamic balance in the distribution of salts across different regions and depths [10].

1.2 Principle of Soil Moisture and Ocean Salinity (SMOS) Mission

Since its launch on November 2, 2009, the Soil Moisture and Ocean Salinity (SMOS) mission has pioneered the comprehensive observation of global soil moisture and sea surface salinity (SSS), utilising the Microwave Imaging Radiometer using Aperture Synthesis (MIRAS) as its singular onboard instrument. Operating in a sun-synchronous polar orbit with altitudes ranging from 761.3 km to 788.4 km and an inclination of 98.42°, SMOS achieves a repeat cycle of 149 days, subdivided into an 18-day sub-cycle [11]. Notably, SMOS attains an impressive accuracy of 0.1 practical salinity scale unit over open ocean expanses measuring 200×200 km^2, facilitating the study of ocean circulation patterns, global freshwater dynamics, and climate impacts [12]. MIRAS, functioning at the L-band frequency of 1.413 GHz, employs a 2-D interferometric polarimetric radiometric approach, synthesising signals from its array of 69 receivers to generate precise images. With a swath coverage of approximately 1000 km and spatial resolution varying between 35 and 50 km, SMOS data aids in diverse applications, including weather forecasting, climate studies, and environmental monitoring initiatives [13].

1.3 Seasonal Variability in Malaysian Waters

Monsoons, a seasonal phenomenon prevalent across much of South and East Asia, including Malaysia, are driven by the movement of air masses between land and sea and are governed by density and pressure differentials [14]. During the cold season, denser, high-pressure air over land leads to winds blowing from land to sea, while in the hot season, the reverse occurs, resulting in monsoonal shifts in wind direction. In summer, the Asian continent's high temperatures create a low-pressure zone, causing the southwest monsoon, which brings moisture-laden air from the Indian Ocean to the land, manifesting in Malaysia from May to August, ushering in a slightly wet season. Conversely, during winter, high pressure over Asia triggers the northeast monsoon, characterised by rain in Malaysia from November to February, as winds from the South China Sea combine with cold air from Siberia. Transition monsoons, occurring in March-April (First-inter monsoon) and September-October (Second-inter monsoon), involve slow and fluctuating wind movements between the northeast and southwest regions, marking brief periods of transition [15].

2 Data and Methods

The seasonal variability of this study utilises the SSS data extracted from the online dissemination service of the European Space Agency (ESA). The observation of SSS from the SMOS mission is done on a daily basis. Two years of data sets from 1st January 2022 until 31st December 2023 retrieved covered the study area bounded from 0° N to 14° N and 95° E to 126° E, as depicted in Fig. 1. The available data is based on the SMOS tracks over the study area, disseminated in the form of Network Common Data Form (NetCDF), and readable using the programming language. MATLAB software is utilised to develop code for data extraction, management, and analysis.

Prior to seasonal assessment, a gridding process using the geostatistical interpolation technique of Kriging is performed. This method is useful to deal with spatially distributed

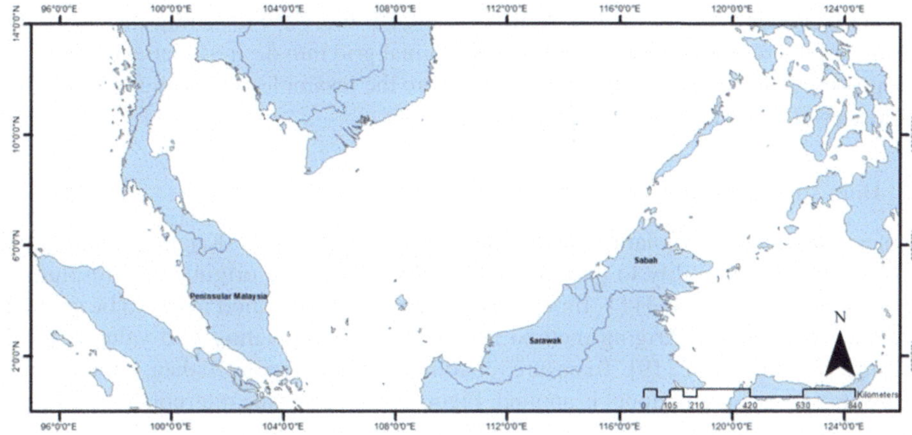

Fig. 1. The study area encompasses the South China Sea, Malacca Straits, Sulu Sea, and Celebes Sea.

data points to estimate values at unmeasured locations within the study area [16]. For validation purposes, this study employed external validation, which involves confirming whether the SSS data used in this study supports the results of a prior study. The external validation is performed by considering the availability of in-situ observations from other sources, such as buoys. However, the lack of in-situ data availability due to the sparse buoy network and disengagement of data sharing for marine authorities made this study employ external validation compared to performing the ground-truthing process. After that, seasonal variability of SSS is performed to examine the seasonal effect of SSS on Malaysian waters.

2.1 Data Extraction and Preprocessing

As mentioned before, the SSS data is an open data type which can be downloaded through the ESA online dissemination service (https://smos-diss.eo.esa.int/oads/access/). ESA provides different types of data, which are categorised into two levels: scientific data Level 1 and Level 2. Level 1 data is designed for scientific users who need to work with calibrated MIRAS instruments measurements. In contrast, Level 2 data products are designed for scientific and operational users who need to work with geo-located soil moisture and sea surface salinity as retrieved from the Level 1 dataset [17]. This study utilised the Level 2 Ocean Salinity (OS) product, which comprises the sea surface salinity measurements geo-located in grid system ISEA 4H9. The extracted Level 2 data used in this study is swath-based sea surface salinity with land-sea contamination correction and is termed SSS_corr. The corrected salinities data is almost cancelling the global mean bias in near-to-coast regions within the range of more than 40 km and less than 800 km [18].

The retrieved data is large; thus, comprehensive managing is performed to ensure the data is extracted inside the study area boundary. The MATLAB software is used to handle data. A series of MATLAB codes are developed to manage the big data, beginning with (1) Extracting SSS_corr data from the NetCDF file, (2) Subsetting the SSS data into area of interest, and (3) Resampling the original grid into desired spatial resolution. This process requires thorough planning to ensure the resample data is aligned with the boundaries of the study.

2.2 Data Interpolation

As mentioned, the interpolation method used in this study is the Kriging method due to its flexibility and ability to give accurate gridded data. The original grid resolution for the SSS_corr data is 0.3111° times 0.3131°. The kriging method needs to be applied accurately because the Kriging method is capable of extrapolating grid value beyond the range of data's value [19]. To ensure reliability of the interpolated data, a linear with an anisotropy variogram model is applied. Figure 2 represents the interpolated SSS data from the Kriging technique and visualised by ArcMap 10.3 software.

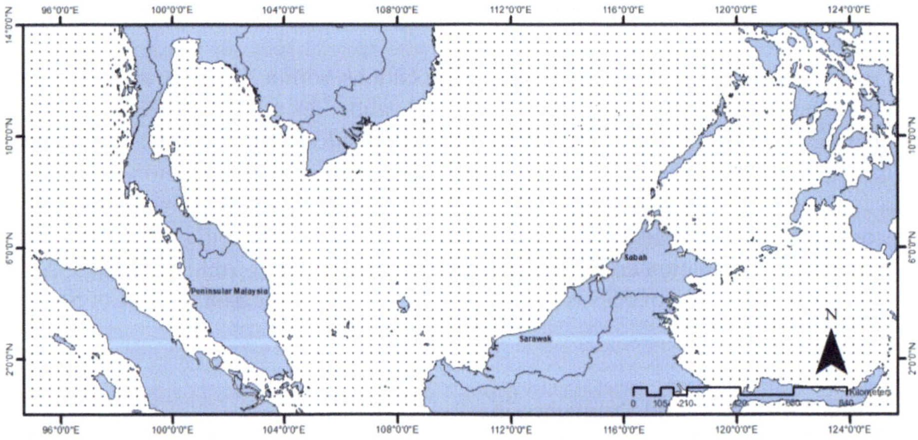

Fig. 2. The gridded Sea Surface Salinity (SSS) data, generated using the Kriging technique.

2.3 Seasonal Variability Assessment

In weather-dependent countries like Malaysia, the monsoon season can have a significant impact on various weather parameters, including sea surface salinity. Malaysia experiences two main monsoon seasons—the Southwest Monsoon (May to August) and the Northeast Monsoon (November to February) [6]. Furthermore, the salinity in a particular region can also be changed by other factors such as local geography, ocean currents, and human activities [20]. Besides, the changes in SSS in particular areas also vary due to precipitation, rainfall, and temperature [21]. This paper highlights the changes in the sea surface salinity due to the seasonal variability experienced in Malaysian waters. Studying the changes in SSS in Malaysia is essential for several reasons, as it provides valuable insights into various environmental and climatic aspects. For example, monitoring changes in salinity helps in understanding the climatic phenomena, which are crucial for predicting and managing the impacts of climate change in coastal agriculture due to salinity intrusion [22]. The assessment of seasonal variability is performed by developing a programming code using MATLAB software. Apart from data extraction and management, this study applied the MATLAB functions for data visualisation by plotting the interpolated SSS data. This process is crucial for identifying the trends and the SSS variations over time.

3 Results and Discussion

In this section, the outcomes of the investigation into sea surface salinity integrate essential elements of external validation and the intricate nuances of seasonal variability. Delving into the seasonal variability of sea surface salinity reveals distinct patterns, unravelling the dynamic interplay between monsoon season and sea surface salinity over Malaysian waters.

3.1 Seasonal Variability of Sea Surface Salinity Over Malaysian Waters

As mentioned, Malaysia experienced two major monsoons: the southwest monsoon and the northeast monsoon. During the Southwest monsoon (May to August), Malaysia typically experiences dry and hot weather, with winds blowing from the southwest direction [6, 14]. This season is associated with lower precipitation in most parts of the country. On the other hand, the Northeast Monsoon brings heavy rainfall as winds blow from the northeast, bringing moisture-laden air from the South China Sea [6, 23].

During the southwest monsoon, reduced precipitation leads to less freshwater input into the sea, causing an increase in sea surface salinity in the coastal area. Besides that, the slightly wet weather could increase evaporation due to higher temperatures, which can also contribute to higher salinity levels [24]. Figure 3 represents the variations of sea surface salinity over the Malaysian waters during the Southwest monsoon (May to August).

Compared to the Southwest monsoon, the sea surface salinity levels in the coastal area during the Northeast monsoon are expected to be lower, as shown in Fig. 4. This is due to the Northeast monsoon bringing heavy rainfall to the region and diluting the salinity at the surface of the sea. Heavy rainfall also increases river discharge, carrying freshwater from the land into the ocean and further diluting the salinity of the surface water.

3.2 External Validation

Based on the extracted SSS data from SMOS, they are congruent with the earlier research. A study by Roseli et al. [25] presented seasonal variability of salinity data (March to May) from the global oceanic prediction model generated by HYCOM in the Malacca Straits and South China Sea. Based on Roseli et al. [12] study showed that the average value of modelled SSS (March to April) in Malacca Straits and South China Sea ranges from 32 PSU to 35 PSU, as shown in Fig. 5 (a). Meanwhile, the results of the average

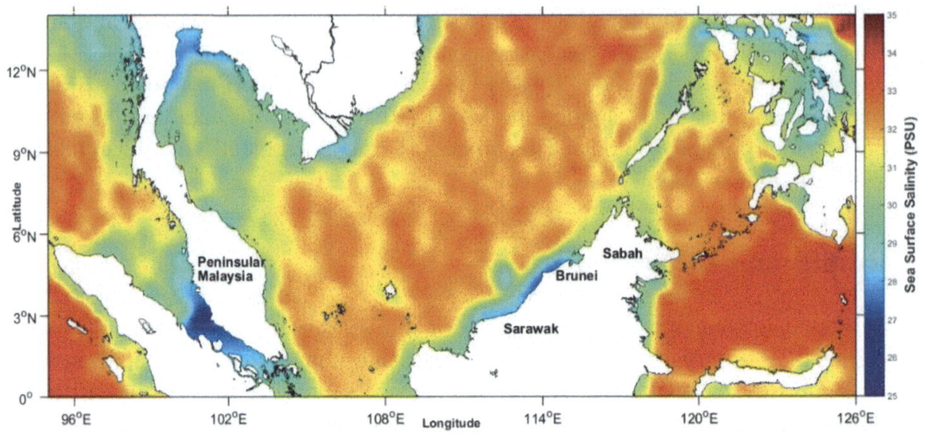

Fig. 3. The average seasonal variability of Sea Surface Salinity (SSS) during the Southwest monsoon (May–Aug) of 2022–2023.

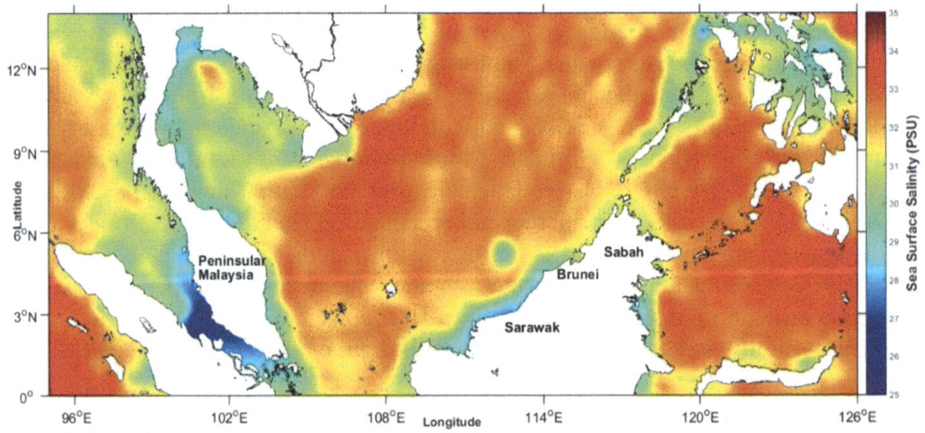

Fig. 4. The average seasonal variability of Sea Surface Salinity (SSS) during the Northeast monsoon (Nov–Feb) of 2022–2023

SSS (March to May) of this study using the SSS data from SMOS also present SSS values ranging up to 35 PSU, as depicted in Fig. 5 (b).

In this study, the SSS observed by SMOS revealed a minimum value of 25 PSU in the Malacca Straits (as in Fig. 5 (b)). According to Refs. [26, 27], the ocean salinity observed by SMOS considers additional factors such as the water cycle and sea surface temperature when observing sea surface salinity (SSS). This integration makes SMOS salinity data considerably more reliable than model data. According to Ref. [26], the SMOS SSS errors are small in the tropic regions. Mu et al. [28] demonstrated that bias and root mean square error (RMSE) in SMOS sea surface salinity (SSS) compared with Argo observations decreased in the South China Sea. The values were reduced from 0.155

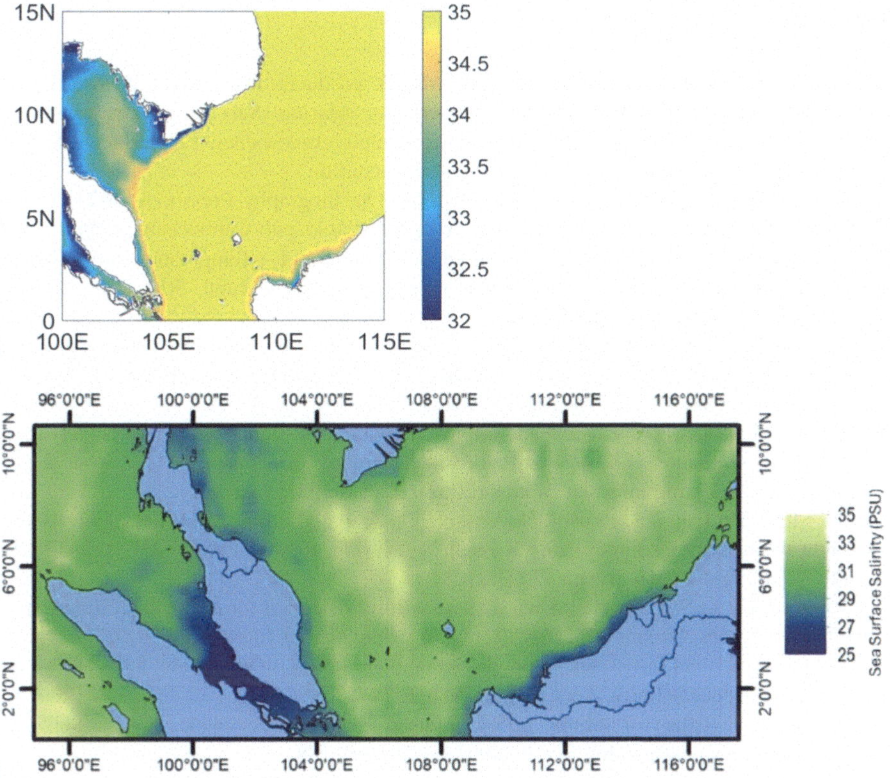

(a)

(b)

Fig. 5. Figure (a) depicts the seasonal variability of salinity (March to April) from model data (HYCOM) as presented by [12], while Figure (b) portrays the seasonal variability (March to April) of Sea Surface Salinity (SSS) data from SMOS, utilised in this study.

PSU and 0.415 PSU to − 0.003 PSU and 0.112 PSU, respectively. The assimilation of corrected SMOS SSS effectively enhanced upper ocean salinity simulation.

4 Conclusion

In conclusion, the positive correlation and acceptable RMSE values have proved that the measurement of sea surface salinity from the SMOS mission is reliable for seasonal variability studies. The climate factor of both rainfall and river discharge presents a better understanding of the influence of seasonal variability towards sea surface salinity. Both heavy rainfall and high river discharge during the Northeast Monsoon (November-February) and less during the Southwest Monsoon (May-August). This study is critical to providing better ocean dynamic information for the marine sectors, such as fisheries, tourism, renewable energy, and oil and gas activity.

Acknowledgements. This work was funded by the Ministry of Higher Education under the Fundamental Research Grant Scheme (FRGS/1/2023/WAB05/UTM/02/1).

References

1. Copernicus: New Sea Surface Salinity (SSS) products from the Soil Moisture and Ocean Salinity (SMOS) Satellite. Copernicus Marine Service. Accessed: 07 Mar 2024. [Online]. Available: https://marine.copernicus.eu/news/new-sea-surface-salinity-sss-products-soil-moisture-and-ocean-salinity-smos-satellite
2. Webb, P.: 5.3 salinity patterns. In: Introduction to Oceanography, Pressbook, 2023. Accessed: 07 Mar 2024. [Online]. Available: https://rwu.pressbooks.pub/webboceanography/
3. Aletheia, A., Manson, A., Saleh, E., Mujahid, A., Juneng, L.: Temperature and salinity profiling analysis off Sarawak waters, Malaysia. Trans. Sci. Technol. **5**(3), 216–220 (2018). [Online]. Available: http://transectscience.org/
4. Reba, M.N.M., Rosli, A.Z., Rahim, N.A.: Soil moisture ocean salinity (SMOS) salinity data validation over Malaysia coastal water. IOP Conf. Ser. Earth Environ. Sci. **18**, 012112 (2014). https://doi.org/10.1088/1755-1315/18/1/012112
5. Ramli, M.R.M.: Physical oceanography activities in Malaysian waters. Hydro Int. [Online]. Available: https://www.hydro-international.com/content/article/physical-oceanography-activities-in-malaysian-waters (2002). Accessed: 07 Mar 2024
6. Uti, M.N., Md Din, A.H., Yusof, N., Yaakob, O.: A spatial-temporal clustering for low ocean renewable energy resources using K-means clustering. Renew. Energy **219**, 119549 (2023). https://doi.org/10.1016/j.renene.2023.119549
7. Boutin, J. et al.: Satellite-based sea surface salinity designed for ocean and climate studies. J. Geophys. Res. Oceans **126**(11) (2021). https://doi.org/10.1029/2021JC017676
8. Ambulkar, A.: "Salinity," Encyclopedia Britannica. [Online]. Available: https://www.britannica.com/science/salinity. Accessed: 08 Mar 2024
9. Obianyo, J.I.: Effect of salinity on evaporation and the water cycle. Emerg. Sci. J. **3**(4), 255–262 (2019). https://doi.org/10.28991/esj-2019-01188
10. Olson, S., Jansen, M.F., Abbot, D.S., Halevy, I., Goldblatt, C.: The Effect of Ocean Salinity on Climate and Its Implications for Earth's habitability. Geophys. Res. Lett. **49**(10) (2022). https://doi.org/10.1029/2021GL095748
11. ESA: Types of Orbit. The European Space Agency. [Online]. Available: https://www.esa.int/Enabling_Support/Space_Transportation/Types_of_orbits. Accessed: 06 Aug 2023
12. ESA: SMOS performance specification. Earth Online. [Online]. Available: https://www.eoportal.org/satellite-missions/smos#mission-capabilities. Accessed: 19 Jan 2023
13. Martín-Neira, M., et al.: Microwave interferometric radiometry in remote sensing: an invited historical review. Radio Sci. **49**(6), 415–449 (2014). https://doi.org/10.1002/2013RS005230
14. Uti, M.N., Din, A.H.M., Omar, A.H.: Assessment of seasonal variability for wind speed and significant wave height using satellite altimeter over Malaysian seas. In: The International Archives of the Photogrammetry, Remote Sensing and Spatial Information Sciences, vol. XLII-4/W9, pp. 153–158 (2018). https://doi.org/10.5194/isprs-archives-XLII-4-W9-153-2018
15. Masseran, N., Razali, A.M.: Modeling the wind direction behaviors during the monsoon seasons in Peninsular Malaysia. Renew. Sustain. Energy Rev. **56**, 1419–1430 (2016). https://doi.org/10.1016/j.rser.2015.11.040
16. Caloiero, T., Pellicone, G., Modica, G., Guagliardi, I.: Comparative analysis of different spatial interpolation methods applied to monthly rainfall as support for landscape management. Appl. Sci. **11**(20), 9566 (2021). https://doi.org/10.3390/app11209566
17. ESA: SMOS L1 and L2 Science Data: How to Access Data," Earth Online. [Online]. Available: https://earth.esa.int/eogateway/catalog/smos-science-products. Accessed: 5 Mar 2024
18. ARGANS: SMOS L2 OS Algorithm Theoretical Baseline Document. United Kingdom (2021)

19. SURFER: A Basic Understanding of Surfer Gridding Methods—Part 1. GOLDENSOFTWARE. [Online]. Available: https://support.goldensoftware.com/hc/en-us/articles/231348728-A-Basic-Understanding-of-Surfer-Gridding-Methods-Part-1. Accessed: 05 Mar 2024
20. Röthig, T., et al.: Human-induced salinity changes impact marine organisms and ecosystems. Glob. Chang. Biol. **29**(17), 4731–4749 (2023). https://doi.org/10.1111/gcb.16859
21. Liu, F., Zhang, H., Ming, J., Zheng, J., Tian, D., Chen, D.: Importance of precipitation on the upper ocean salinity response to Typhoon Kalmaegi (2014). Water (Basel) **12**(2), 614 (2020). https://doi.org/10.3390/w12020614
22. Ullah, A., Bano, A., Khan, N.: Climate change and salinity effects on crops and chemical communication between plants and plant growth-promoting microorganisms under stress. Front Sustain. Food Syst. **5** (2021). https://doi.org/10.3389/fsufs.2021.618092
23. Krishnamurti, T.N., Smith, Philip, J., Gentilli, J.: Malaysian-Australian monsoon. Encyclopedia Britannica. [Online]. Available: https://www.britannica.com/science/Malaysian-Australian-monsoon. Accessed: 05 Mar 2024
24. Du, Y., Zhang, Y., Shi, J.: Relationship between sea surface salinity and ocean circulation and climate change. Sci. China Earth Sci. **62**(5), 771–782 (2019). https://doi.org/10.1007/s11430-018-9276-6
25. Roseli, N.H., Maisarah, N., Shakila, N., Wahid, A., Irwan, M.S.: Temperature and salinity characteristics in the Southern South China Sea and Strait of Malacca. Universiti Malaysia Terengganu J. Undergraduate Res. **4**(1), 99–116 (2022). https://doi.org/10.46754/umtjur.v4i1.264
26. Jian, C., Ren, Z., Luhua, W., Gongjie, W.: Performance evaluation of SMOS sea surface salinity observations in retrieving salinity profiles. In: Proceedings of the 2013 International Conference on Remote Sensing, Environment and Transportation Engineering, Paris, France. Atlantis Press (2013). https://doi.org/10.2991/rsete.2013.164
27. Berger, M., et al.: Measuring Ocean Salinity with ESA's SMOS Mission—Advancing the Science," Aug. 2002. [Online]. Available: http://www.esa.int/livingplanet. Accessed: 06 Mar 2024
28. Mu, Z., Zhang, W., Wang, P., Wang, H., Yang, X.: Assimilation of SMOS sea surface salinity in the regional ocean model for South China Sea. Remote Sens. (Basel) **11**(8), 919 (2019). https://doi.org/10.3390/rs11080919

Investigation of Long-Term Sea Level Variation from Satellite Altimeter and Tide Gauge in Southeast Asia Region

Muhammad Luqman Ahmad Affandi(✉), Ami Hassan Md Din, Mohammad Hanif Hamden, and Abd Wahid Rasib

Geospatial Imaging and Information Research Group (GI2RG), Faculty of Built Environment and Surveying, Universiti Teknologi Malaysia, 81310 Johor Bahru, Johor, Malaysia
mluqman35@graduate.utm.my, amihassan@utm.my

Abstract. The use of satellite altimeters has revolutionised our ability to monitor and understand long-term sea level variations. With advancements in technology, satellite altimetry has become an essential tool for studying global sea level anomalies. The research leverages the capabilities of satellite altimeters, providing global coverage, and combines these observations with the localised and historical records obtained from tide gauges. The synergy of these two datasets enables the identification of trends, patterns, and anomalies in sea level behaviour over time. This method has not only provided us with long-term data but has also been proven to complement and enhance the accuracy of in-situ observations from tide gauges. The data obtained from satellite altimeters, known as Sea Level Anomaly (SLA), has been shown to be in good agreement with tide gauge measurements. It is evident that the sea level anomaly and tidal anomaly exhibit good agreement with a high correlation (>0.89). The findings of this exploration contribute valuable insights to the ongoing discourse on climate change and its repercussions on sea level dynamics. The combined use of satellite altimetry and tide gauge data enhances our comprehension of long-term sea level variations but also provides a foundation for predicting future trends and formulating informed strategies for coastal adaptation and mitigation.

Keywords: Satellite altimeter · Tide gauge · Sea level anomaly · Southeast Asia · Climate change

1 Introduction

Exploring the dynamics of long-term sea level variation is crucial for understanding the impacts of climate change and implementing effective coastal management strategies. According to Ref. [1], sea level variations are influenced by a combination of factors such as thermal expansion of seawater, melting of ice caps and glaciers, and changes in ocean circulation patterns. One of the most effective methods for studying long-term sea level variation is through the use of satellite altimeters and tide gauges. Satellite altimeters provide a global perspective by measuring the height of the sea surface from

A. Abdul-Rahman et al. (Eds.): GeoWeek 2024, LNGC, pp. 172–184, 2025.
https://doi.org/10.1007/978-3-031-86654-8_13

space, allowing for a comprehensive analysis of sea level changes on a large scale [2]. On the other hand, tide gauges provide local and historical data, offering insights into long-term trends and variations in sea level at specific coastal locations.

However, accurate sea level measurements are limited. Only satellite-derived altimeter data sets from late 1992 are available, which are insufficient to address the decadal variability. Acquiring credible sea-level datasets for analysis, particularly those predating the 1950s, is challenging [3]. Consequently, tide gauge stations located in coastal regions serve as the primary source of long-term sea level measurements [4]. Despite their limited distribution, tide gauges offer valuable long-term data for analysing sea level rise trends.

Southeast Asia boasts a complex and varied coastline [5]. The countries of the SEA are surrounded mainly by the sea and are vulnerable to rising sea levels and coastal flooding. However, the study interpretation of sea level regionally for the SEA is still minimal compared to globally and locally. Most of the SEA countries are located on the same continental and surrounded by archipelagos (extensive groups of islands), particularly Indonesia and the Philippines, which have the most islands in their regions. Most countries in SEA are encircled by oceans, with massively populated lowlands and coastal areas.

Sea level rise holds great significance due to its potential to result in significant environmental, social, and economic repercussions. As global temperatures continue to rise, the melting of glaciers and ice sheets contributes to the elevation of sea levels [6]. This occurrence directly endangers coastal communities, low-lying islands, and vulnerable ecosystems. The consequences of sea level rise encompass heightened coastal erosion, land inundation, intrusion of saltwater into freshwater sources, loss of coastal habitats, and intensified storm surges. These outcomes can lead to population displacement, livelihood loss, infrastructure damage, and disruptions to ecosystems. Consequently, it is imperative to gain a precise understanding of sea level rise and take appropriate actions to plan and implement adaptive measures that safeguard coastal regions, ensuring the sustainability and well-being of communities and ecosystems dependent on them.

This study leverages both tide gauge stations and multi-satellite altimetry missions across Southeast Asia to investigate temporal changes in sea level and to discern patterns of relative and absolute sea level anomalies. Subsequently, the accuracy of satellite altimetry-derived sea level anomalies is assessed by comparison with in-situ tide gauge data. This validation process involves calculating correlation coefficients and root mean square errors.

2 Methodology

In this section, the utilisation of tide gauge data and data processing strategies for multi-mission satellite altimeters for sea level trends are discussed. In this study, we use recorded sea level data around the SEA region. These include two different types of instruments to get the sea level data: satellite altimeter-derived sea level anomaly and tide gauge-derived tidal anomaly. This part will be discussed in the subsection below.

2.1 Tide Gauge Data Acquisition

In this study, we used thirty (30) tide gauges in the SEA region are used to investigate sea level trends. The tidal data are retrieved from The Permanent Service for Mean Sea Level (PSMSL) website (http://www.psmsl.org/). Determining mean sea level solely from tide gauge data is complex due to the influence of local and regional factors like winds and ocean currents. Figure 1 outlines the process of calculating relative sea level anomalies from these tide gauge measurements, which is detailed in the following section.

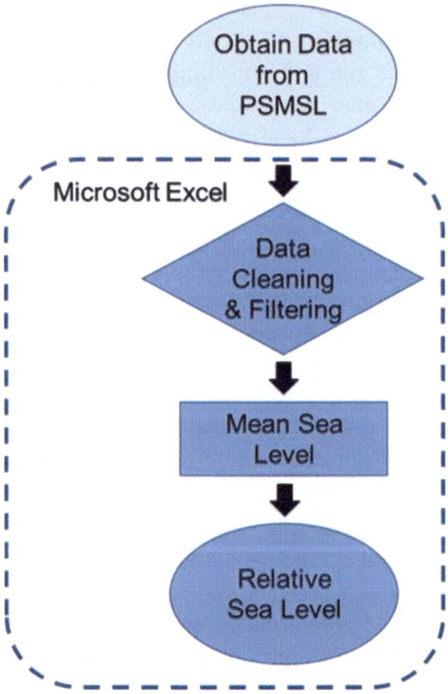

Fig. 1. Calculating relative sea level anomaly from tide gauge data.

Tidal Data Compilation and Filtering. The monthly-averaged tide gauge records obtained through downloads are referenced to the Revised Local Reference (RLR) for measurements. The data obtained from the Permanent Service for Mean Sea Level (PSMSL) is transferred to Microsoft Excel for data cleaning and filtering purposes. This step aims to remove outliers, data errors, and miscalculated values. Any data points that contain the value −9999, which is indicative of missing or erroneous data, are excluded from further analysis. Instead of using −9999, these data points are replaced with NaN (Not a Number) to prevent miscalculations during subsequent operations, particularly when calculating the mean sea level. By performing data cleaning and filtering, the resulting dataset becomes more reliable and suitable for accurate calculations and referencing the mean sea level.

Tidal Anomalies. Calculating relative sea level anomalies from tidal data involves multiple steps, as outlined in Eqs. 1 and 2. It is essential to convert raw tidal data into tidal anomalies to account for inconsistencies in mean sea level or mean sea surface. Subsequently, to evaluate the agreement between tide gauge and satellite altimetry data, the correlation between their respective tidal and sea level anomalies is calculated. This analysis aids in visualising both absolute and relative sea level patterns. The specific equations used for computing tidal anomalies are detailed below [7]:

$$\underline{X}_{MSL} = \frac{\sum_{i=n}^{n} AMSL_i}{n} \tag{1}$$

$$Tidal_{SLA} = AMSL_i - \underline{X}_{MSL} \tag{2}$$

where,

$\underline{X}_{MSL}$: Average MSL up to 2020 at the tide gauge stations.
n: Observation Count.
$AMSL_i$: Monthly sea level for every observation.
$Tidal_{SLA}$: Tidal Anomaly with respect to MSL.

2.2 Satellite Altimetry Data Acquisition

The Radar Altimeter Database System (RADS) was utilised to extract Sea Level Anomalies (SLAs) for this study. RADS, developed by the Delft Institute for Earth-Oriented Space Research (DEOS), is a comprehensive database housing validated altimetry products [8]. Users can select appropriate corrections for their specific needs from a range of options curated by the global scientific community. Sea level data was collected from ten different satellite altimetry missions: TOPEX, Jason-1, Jason-2, Jason-3, ERS-1, ERS-2, Envisat, CryoSat-2, SARAL, and Sentinel-3A. Satellite altimetry data from January 1993 to December 2018, spanning 26 years, is employed in this study. Since the life span of each satellite altimeter is different, implementing ten multi-mission satellites indirectly provides continuous data retrieval with denser coverage of monthly sea level data, which will improve sea level rise trends. Figure 2 shows the combination of 10 multi-mission satellite altimeter tracks.

Data Filtering and Gridding. RADS processing enhances data quality through a distance-weighted gridding method. This technique prioritises information preservation while minimising the impact of data gaps between satellite tracks. By assigning greater weight to data points closer to the grid center and lesser weight to those farther away, the weighting function (Eq. 3) optimises data distribution [8]:

$$F_W(R) = e^{\frac{r^2}{\sigma^2}} \tag{3}$$

The Gaussian weighting function, $F_W(R)$, is employed for data processing, where 'e' represents the station's unit vector, 'σ' (sigma) determines the filter's smoothing effect, and 'r' signifies the spatial relationship between raw data points and grid locations is crucial for the subsequent processing steps. A Gaussian weighting algorithm is employed to distribute these data points into designated grid cells while filtering out noise and irregularities. This process ensures that the resulting gridded dataset accurately represents the underlying sea level anomalies, excluding any extraneous or sensitive information.

Daily altimetry data from Topex and ERS class are gridded to generate SLAs over bins using Gaussian weighting functions. A square grid with a specific block dimension was used to categorise the data. To optimise SLA estimates, a 0.25° by 0.25° grid size was generated using the Inverse Distance Weighting (IDW) method for altimetry data processing. Although gridding is not a correction technique, it is critical to collect, enhance, and analyse RADS data.

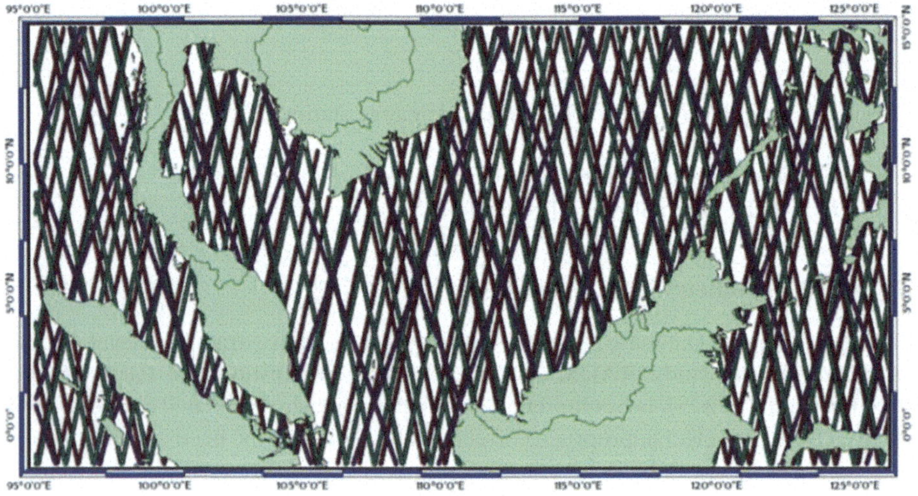

Fig. 2. Combination of multi-mission satellite altimetry tracks.

3 Results and Discussion

The result is explained in a clustering type, where the data points are grouped together based on their characteristics or attributes. This helps to identify patterns, relationships, and distinct clusters within the dataset. This visual representation provides valuable insights into the data and facilitates further analysis. This section analyses variations in relative and absolute sea level anomalies from the initial establishment of tide gauges until 2020. Additionally, the latter part of this section will discuss the validation of data by comparing sea level anomaly with tidal anomaly.

3.1 Relative Sea Level Anomaly

For instance, Fig. 3 presents sea level anomaly data obtained from three tide gauge stations situated on the eastern coast of Borneo: Tawau, Lahad Datu, and Sandakan. These stations face the Sulu and Celebes Seas. As shown in Fig. 3, the sea level variations from the three tide gauges also indicate a similar pattern. Intriguingly, the data reveal a significant decrease in sea level during 1997. This coincides with the well-documented strong El-Niño event of that period. El-Niño's influence on the west coast of Peninsular Malaysia and the east coast of Borneo is evident, as it typically leads to a decrease in sea level in these regions. Conversely, the east coast of Borneo appears to exhibit a contrasting pattern during strong La-Niña events. The time series indicated a rise in sea level during 2010–2011, which aligns with the occurrence of a prominent La-Niña event. La-Niña is known to induce opposite effects compared to El-Niño, potentially causing sea level rise in specific regions.

The South China Sea is a major maritime corridor surrounded by most Southeast Asian countries. Tidal anomaly patterns from coastal tide gauge stations within this sea exhibit strong similarities, as depicted in Fig. 4. Predominantly influenced by diurnal tides, as documented by [9], the South China Sea experiences notably larger tidal ranges compared to other regions. Interestingly, the impact of El Niño and La Niña on sea level fluctuations appears to be muted in the open waters of the South China Sea, as suggested by Ref. [10].

Figure 5 illustrates the sea level characteristics of the Gulf of Thailand, as determined by data from the Koh Lak and Fort Phrachula tide gauge stations. In contrast to the Gulf of Thailand, the Andaman Sea's sea level pattern is represented by the Ko Tapha Noi tide gauge station. The tidal regime in the Gulf of Thailand is characterised by a mix of diurnal and mixed tides, while the Andaman Sea predominantly experiences semi-diurnal tides [11]. The tidal dataset contains a data gap in 2004, attributed to the devastating Sumatra earthquake, which triggered significant land subsidence [12]. Additionally, the Fort Phrachula tide gauge station in Bangkok has recorded a sudden increase in sea level. This anomaly is linked to land subsidence caused by excessive groundwater extraction in Thailand [13, 14].

The Philippines stands out as an archipelago with a widespread network of tide gauge stations [15]. Three tide gauge stations were strategically chosen for this study to represent the Philippine archipelago: Manila, Cebu, and Legazpi, situated along the South China Sea, Sulu Sea, and Pacific Ocean, respectively. Over the past 70 years, these stations have recorded tidal variations ranging from −0.5 to 1 m, as illustrated in Fig. 6. While Cebu and Legazpi exhibit comparable tidal anomaly patterns, Manila has displayed a distinct upward trend since 1970. This divergence is likely due to land subsidence caused by reclamation activities and increased river discharge [13, 14].

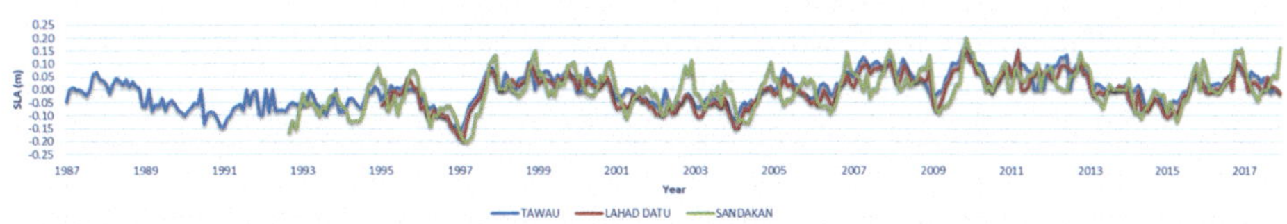

Fig. 3. Tidal variations derived from tide gauge stations located near the Sulu and Celebes Seas

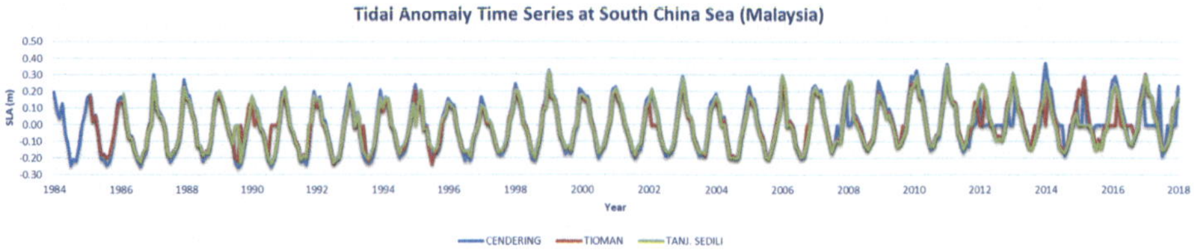

Fig. 4. Tidal variations derived from tide gauge stations located near the South China Sea (Malaysian region).

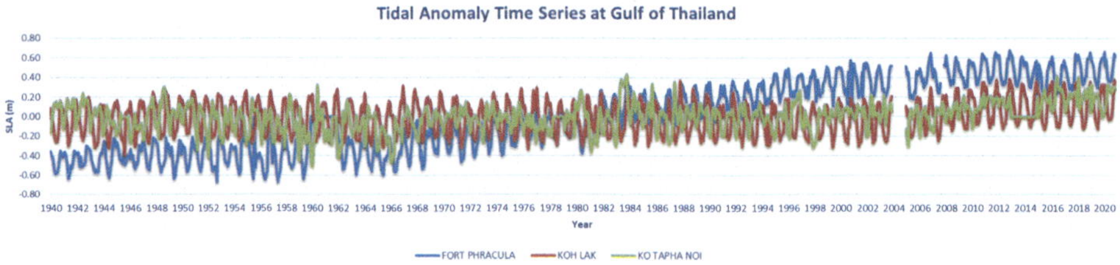

Fig. 5. Tidal variations derived from tide gauge stations located near the Gulf of Thailand.

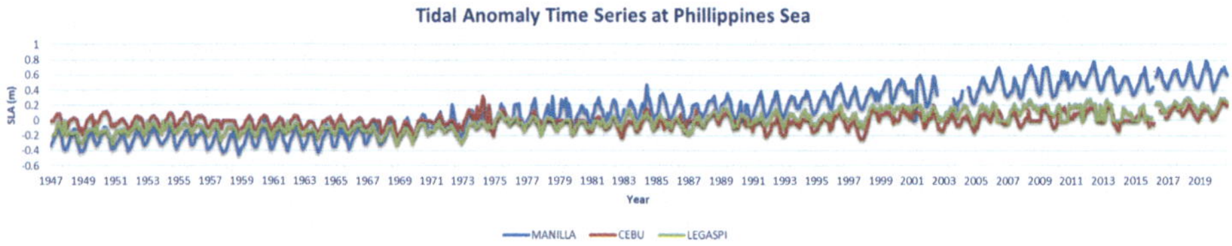

Fig. 6. Tidal variations derived from tide gauge stations located near the Phillipines sea.

3.2 Absolute Sea Level Anomaly

Absolute sea level anomalies can be influenced by various factors, including interannual climate patterns like El Niño and La Niña, as well as long-term climate change. Figure 7 presents a spatial representation of average absolute sea level anomalies relative to the mean sea surface across Southeast Asia. The calculated anomalies, ranging from −0.07 to 0.12 m, exhibit significant heterogeneity across the region. These fluctuations are primarily driven by factors such as water temperature, salinity, coastal currents, wind speed, and atmospheric pressure [16].

The spatial distribution of sea level anomalies, as depicted in Fig. 7, reveals distinct patterns. Dark blue and red tones highlight areas with elevated sea level anomalies, notably along the northern coast of the Gulf of Thailand and parts of the northern and southern Philippines. Major cities like Bangkok and Manila, situated in low-lying coastal zones, are particularly vulnerable to these higher sea levels [17]. Additionally, the western coast of Sabah in the Celebes and Sulu Seas also experiences elevated sea level anomalies, corroborating findings from previous studies [10].

Figure 7 illustrates the sea level anomaly patterns in the Indonesian region, encompassing the Karimata Strait, Java Sea, and Makassar Strait. Dominated by blue and green

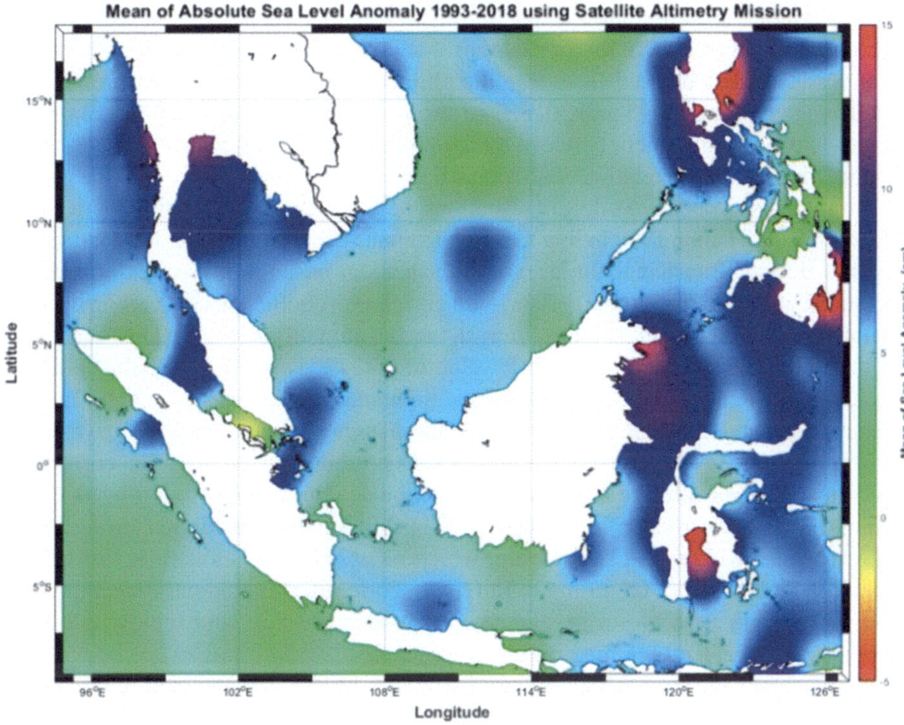

Fig. 7. Average Sea Level Anomaly (SLA) calculated from satellite altimeter data spanning the period between 1993 and 2018 across Southeast Asian waters.

hues, the colour scheme indicates relatively small sea level anomalies ranging from 0.25 to 10.0 cm. The SLA patterns observed in the South China Sea and Malacca Strait align with findings from previous studies [18]. However, a distinct yellow colouration in the southern Malacca Strait suggests potential data limitations due to sparse satellite altimetry coverage in this region [15]. The overall sea level anomaly patterns in Southeast Asia are likely influenced by a combination of factors, including El Niño-Southern Oscillation (ENSO) events, global warming, human activities, and ice sheet melting.

3.3 Data Validation: Comparing Sea Level and Tidal Anomalies

Before analysing long-term sea level rise trends, it is crucial to validate the accuracy of sea level anomaly data obtained from the Radar Altimeter Database System (RADS). This involves comparing the altimetry-derived data with in-situ measurements from tide gauge stations. Previous sections have independently explored sea level variability based on both datasets. Correlation coefficients were computed to assess the relationship between these datasets. The findings of this correlation analysis are presented in this section.

The R-values are computed at the tide gauge points encompassing the areas of interest. Figure 8 presents a Taylor diagram that compares the variability and correlation of sea level data derived from satellite altimetry and tide gauge measurements at Pulau Tioman between 1993 and 2018. The strong correlation coefficient (R) of 0.9701 indicates excellent agreement between the two datasets.

Table 1 shows the correlation coefficients between satellite altimetry and tide gauges SLA computed at respective tide gauge stations. Tide gauge stations in the South China Sea and Gulf of Thailand, including Geting, Tioman, Bintulu, Ko Lak, Ko Mattaphon, Vung Tau II, and Kota Kinabalu, as well as Sandakan in the Sulu Sea, exhibit strong correlations with satellite altimetry data, with correlation coefficients exceeding the 0.80 thresholds suggested by Ref. [19]. In contrast, the Tawau station in the Celebes Sea shows a slightly weaker correlation, with an R-value of 0.7242. This is because the station is located in a semi-enclosed area located at the Sulu and Celebes seas.

Thirty tide gauge stations with complete data spanning from January 1993 to December 2018 were selected for correlation analysis, as presented in Table 1. The correlation coefficients between satellite altimetry and tide gauge data ranged from a weak correlation of 0.5155 to a strong correlation of 0.9734. Most stations exhibited strong agreement, except for Kukup and Lahat Datu, which recorded correlation values of 0.5992 and 0.5475, respectively. According to Ref. [7], a correlation coefficient exceeding 0.80 is considered indicative of strong agreement between absolute and relative sea level anomalies, essential for obtaining high-quality results.

Table 1 shows the Root Mean Square Error (RMSE) values obtained in this study, ranging from 0.03 m to 0.37 m. Lower RMSE values indicate better agreement between the altimetry data and in-situ tidal data. For example, a Tioman tide gauge station has an RMSE of 0.03 m and a correlation coefficient (R) of 0.97 m, signifying excellent agreement. Furthermore, interpreting the RMSE alongside the correlation coefficient, which is typically above 0.8 in this case, provides a comprehensive evaluation of altimetry performance. These insights are crucial for improving the accuracy and reliability of future sea level studies.

Table 1. The agreement between satellite altimetry and tide gauge data

Location	Tide gauge station	R-value (correlation)	RMSE (m)
Peninsular Malaysia	Lumut	0.84	0.05
	Langkawi	0.89	0.05
	Kukup	0.70	0.06
	Tanjung Keling	0.96	0.04
	Geting	0.97	0.04
	Tanjung Sedili	0.94	0.04
	Pelabuhan Klang	0.66	0.06
	Tioman	0.97	0.03
	Cendering	0.96	0.04
	Johor Bahru	0.89	0.05
Borneo Sabah Sarawak	Bintulu	0.91	0.03
	Kota Kinabalu	0.95	0.02
	Kudat	0.96	0.04
	Tawau	0.72	0.08
	Labuan	0.92	0.03
	Lahad Datu	0.52	0.14
	Sandakan	0.84	0.06
	Miri	0.89	0.06
Thailand	Fort Phrachula	0.82	0.37
	Ko Lak	0.95	0.05
	Komattaphon	0.85	0.12
	Ko Taphao Noi	0.86	0.06
	Ko Si Chang	0.81	0.06
Vietnam	Vung Tau	0.94	0.05
	Danang	0.90	0.07
	Hondau	0.72	0.08
Philippine	Manila	0.76	0.35
	Cebu	0.76	0.05
	Legaspi	0.90	0.10
	Puerto	0.89	0.03

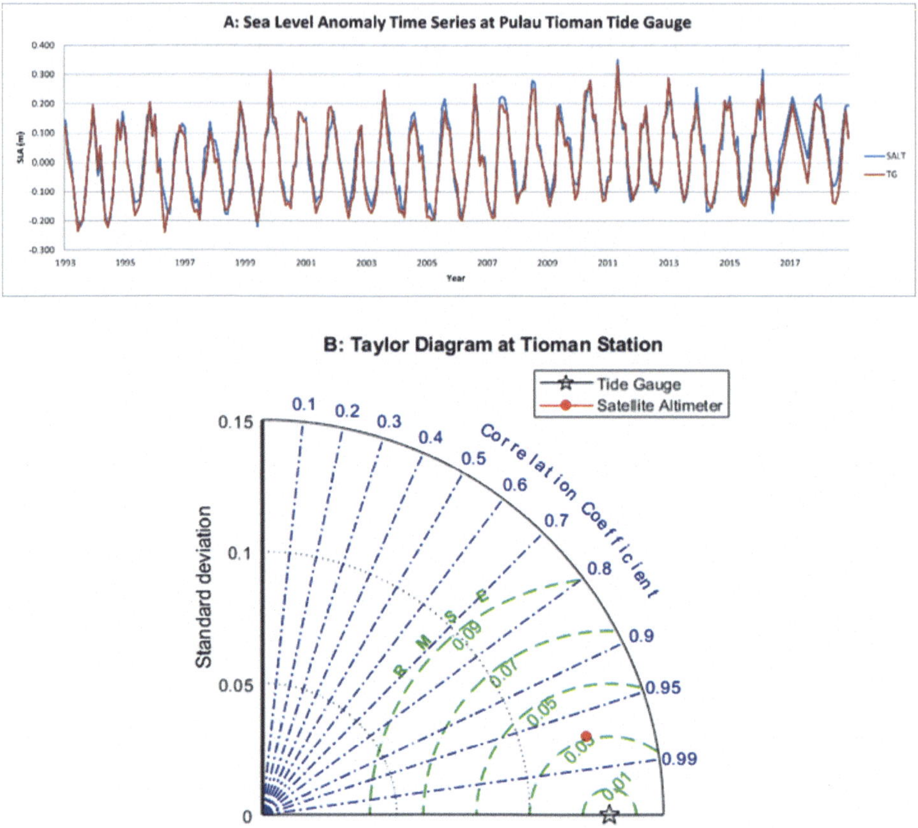

Fig. 8. The altimetry and tide gauge SLA time series and the correlation coefficient (R) at Pulau Tioman tide gauge station.

4 Conclusion

Sea level rise, primarily driven by thermal expansion, melting ice, and glaciers, poses a significant threat to coastal ecosystems and economies, including the lucrative ecotourism sector. These ecosystems may require millennia to recover from the destabilising effects of rising seas. Human activities exacerbate this issue. Southeast Asia, in particular, is highly vulnerable to rising sea levels. This study demonstrates strong agreement between satellite altimetry and tide gauge data for sea level anomalies across the region, with correlation coefficients consistently exceeding 0.8. These findings provide a valuable foundation for developing and implementing effective coastal management strategies. Such measures may include coastal defences, beach nourishment, offshore barriers, flood gates, and mangrove restoration to mitigate the adverse impacts of rising sea levels in Southeast Asia.

Acknowledgment. This project is funded by the Ministry of Higher Education (MOHE) Malaysia under the Universiti Teknologi Malaysia Fundamental Research Grant (UTMFR) (UTM Vote Number: Q.J130000.3852.22H75).

References

1. Khojasteh, D., Haghani, M., Nicholls, R.: The evolving landscape of sea-level rise science from 1990 to 2021. https://doi.org/10.21203/rs.3.rs-2043042/v1
2. Dettmering, D., Passaro, M., Braun, A.: Editorial for special issue 'advances in satellite altimetry and its application.' Remote Sens (Basel) **11**(24), 10–12 (2019). https://doi.org/10.3390/rs11242913
3. Affandi, M.L.A., Din, A.H., Wahid, A.: A short review on causes of sea level rise for climate monitoring a short review on causes of sea level rise for climate monitoring (2022). https://doi.org/10.1088/1755-1315/1051/1/012003
4. Deepa, J.S., Gnanaseelan, C.: The decadal sea level variability observed in the Indian Ocean tide gauge records and its association with global climate modes. Glob. Planet Change **198**, 103427 (2021). https://doi.org/10.1016/j.gloplacha.2021.103427
5. Leinbach, T.R., Frederick, William, H.: "Southeast Asia." Encyclopedia Britannica. 10 Nov 2020. [Online]. Available: https://www.britannica.com/place/Southeast-Asia. Accessed: 19 Aug 2022
6. Affandi, M.L.A.: Estimating sea level rise trends and projections in the Southeast Asia region using robust fit regression technique. Master Thesis, Universiti Teknologi Malaysia, Johor Bahru, 2023
7. Din, W.A.W., A.H.M., Zulkifli, N.A., Hamden, M.H., Aris: ScienceDirect sea level trend over Malaysian seas from multi-mission satellite altimetry and vertical land motion corrected tidal data. Adv. Space Res. **63**(11), 3452–3472 (2019). https://doi.org/10.1016/j.asr.2019.02.022
8. Ami Hassan, M.D.: Sea Level Rise Estimation and Interpretation in Malaysian Region Using Multi-Sensor Techniques. Universiti Teknologi Malaysia, Skudai (2014)
9. Yan, T., Qi, Y., Jing, Z., Cai, S.: Seasonal and spatial features of Barotropic and Baroclinic tides in the Northwestern South China Sea. J. Geophys. Res. Oceans **125**(1). https://doi.org/10.1029/2018JC014860
10. Hamid, A.I.A., Din, A.H.M., Hwang, C., Khalid, N.F., Tugi, A., Omar, K.M.: Contemporary sea level rise rates around Malaysia: altimeter data optimisation for assessing coastal impact. J. Asian Earth Sci. **166**(October 2017), 247–259 (2018). https://doi.org/10.1016/j.jseaes.2018.07.034
11. Pongsiri, N., McNeil, R., Chuai-Aree, S.: Trends in tidal levels and mean sea level in the Gulf of Thailand. Ocean Sci. J. **55**(4), 495–503 (2020). https://doi.org/10.1007/s12601-020-0039-1
12. Sojisuporn, P., Sangmanee, C., Wattayakorn, G.: Recent estimate of sea-level rise in the Gulf of Thailand ocean forecasting and marine disasters mitigation for Southeast Asia seas view project. [Online]. Available: www.mijst.mju.ac.th
13. Adebisi, N., Balogun, A.L., Min, T.H., Tella, A.: Advances in estimating sea level rise: a review of tide gauge, satellite altimetry and spatial data science approaches. Ocean Coast Manag. **208**(April), 105632 (2021). https://doi.org/10.1016/j.ocecoaman.2021.105632
14. PSMSL: The Permanent Service For Mean Sea Level (2021)
15. Affandi, M.L.A., Din, A.H.M., Rasib, A.W.: Sea level rise estimation and projection from long-term multi-mission satellite altimetry and tidal data in the Southeast Asia region. Int. J. Remote Sens. (2024). https://doi.org/10.1080/01431161.2023.2297179

16. Khalid, N.F.: Coastal inundation risk assessment using multi-mission satellite altimeter and digital elevation models. Universiti Teknologi Malaysia, Johor Bahru (2018)
17. IPCC: Climate Change 2021 The Physical Science Basis Working Group I Contribution to the Sixth Assessment Report of the Intergovernmental Panel on Climate Change (2021)
18. Zulkifli, N.A.: Vertical Land Motion Estimation Derived From Spase-based Geodetic Technique. PhD Thesis, Universiti Teknologi Malaysia, Johor Bahru (2022)
19. Schober, P., Schwarte, L.A.: Correlation coefficients: appropriate use and interpretation. Anesth. Analg. **126**(5), 1763–1768 (2018). https://doi.org/10.1213/ANE.0000000000002864

A Review of Global Ocean Tide Models for Tidal Prediction to Support Hydrographic Separation Model

Fatin Nabihah Syahira Ridzuan(✉), Mohammad Hanif Hamden, and Ami Hassan Md Din

Geospatial Imaging and Information Research Group (Gi2RG), Faculty of Built Environment and Surveying, Universiti Teknologi Malaysia, 81310 Johor Bahru, Johor, Malaysia
fnsyahira2@graduate.utm.my, mohammad.hanif@utm.my

Abstract. The ocean tide model is critical for determining accurate vertical datum transformations with high accuracy and resolution. The advent of satellite altimeters has increased the accuracy of tidal predictions for the open ocean by around 2 cm. However, there are significant variations in shallow water and coastal zones, which could correspond to the complex hydrodynamic systems and a lack of high-quality satellite altimetry data. Hence, an overview of the global tide model is performed to evaluate the performance of global ocean tide models in open ocean and coastal areas. This paper attempts to review the performance of several tide models based on the previous comprehensive tests. Such an attempt foresees the potential application of local hydrodynamic models for accurate tidal prediction and could provide an opportunity to produce a seamless hydrographic separation model in coastal and offshore areas, especially in Malaysian waters.

Keywords: Global ocean tide · Tidal prediction · Separation model

1 Introduction

Ocean tides refer to the cyclical rise and fall of sea levels. Accurate tide measurements and forecasts are essential for determining vertical datum transformations and correcting sea level data from various sources [1]. Continuous tide monitoring through tide gauge stations can yield highly precise, high-frequency, and verifiable tidal data at the millimetre level. However, the effectiveness of these measurements is limited by the sparse distribution of tide gauges along the coast. The introduction of satellite altimeters has improved the accuracy of tidal predictions in the open ocean by approximately 2 cm [2]. Nevertheless, in shallow-water areas, particularly near coastlines, the accuracy of altimetry data diminishes due to various challenges, leading to a rapid decline in data reliability as measurements approach the shore [3]. The accumulation of tide observations from tide gauges and the widespread use of satellite altimeters have significantly advanced the development of tide models to address these issues [4]. Additionally, tide gauge stations have been expanded to serve as ground-truth data for assessing the accuracy of such models and tidal constituents [5–7].

A. Abdul-Rahman et al. (Eds.): GeoWeek 2024, LNGC, pp. 185–192, 2025.
https://doi.org/10.1007/978-3-031-86654-8_14

Advances in global ocean tide modelling have enabled the identification of tidal phenomena in the open sea [8, 9]. Schwiderski's (SCHW) model was a pioneering effort in this domain, rigorously grounded in oceanic dynamics and incorporating extensive tide gauge data within a hydrodynamic model [10, 11]. Subsequent models have refined these foundations, leading to continuous improvements [2]. A comprehensive evaluation by Shum et al. [5] highlighted significant advancements, especially in shallow and deep ocean regions. Recent in-depth assessments of state-of-the-art altimeter-based ocean tide models by Stammer et al. [2], Fu et al. [7], and Sun et al. [4] have revealed disparities between model accuracy in open ocean versus coastal and shallow water areas, particularly in terms of tidal constituent estimates and predicted tidal heights. A comparative analysis by Stammer et al. [2] further emphasised the superior accuracy of assimilative models over their non-assimilative counterparts.

This paper aims to evaluate the performance of various tide models based on previous comprehensive assessments. The remainder of the paper is structured as follows: An overview of the global ocean tide model is provided in Sect. 2, and Sect. 3 provides a discussion on the performance of tide models. A summary and an outlook are provided in Sect. 4.

2 Overview of Ocean Tide Model

Tide models are primarily developed through two distinct methodologies: empirical and assimilation approaches. Empirical models construct predictive frameworks solely based on observational data, such as satellite altimetry and tide gauge measurements. These models excel at capturing large-scale tidal patterns but might struggle to represent complex coastal and shallow-water dynamics accurately. Prominent examples of empirical models include the DTU, OSU, GOT, and EOT series. In contrast, assimilation models combine observational data with hydrodynamic models to improve tidal predictions. These models offer a more comprehensive representation of tidal phenomena by incorporating physical processes and boundary conditions. However, they require careful calibration and validation. Notable assimilation models include the HAMTIDE, FES, and TPXO series. While both approaches have their strengths and weaknesses, the optimal choice of the tide model depends on the specific application and geographical region.

The TPXO8 inverse tide model, developed by Egbert and Erofeeva at Oregon State University in 2002, represents a significant advancement in tidal modelling. It incorporates an efficient computational approach [12] and tools for grid generation, tidal forcing, boundary conditions, error estimation, and data assimilation from multiple satellite altimeter and tide gauge missions. The model incorporates 13 distinct tidal components and employs a highly detailed grid spacing of $1/30^\circ$ for the nine dominant tidal patterns. A less detailed grid spacing of $1/6^\circ$ is used for the remaining four tidal components, particularly in coastal regions, to enhance efficiency.

The DTU10 global empirical model, developed by Cheng and Andersen [13], utilised 17 years of data from the T/P, Jason-1, and Jason-2 missions, along with observations from ERS2 and Envisat. The multi-mission altimeter tide residuals were processed using the response method [14, 15], and the along-track data was interpolated onto the FES2004 grid using the dynamic interpolation method [16]. The DTU16 global ocean tide model was created by estimating along-track residual tides using the response analysis method in FES2012 [13]. Using a dynamic interpolation technique, the cosine and sine components were spatially adjusted to fit the FES2012 grid (resolution of 1/16° by 1/16°). This process accounted for the varying lengths of diurnal and semidiurnal tidal waves and incorporated water depth information from the DTU10 bathymetry model.

Compared to DTU10, DTU16 leverages a substantially larger dataset by incorporating two years of Jason-1 tandem mission data and another five years of Jason-2 mission data. This extended time series, spanning over 23 years of Topex-Jason-1/2 data since 1993, enables a more accurate differentiation between the M2 and S2 tidal constituents by overcoming their three-year aliasing period. In the Arctic Ocean, including AltiKa data, with its 35-day repeat cycle similar to Envisat, enhances the model's precision in high-latitude regions.

The OSU12 is an empirical global tidal model developed at Ohio State University using an advanced method for representing tidal patterns [17]. As a fully empirical tide model, it employs least squares collocation to interpolate data onto a 0.25° × 0.25° grid, using satellite altimetry data from T/P, Jason-1, Jason-2, Envisat, and GFO. The covariance matrix used in this interpolation varies by location and depends on ocean depth. The model provides harmonic constituents for diurnal and semidiurnal tides only.

HAMTIDE is a global assimilation ocean model developed by Taguchi et al. [18], utilising linearised hydrodynamic equations on a global scale. The HAMTIDE series tide models have integrated time series data from T/P and Jason-1 over 15 years, with a grid resolution 1/8°. The latest iteration, HAMTIDE12, employs a straightforward harmonic time dependence of the variables and linearised tidal hydrodynamic equations.

The GOT series, developed by NASA's Goddard Space Flight Center, are empirical tide models derived from analysing multiple satellite altimeter datasets. Building upon the methods of Schrama and Ray [19] and Ray [20], these models employ a comparative approach, analysing the discrepancies between observed and predicted sea levels from a combination of global, regional, and local tide models. In deep ocean areas, dominant tidal components are determined through standard harmonic analysis, while less significant constituents are estimated using admittance relationships. The resulting amplitude and phase data are provided on a 0.5° global grid, suitable for deep ocean predictions but less adequate for shelf seas and coastal waters.

The GOT4.8 model is based on an empirical harmonic analysis of satellite altimetry data, integrating both global and regional models. It primarily uses data from the TOPEX/Poseidon satellite for deep ocean areas between latitudes ± 66°, while GEOSAT Follow-On, ERS-1, and ERS-2 satellite data are used for shallow seas and higher latitudes [7]. GOT4.8 represents an improvement over its predecessor, GOT4.7, by rectifying a minor error in the S2 atmospheric tide component, which in turn enhances the accuracy of the S2 ocean tide solution. GOT4.10 shares similarities with versions 4.7, 4.8, and 4.9 but exclusively relies on data from Jason-1 and Jason-2 satellites, excluding Topex/Poseidon data. Notably, the dry-tropospheric correction applied to Jason's satellite data does not contain the S2 air-tide error in the original Topex/Poseidon data.

The Finite Element Solution (FES) tide model is a sophisticated computer simulation that accurately predicts ocean tides. It combines tide gauge observations with satellite data from multiple missions, including Topex/Poseidon, Jason-1, Jason-2, and European satellites like ERS-1, ERS-2, and Envisat. This comprehensive dataset allows the model to produce highly detailed and reliable tidal information. The latest version, FES2014, employs a global finite element mesh with higher resolution in coastal and shallow-water regions, solving shallow-water hydrodynamic equations in a spectral configuration [22]. The FES2014 dataset, consisting of approximately 2.9 million nodes, benefits from extensive altimeter time series, tide gauges, enhanced modelling, and advanced data assimilation techniques, resulting in an accurate solution [22, 23]. It includes 34 tidal constituents with a gridded resolution of 1/16°. The primary advantage of the FES2014 tidal database is its ability to generate tidal time series for arbitrary periods globally without requiring complex direct satellite altimeter data analysis.

EOT11a is a global tidal model developed in 2011 [24]. Unlike traditional tidal models, EOT11a relies on an "empirical" approach. It does not simulate the physical processes driving tides but analyses actual satellite tidal observations. Specifically, EOT11a utilises data from multiple satellite altimeter missions, including T/P, ERS-2, Envisat, Jason-1, and Jason-2. The data spans a period between September 1992 and April 2010. By analysing these residual tidal observations (the leftover signal after removing other oceanic influences), EOT11a provides a global picture of tidal patterns across the Earth's oceans.

EOT20 is the latest in a series of empirical ocean tide models created using tidal residue analysis of data from multiple satellite altimetry missions. This model provides detailed information about the size and timing of 17 tidal components on a global scale, with a high level of precision. It incorporates data from seven primary satellite missions and four extended missions. Due to the addition of more missions and more recent satellite altimetry data, the EOT20 model exhibits notable improvements over the previous global model iteration (EOT11a), especially in the coastal and shelf regions of the ocean [25].

Table 1 summarises the global tide model characteristics of TPXO8, DTU10, OSU12, HAM12, GOT4.8, FES2014, GOT4.10, DTU16 and EOT20. "E" indicates an empirical adjustment to a prior model, while "A" denotes assimilation into a hydrodynamic model.

3 Discussion

The performance of ocean tide models is highly variable across different geographic regions. As illustrated in Fig. 1, the world's oceans are vast and diverse, with distinct characteristics in terms of depth, bathymetry, and hydrodynamic processes. While providing a general overview of tidal patterns, global tide models often struggle to represent the complex tidal dynamics in specific areas accurately. Shallow water environments and coastal regions pose particular challenges for these models due to the intricate interplay between ocean tides and local topography. Factors such as continental shelves, islands, and estuaries significantly influence tidal propagation and amplitude. Consequently, global tide models often exhibit reduced accuracy in these areas, necessitating the use of regional or localised models for more precise tidal predictions.

Tide models in the Arctic and Antarctic regions often exhibit lower accuracy than other areas due to various factors [4]. The primary challenges in developing accurate

Table 1. Spatial and modelling approach.

Tide models	Time/Year	Resolution	Type
TPXO8	2002	1/30° × 1/30°	A
DTU10	2010	1/8° × 1/8°	E
EOT11a	2011	1/8° × 1/8°	E
OSU12	2012	1/4° × 1/4°	E
HAM12	2014	1/8° × 1/8°	A
GOT4.8	2013	1/2° × 1/2°	E
FES2014	2016	1/16° × 1/16°	A
GOT4.10	2017	1/2° × 1/2°	E
DTU16	2017	1/16° × 1/16°	E
EOT20	2021	1/8° × 1/8°	E

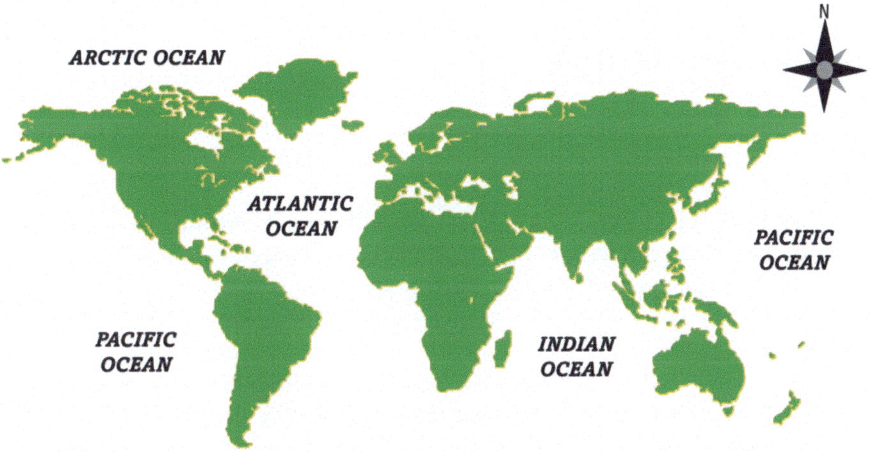

Fig. 1. Five Oceans of the World

tide models for the Arctic and Antarctic are the scarcity of high-precision altimetry data and the limitations of satellite coverage, especially from the TOPEX/Poseidon series. The presence of extensive sea ice and large ice shelves significantly hinders the ability of satellite altimeters to measure sea levels accurately, mainly due to seasonal ice variations. Nevertheless, because of the severe climate in the polar seas, there are also few available measured data sources, such as tide stations. Consequently, there is insufficient data to model or constrain ocean tides in polar seas, and model consistency is poor. Stammer et al. [2] and Lei et al. [26] concur that TPXO8 is the most accurate tidal model for the Antarctic region, with the latter study quantifying its accuracy at 7.52 cm through comparison with in-situ tidal measurements. Sun et al. [4] found that the EOT20 model exhibits significantly improved accuracy in the Antarctic Peninsula, with root mean square errors of 8.29 and 7.46 cm. This suggests that tidal predictions in Antarctica are steadily advancing.

The accuracy and performance of tidal models exhibit significant spatial variability, influencing the precision of tidal height predictions [17]. While models such as DTU10, FES2014, HAMTIDE12, OSU12, and TPXO8 demonstrate high precision in deep-water regions of the South China Sea, substantial discrepancies arise in shallow waters and coastal zones due to variations in satellite data used for model development. FES2014 outperforms other models and the tide gauge with an RSS value of 9.35 cm, although HAMTIDE's lower accuracy in the Strait of Malacca affects its shallow-water performance. Moreover, the optimal tidal model varies regionally. FES2014 excels in southeastern and northeastern China, while EOT11a is more suitable for southern coastal stations. SCHW demonstrates superiority in northern China and inland areas despite its lower resolution than newer models like FES2014 and DTU10. These newer models, incorporating satellite altimetry and improved bathymetry, do not consistently surpass SCHW in inland regions. Overall, determining the best ocean tide model is challenging due to regional variations and the complex interplay of factors influencing tidal constituents.

4 Summary and Outlook

Previous studies have highlighted the superior accuracy of global ocean tide models in deep-water environments compared to their performance in shallow water areas, especially in regions characterised by complex hydrodynamic processes. This disparity underscores the challenges in accurately modelling tidal constituents in coastal and offshore regions. However, recent satellite technology advancements have improved precision in deriving tidal constituents for shallow water regions. This development offers exciting possibilities for enhancing our understanding of oceanic phenomena in these areas.

It is conceivable to develop highly accurate local hydrodynamic models capable of precise tidal prediction by leveraging these improved satellite-derived tidal constituents. This would particularly benefit coastal and offshore regions, such as Malaysia, where complex hydrographic conditions prevail. Ultimately, integrating these local models into a cohesive hydrographic separation model could provide valuable insights for various applications, including marine transportation, coastal engineering, and environmental management.

Acknowledgement. The authors would like to acknowledge the financial support from the Ministry of Education Malaysia under the Fundamental Research Grant Scheme (FRGS) (FRGS/1/2023/WAB05/UTM/02/1) and Universiti Teknologi Malaysia for funding under UTM Encouragement Research (UTMER) (Q.J130000.3852.31J46).

References

1. Keysers, J.H., Quadros, N.D., Collier, P.A.: Vertical datum transformations across the Australian littoral zone. J. Coastal Res. **31**(1), 119–128 (2015)
2. Stammer, D., et al.: Accuracy assessment of global barotropic ocean tide models. Rev. Geophys. **52**(3), 243–282 (2014)
3. Laignel, B., Vignudelli, S., Almar, R., Becker, M., Bentamy, A., Benveniste, J., Birol, F., Frappart, F., Idier, D., Salameh, E., Passaro, M.: Observation of the coastal areas, estuaries and deltas from space. Surv. Geophys. 1–48 (2023)
4. Sun, W., Zhou, X., Zhou, D., Sun, Y.: Advances and accuracy assessment of ocean tide models in the Antarctic Ocean. Front. Earth Sci. **10**, 757821 (2022)
5. Shum, C.K., et al.: Accuracy assessment of recent ocean tide models. J. Geophys. Res. Oceans **102**(C11), 25173–25194 (1997)
6. Seifi, F., Deng, X., Baltazar Andersen, O.: Assessment of the accuracy of recent empirical and assimilated tidal models for the Great Barrier Reef, Australia, using satellite and coastal data. Remote Sens. **11**(10), 1211 (2019)
7. Fu, Y., Feng, Y., Zhou, D., Zhou, X., Li, J., Tang, Q.: Accuracy assessment of global ocean tide models in the South China Sea using satellite altimeter and tide gauge data. Acta Oceanol. Sin. **39**, 1–10 (2020)
8. Zahel, W.: Modeling ocean tides with and without assimilating data. J. Geophys. Res. Solid Earth **96**(B12), 20379–20391 (1991)
9. Zahran, K.H., Jentzsch, G., Seeber, G.: Accuracy assessment of ocean tide loading computations for precise geodetic observations. J. Geodyn. **42**(4–5), 159–174 (2006)
10. Schwiderski, E.W.: Ocean tides, part I: global ocean tidal equations. Mar. Geodesy **3**(1–4), 161–217 (1980)
11. Tan, H., et al.: Evaluation of global ocean tide models based on tidal gravity observations in China. Geodesy Geodyn. **12**(6), 451–458 (2021)
12. Egbert, G.D., Bennett, A.F., Foreman, M.G.: TOPEX/POSEIDON tides estimated using a global inverse model. J. Geophys. Res. Oceans **99**(C12), 24821–24852 (1994)
13. Cheng, Y., Andersen, O.B.: Towards further improving DTU global ocean tide model in shallow waters and Polar Seas. In: OSTST, Poster in: Proceedings of the Ocean Surface Topography Science Team (OSTST) Meeting, Miami, FL, USA, pp. 23–27 (2017)
14. Munk, W.H., Cartwright, D.E.: Tidal spectroscopy and prediction. Philosophical transactions of the royal society of London. Series A, Math. Phys. Sci. **259**(1105), 533–581 (1966)
15. Groves, G.W., Reynolds, R.W.: An orthogonalised convolution method of tide prediction. J. Geophys. Res. **80**(30), 4131–4138 (1975)
16. Lyard, F., Lefevre, F., Letellier, T., Francis, O.: Modelling the global ocean tides: modern insights from FES2004. Ocean Dyn. **56**, 394–415 (2006)
17. Fok, H.S.: Ocean tides modeling using satellite altimetry. Doctoral dissertation, The Ohio State University (2012)
18. Taguchi, E., Stammer, D., Zahel, W.: Inferring deep ocean tidal energy dissipation from the global high-resolution data-assimilative HAMTIDE model. J. Geophys. Res. Oceans **119**(7), 4573–4592 (2014)
19. Schrama, E.J.O., Ray, R.D.: A preliminary tidal analysis of TOPEX/POSEIDON altimetry. J. Geophys. Res. Oceans **99**(C12), 24799–24808 (1994)
20. Ray, R.D.: A global ocean tide model from TOPEX/POSEIDON altimetry: GOT99. 2. National Aeronautics and Space Administration, Goddard Space Flight Center (1999)
21. Ray, R.D.: Precise comparisons of bottom-pressure and altimetric ocean tides. J. Geophys. Res. Oceans **118**(9), 4570–4584 (2013)

22. Carrère, L., Lyard, F., Cancet, M., Guillot, A., Picot, N.: May. FES 2014, a new tidal model—validation results and perspectives for improvements. In: Proceedings of the ESA Living Planet Symposium, pp. 9–13 (2016)
23. Lyard, F.H., Allain, D.J., Cancet, M., Carrère, L., Picot, N.: FES2014 global ocean tide atlas: design and performance. Ocean Sci. **17**(3), 615–649 (2021)
24. Savcenko, R., Bosch, W.: EOT11a-empirical ocean tide model from multi-mission satellite altimetry. In: DGFI Report No. 89 (2012)
25. Hart-Davis, M.G., Piccioni, G., Dettmering, D., Schwatke, C., Passaro, M., Seitz, F.: EOT20: a global ocean tide model from multi-mission satellite altimetry. Earth Syst. Sci. Data **13**(8), 3869–3884 (2021)
26. Lei, J., Li, F., Zhang, S., Ke, H., Zhang, Q., Li, W.: Accuracy assessment of recent global ocean tide models around Antarctica. Int. Arch. Photogramm. Remote. Sens. Spat. Inf. Sci. **42**, 1521–1528 (2017)

Assessment of the Optimal Hydrodynamic Model for Malaysian Coastal Waters

Nur Surayatul Atikah Alihan[1(✉)], Ami Hassan Md Din[1], Mohammad Hanif Hamden[1], and Dudy Darmawan Wijaya[2,3]

[1] Geospatial Imaging and Information Research Group (Gi2RG), Faculty of Built Environment and Surveying, Universiti Teknologi Malaysia, 81310 Johor Bahru, Johor, Malaysia
nsatikah2@live.utm.my, amihassan@utm.my

[2] Geodesy Research Group, Faculty of Earth Sciences and Technology, Institute ofTechnology Bandung, Bandung, Indonesia

[3] Geodetic Science, Engineering and Innovation Research Group, Faculty of Earth Science and Technology, Institute of Technology Bandung, Bandung, Indonesia

Abstract. Coastal zones serve as crucial interfaces where terrestrial and marine systems converge, substantially impacting economies, ecosystems, and overall societal well-being. Regarding tidal dynamics, the accuracy of the tidal model experiences pronounced deterioration in proximity to coastal regions due to the intricate tidal interactions arising from hydrodynamical interactions. This accuracy degradation stems from many factors, including the effects of shallow water, tidal amplification, local topography, tidal resonance and the complexities introduced by tidal currents. Therefore, this study aims to assess the optimal hydrodynamic model in Malaysian coastal waters. In this study, the Tide Model Driver (TMD) is utilised to retrieve the hydrodynamic model data from the Indian Ocean model, TPXO9 model and TPXO9-Atlas model. Incorporating ground-based tidal data in the assessment contributes to the overall robustness of the analysis. This evaluation reveals that the TPXO9 model demonstrates superior performance in the context of Malaysian coastal waters, with an RMSmisfit below 10 cm for primary tidal constituents. This research is beneficial for coastal management, environmental protection, and diverse applications that necessitate a comprehensive understanding of coastal dynamics in the maritime regions of Malaysia.

Keywords: Coastal zones · Tidal dynamics · Hydrodynamic models · Malaysian coastal waters

1 Introduction

Ocean tides are described as the fluctuation of the sea level driven by the gravitational attraction between the earth and celestial bodies, specifically the moon and the sun [1–3]. The predictable movements of the solar system influence tidal patterns, allowing for the anticipation of variations in tide height and timing. As the tide changes, substantial volumes of water advance and recede along the coastline, causing tidal currents that shape oceanic circulation. Usually, ocean tides were measured using tide gauges along

A. Abdul-Rahman et al. (Eds.): GeoWeek 2024, LNGC, pp. 193–202, 2025.
https://doi.org/10.1007/978-3-031-86654-8_15

the coastlines and either tide gauges or bottom pressure recorders at islands before the satellite altimeter era. As a conventional method for determining tidal height, tide gauge stations can provide accurate, verifiable and high-frequency tidal information. However, the sparse distribution of tidal stations in Malaysia is still inadequate in providing tidal information, especially in open seas. Due to the localised nature of tidal characteristics, which are approximately within tens of kilometres, there arises a challenge in obtaining accurate tidal information for areas located beyond the vicinity of tide gauge stations [4].

The advent of satellite altimeters introduces the technique of estimating ocean parameters that can be further studied in a synoptic manner. One of the prominent benefits of using a satellite altimeter is its ability to observe the global and regional ocean tides in the offshore area, as the availability of tide gauge stations is scarce along the coastal area. Nonetheless, as the satellite altimeter approaches coastal areas, the accuracy of sea level measurements significantly diminishes due to substantial variations in ocean conditions throughout time and space [5–7]. Abdullah (2018) highlights two major challenges of employing satellite altimeters in coastal areas. Firstly, waveform distortion due to land contamination effects poses a significant issue. Secondly, accurate geophysical corrections for sea level retrieval are required, such as ionospheric and tropospheric correction models. The reliability of satellite altimetry data degrades when the sea floor topography abruptly becomes shallow and significant surface changes occur rapidly between land and ocean [8].

The consistent oscillations of the sea surface induced by ocean tides have captivated scientists for centuries due to their impact on oceanic processes. The advancement of satellite altimeters allows researchers to broaden their knowledge to generate the ocean tide model. Precise ocean tide models offer sea level corrections applicable to various measurements, including bathymetry computation and vertical datum determination [9]. There are three types of categories in tidal modelling: hydrodynamic model, assimilation model and empirical model [10, 11]. Hydrodynamic models are derived numerically from the Laplace tidal equation related to bottom topography, advection, bottom friction, tidal loading and self-attraction, which delineate the variability of the ocean as a result of tidal forcing [11, 12]. The model's accuracy significantly relies on bathymetry and tide gauge data quality. The assimilation model is the hydrodynamic approach, which proposes the integration of satellite altimeter data with or without tide gauge data into a hydrodynamic model to compute ocean tides, with some cases incorporating both types of data. This model is alternatively known as the barotropic hydrodynamic model, which is constrained by assimilated tide information [13]. The empirical model is contrary to the hydrodynamic and assimilation models since this approach is free from external effects due to hydrodynamical influences. This model adopts purely satellite altimeter data as the sole data source to quantify the ocean tides [11, 12, 14]. According to Fu et al. (2020), the accuracy of the global ocean tide models varies significantly between shallow waters of coastal areas and deep ocean areas. Therefore, this study evaluates the accuracy of three assimilation tidal models in the coastal areas of Malaysian waters. The optimal hydrodynamic model is determined through a statistical comparison between the derived tidal constituents from the hydrodynamic model and the ground truth tidal data acquired from the nearest coastal tide gauge stations. The closest hydrodynamic point to the tide gauge stations is validated using eight tidal constituents, encompassing primary diurnal and semi-diurnal tide components: K1, M2, S2, O1, N2, K2, P1, and Q1. The findings contribute to an extensive comprehension of the coastal dynamics within the maritime regions of Malaysia.

2 Materials and Methods

2.1 Research Domain and Data Sources

A total of 377 hydrodynamic points was established in Peninsular Malaysia, complemented by 491 hydrodynamic points in East Malaysia. These points are strategically located within the designated buffer region with a spacing grid of 0.09°, ranging from 0 to 20 km from the Malaysian coastline. This study uses the TPXO model to determine the optimal fit of the assimilation into a barotropic hydrodynamic model within the Malaysian coastal region. TPXO comprises a series of fully global ocean tide models, which, in a least-squares sense, optimally fit the Laplace Tidal Equations and altimetry data. Each subsequent model in the TPXO series is built upon updated bathymetry and incorporates more extensive data than the preceding versions [15]. This study employed the Indian Ocean regional ocean tide model, the TPXO9 model and the latest version, the TPXO9-Atlas model, spanning from 1993 to 2022.

The Oregon State University (OSU) is credited with developing the inverse tide model of the Indian Ocean, TPXO9 and TPXO9-Atlas. These models were acquired using OSU Tidal Inversion Software (OTIS), which integrates satellite altimeters with or without tidal data into a hydrodynamic model, including global shallow-water and inverse barotropic tidal models. Indian Ocean, a regional model with 1/12° resolution, is incorporated into a local inverse solution. In contrast, TPXO9 and TPXO9-Atlas represent the most recent global models provided by Oregon State University (OSU), best fit in the Laplace tidal equations and are based on updated bathymetry data, which incorporate assimilation data of hydrodynamic-based models. TPXO9-Atlas offers a comprehensive global solution featuring a 1/30° resolution, obtained by integrating a 1/6° base global solution from TPXO9 and a 1/30° resolution of local solutions for coastal areas worldwide, including the Indian Ocean [13, 15].

This study assessed eight primary tidal constituents (K1, M2, S2, O1, N2, K2, P1, and Q1) from the hydrodynamic models against the nearest coastal tidal data. 21 Malaysian coastal tide gauges acquired from Department of Survey and Mapping Malaysia are utilised to serve as in-situ ground truth data for identifying the optimal fit hydrodynamic model for Malaysian coastal waters.

2.2 Spatial Design of the Buffer Region

The designated buffer region is created using ArcGIS software by adhering to the specified spatial arrangement for the data acquisition employed in this study. The process initiates with forming a spatial grid within the study area, featuring a grid interval of 0.09°, equivalent to 10 km. Subsequently, the buffer region is generated using a multiple-ring buffer in ArcToolbox, with the designated area set extending from 0 to 20 km from the shoreline. The grid points located beyond the buffer region are then truncated to establish a coastal grid, as shown in Fig. 1.

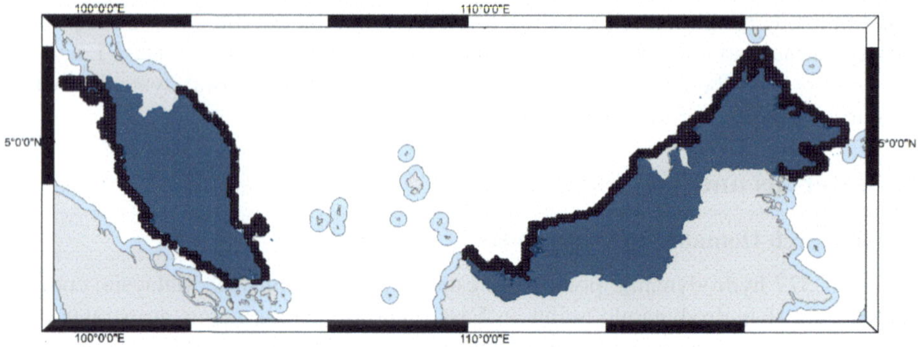

Fig. 1. The coastal grid points are generated within the designated buffer region.

2.3 Hydrodynamic Model Data Processing Via Tide Model Driver (TMD)

Tide Model Driver (TMD) is a MATLAB software tool created by Oregon State University and Earth and Space Research. This toolbox is designed to retrieve the harmonic components in a barotropic tide model and predict the tides. There are two components included in this toolbox. First, a graphical user interface (GUI) streamlines the browsing process by allowing the user to specify the tidal prediction at the chosen time intervals and points of interest. The second component is a module designed for obtaining tide parameters and generating tidal predictions [16]. The TMD software can execute both regional and global ocean tide models from Oregon State University. This model is available for academic research purposes, and registration can be completed at https://www.tpxo.net/. The solution to the extraction problem of tidal components relies on inverse theory. It involves solving the Shallow Water Equations, incorporating an ocean bathymetry model and global coastal and estuary geometries [17].

For this study, TMD is utilised for hydrodynamic model data processing, requiring the height of harmonic coefficient grids data (h_model), velocity data (u_model) and bathymetry grid file (grid_model) to be downloaded in binary format. The Indian Ocean model and TPXO9 provide multi-constituents for h, u and bathymetry grid_model. However, the data downloaded from the TPXO9-Atlas comes in a large file, as the harmonic coefficient grid and velocity data are based on a single tidal constituent. Therefore, the tpxo_atlas2local.m function is used to create a multi-constituents outcut from the single constituent within the specified study area, defined by geographical coordinates between latitude 0° N to 14° N and longitude 95° E to 126° E. The spatial hydrodynamic points shown in Fig. 1, along with start date information, data interval, and total hours, serve as the input data for the TMD software. Afterwards, the tidal constituents from each hydrodynamic model are generated at the designated hydrodynamic points.

2.4 Statistical Data Assessment of Optimal Hydrodynamic Model

The in-situ coastal tide gauges are the ground truth data to validate the hydrodynamic model. The discrepancy of the tidal constituents is evaluated using Root Mean Square misfit (RMSmisfit) as follows [18–20]:

$$RMS_{misfit} = \left\{ \frac{1}{2M} \sum_{k=1}^{M} \left[(A_c \cos\theta_c - A_i \cos\theta_i)^2 + (A_c \sin\theta_c - A_i \sin\theta_i)^2 \right] \right\}^{\frac{1}{2}} \quad (1)$$

where M denotes the number of assessment data used, A_c is amplitude and θ_c is phase derived from the hydrodynamic model. A_i and θ_i represent the amplitude and phase generated by in-situ coastal tidal data. Then, the precision of overall fit and agreement between the hydrodynamic model and coastal tidal data is estimated by Root Sum Square (RSS) as expressed as follows:

$$RSS = \left(\sum_{a=1}^{TC} RMS_a^2 \right)^{\frac{1}{2}} \quad (2)$$

where TC is the tidal constituents and $(RMS_a^2)^{\frac{1}{2}}$ represents the square root of the mean square for each tidal constituent.

3 Results and Discussion

This study uses the tidal observations from 21 Malaysian coastal tide gauge stations to assess the accuracy of the hydrodynamic model's derived tidal constituents. The spatial distribution of the coastal tide gauges is illustrated in Fig. 2, while Table 1 details the distances between each station and the nearest hydrodynamic points.

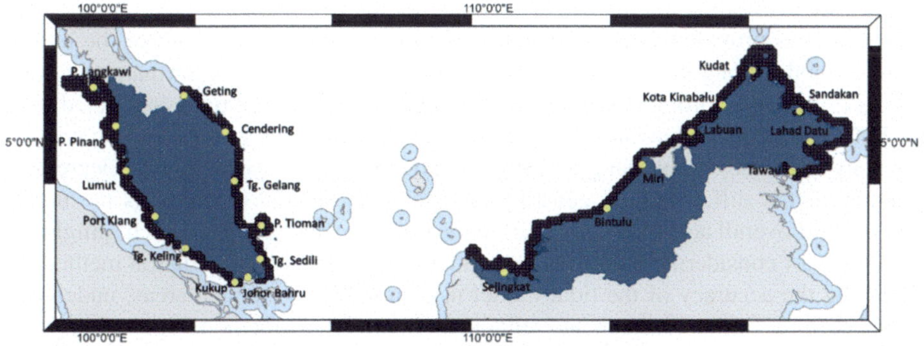

Fig. 2. The spatial distributions of Malaysian coastal tide gauges used for the hydrodynamic model assessment.

Eight major tidal constituents, comprising primary diurnal and semi-diurnal tide components: K_1, M_2, S_2, O_1, N_2, K_2, P_1, and Q_1, are extracted from the TMD and analysed against the derived tidal constituents from the Malaysian tide gauges. This analysis involved a statistical assessment using RMSmisfit and RSS. Table 2 shows the RMSmisfit values between the hydrodynamic model and 21 Malaysian coastal tide gauges stations. The values in bold represent the best accuracy performance of the hydrodynamic model within the coastal area. According to Fu et al. (2020), the tidal variability of the study region influences the misfits between the hydrodynamic model and the observed tidal data. The RMSmisfit values for the dominant diurnal tidal constituent K_1 and O_1 are relatively lower, ranging from 2.2 cm to 3.27 cm, compared to the study by Fu et al. (2020), who reported a wider range of RMSmisfit between the TPXO model and coastal tide gauge with values of 3.82 cm for K_1 and 5.36 cm for O_1 constituents. In addition, the RMSmisfit values for major semi-diurnal M_2 and S_2 are found in the range of 6.88 cm to 15.93 cm, while the remaining tidal constituents, N_2, K_2, P_1, and Q_1, range from 0.92 cm to 6.22 cm.

Overall, the findings indicate that TPXO9 exhibits the best accuracy, with all RMSmisfit values for each tidal constituent below 10 cm. Additionally, TPXO9 shows high precision in its RSS value compared to the other models. Histograms of the RMSmisfit between the hydrodynamic models and 21 coastal tide gauge stations are shown in Fig. 3, providing a visual representation of the discrepancies.

Further assessment in Fig. 4 illustrates the discrepancies between the hydrodynamic model and tidal data for each tide gauge station. Four tide gauge stations, Kukup, Johor Bahru, Sejingkat and Tawau, show low precision in *RSS* values. This discrepancy could be attributed to the complex dynamics of the geographical locations of these tide gauge stations, potentially impacting the accuracy of the hydrodynamic model. Since the hydrodynamic model assimilates data from either solely satellite altimeters or both altimeters and tidal data, the accuracy of the model is limited in complex geographical areas with sparse altimetry data [9, 21].

4 Conclusion

This study has provided valuable insights into the accuracy assessment of the hydrodynamic model along the coastal areas of Malaysian waters. The performance of these models is identified through the analysis of eight major tidal constituents extracted from the TMD and compared against derived tidal constituents from Malaysian tide gauges. The statistical assessment using RMSmisfit and RSS has revealed varying degrees of accuracy across different tidal constituents. Notably, TPXO9 emerged as the model with the best overall accuracy compared to the other models. The findings highlight the importance of considering factors such as tidal range and data assimilation methods in enhancing the accuracy of the tidal model in complex geographical areas, underscoring the need for further research and refinement in modelling methodologies to enhance accuracy in such regions. Moving forward, this research contributes to the ongoing efforts to improve the accuracy of tidal models for better coastal management and decision-making. This study is a stepping stone towards developing more accurate and reliable regional tidal models, ultimately contributing to the sustainable management of coastal environments, disaster preparedness, environmental conservation and various maritime activities that rely on a comprehensive comprehension of tidal behaviour.

Table 1. Distance between Malaysian coastal tide gauges to the nearest hydrodynamic points.

Station	Marker name	Latitude	Longitude	Nearest distance (km)
Bintulu	BINT	3° 15′ 44.00″ N	113° 03′ 50.00″ E	7.52
Cendering	CEND	5° 15′ 54.00″ N	103° 11′ 12.00″ E	6.94
Geting	GETI	6° 13′ 35.00″ N	102° 06′ 24.00″ E	9.47
Johor Bahru	JHBR	1° 27′ 43.20″ N	103° 47′ 31.20″ E	30.25
Kota Kinabalu	KINA	5° 58′ 59.99″ N	116° 04′ 00.00″ E	5.25
Kudat	KUDA	6° 52′ 44.40″ N	116° 50′ 38.40″ E	5.18
Kukup	KUKU	1° 19′ 31.00″ N	103° 26′ 34.00″ E	4.60
Labuan	LABU	5° 16′ 22.80″ N	115° 15′ 00.00″ E	5.86
Lahad Datu	DATU	5° 1′ 8.40″ N	118° 20′ 45.60″ E	8.61
Pulau Langkawi	LGKW	6° 25′ 51.00″ N	99° 45′ 51.00″ E	7.28
Lumut	LUMU	4° 14′ 24.00″ N	100° 36′ 47.99″ E	1.35
Miri	MIRI	4° 24′ 04.00″ N	113° 58′ 27.99″ E	8.27
Pulau Pinang	PINA	5° 25′ 18.00″ N	100° 28′ 00.03″ E	2.46
Port Klang	KELA	3° 02′ 60.00″ N	101° 21′ 29.99″ E	2.29
Sandakan	SAND	5° 48′ 36.00″ N	118° 04′ 02.00″ E	8.26
Sejingkat	SJKT	1° 34′ 58.80″ N	110° 25′ 19.20″ E	16.97
Tawau	TAWA	4° 13′ 58.80″ N	117° 52′ 58.80″ E	2.57
Tanjung Gelang	GELA	3° 58′ 30.00″ N	103° 25′ 48.00″ E	2.77
Tanjung Keling	KELI	2° 12′ 54.00″ N	102° 09′ 11.99″ E	6.09
Tanjung Sedili	SEDI	1° 55′ 54.00″ N	104° 06′ 54.00″ E	4.93
Pulau Tioman	TIOM	2° 48′ 26.00″ N	104° 08′ 24.00″ E	8.15

Table 2. The RMSmisfit and RSS between the hydrodynamic model and 21 coastal tide gauge stations.

Hydrodynamic Model		RMS_{miSfit}(cm)							RSS (cm)
	K_1	O_1	M_2	S_2	P_1	Q_1	N_2	K_2	
Indian ocean	3.26	3.27	15.93	**6.88**	1.08	1.49	3.81	2.10	18.57
TPX09	**3.02**	**2.56**	**9.41**	9.44	**1.07**	**1.01**	**2.44**	3.69	**14.67**
TPX09-atlas	**2.50**	**2.20**	12.66	13.90	**0.92**	**0.95**	2.58	6.22	20.29

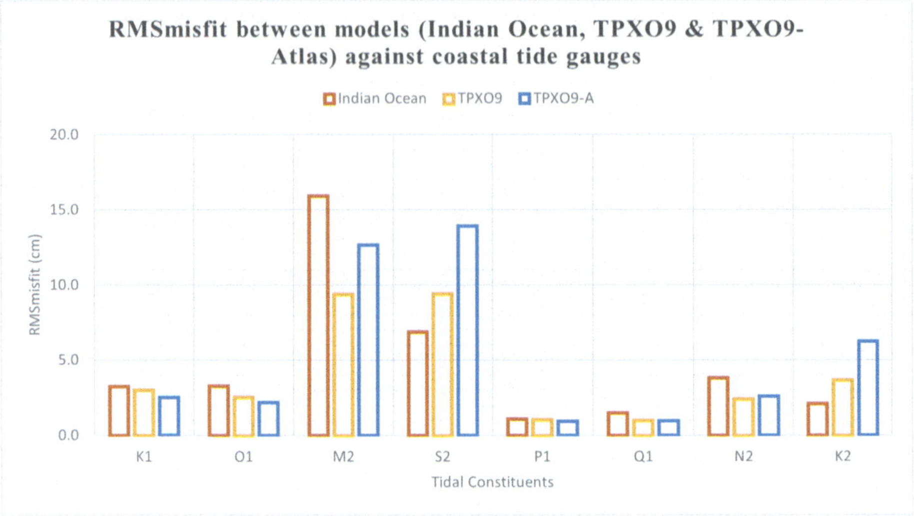

Fig. 3. Histogram of RMSmisfit between hydrodynamic models and coastal tide gauges.

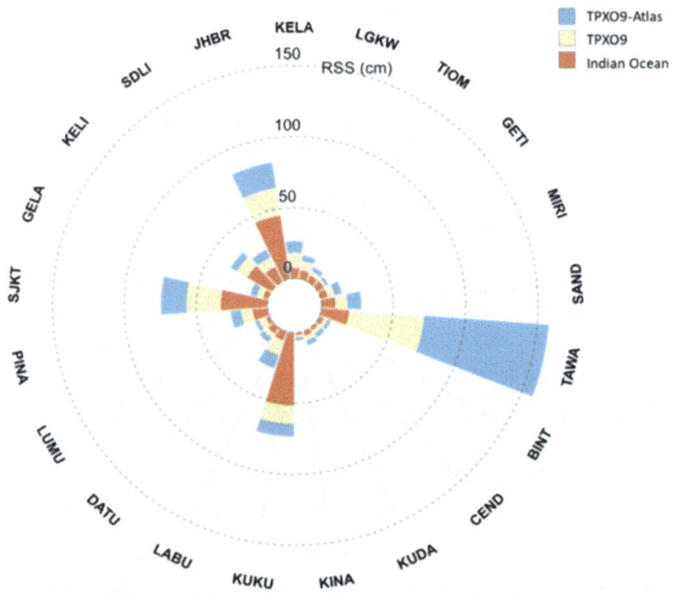

Fig. 4. The discrepancies of the hydrodynamic models at each coastal tide gauge station.

Acknowledgement. The author would like to express their gratitude to Oregon State University and the Department of Survey and Mapping Malaysia for providing valuable barotropic hydrodynamic and coastal tidal data, respectively, which significantly enhanced this study. Additionally, this research was made possible through the support of the UTM Fundamental Research Grant (Q.J130000.3852.22H75).

References

1. Pugh, D.T.: Tides, Surges and Mean Sea-Level. John Wiley & Sons Ltd., Swindon, UK (1996)
2. Gan, M., Pan, H., Chen, Y., Pan, S.: Application of the variational mode decomposition (VMD) method to river tides. Estuar. Coast. Shelf Sci. **261**, 107570 (2021)
3. Wei, Z., Pan, H., Xu, T., Wang, Y., Wang, J.: Development history of the numerical simulation of tides in the East Asian marginal seas: an overview. J. Marine Sci. Eng. **10**(7), 984 (2022)
4. Parvazi, K., Asgari, A.R., Amirisimkooei, J.A.R., Tajfirooz, B.: Determination of difference between datum and reference ellipsoid by using of analysis of altimetry data of TOPEX/Poseidon, Jason-1 and observations of coastal Tide Gauges. J. Geomatic Sci. Technol. **5**(1), 257–269 (2015)
5. Fu, L.L., Cazenave, A.: Satellite Altimetry and Earth Sciences: A Handbook of Techniques and Applications. Academic Press, San Diego, California (2001)
6. Bouffard, J., Vignudelli, S., Cipollini, P., Menard, Y.: Exploiting the potential of an improved multimission altimetric data set over the Coastal Ocean. Geophys. Res. Lett. **35**(10), 1–6 (2008)
7. Abdullah, N.N.: The Optimal Coastal Retracked Sea Levels from Saral/Altika Satellite Altimetry Over the Southeast Asia. Degree of Master of Philosophy (Geomatic Engineering). Faculty Geoinformation and Real Estate, Universiti Teknologi Malaysia, Malaysia (2016)
8. Idris, N.H., Deng, X.: The retracking technique on multi-peak and quasi-specular waveforms for Jason-1 and Jason-2 missions near the Coast. Mar. Geodesy **35**(sup1), 217–237 (2012)
9. Fu, Y., Feng, Y., Zhou, D., Zhou, X., Li, J., Tang, Q.: Accuracy assessment of global ocean tide models in the South China Sea using satellite altimeter and tide gauge data. Acta Oceanol. Sin. **39**, 1–10 (2020)
10. Matsumoto, K., Takanezawa, T., Ooe, M.: Ocean tide models developed by assimilating TOPEX/POSEIDON altimeter data into hydrodynamical model: a global model and a regional model around Japan. J. Oceanogr. **56**, 567–581 (2000)
11. Fok, H.S.: Ocean Tides Modeling Using Satellite Altimetry. Doctor of Philosophy. The Ohio State University (2012)
12. Ainee, A.: Derivation of Tidal Constituents from Satellite Altimetry Data for Coastal Vulnerability Assessment in Malaysia. Degree of Master of Science (Geomatic Engineering). Faculty Geoinformation and Real Estate, Universiti Teknologi Malaysia, Malaysia (2016)
13. Stammer, D., Ray, R.D., Andersen, O.B., Arbic, B.K., Bosch, W., Carrère, L., Cheng, Y., Chinn, D.S., Dushaw, B.D., Egbert, G.D., Erofeeva, S.Y., Fok, H.S., Green, J.A.M., Yi, Y., et al.: Accuracy assessment of global barotropic ocean tide models. Rev. Geophys. **52**(3), 243–282 (2014)
14. Seifi, F., Deng, X., Baltazar Andersen, O.: Assessment of the accuracy of recent empirical and assimilated tidal models for the Great Barrier Reef, Australia, using satellite and coastal data. Remote Sens. **11**(10), 1211 (2019)
15. Egbert, G.D., Erofeeva, S.Y.: Efficient inverse modeling of barotropic ocean tides. J. Atmos. Oceanic Tech. **19**(2), 183–204 (2002)
16. Padman, L., Erofeeva, S., Howard, S.L.: Tide Model Driver (TMD) version 2.5, Toolbox for MATLAB (MATrix LABoratory). Arctic Data Center (2022). https://doi.org/10.18739/A21Z41V08
17. Puntel de Oliveira, E.A., Junqueira, C.P.: On the use of ocean tide model driver, TMD, as a filter to recover reservoir signal from well test pressure history. In: OTC Brasil. OnePetro (2017)
18. Daher, V.B., de Oliveira, C., Vieira Paes, R., França, G.B., Rodrigues Alvarenga, J.B., Galvão Teixeira, G.L.: Extraction of tide constituents by harmonic analysis using altimetry satellite data in the Brazilian coast. J. Atmos. Oceanic Tech. **32**(3), 614–626 (2015)

19. Zhao, H., et al.: Combining the tide gauge stations and GPS/GLONASS observations to validate global and regional ocean tide models around China coast. J. Surv. Eng. **148**(3), 04022004 (2022)
20. Hamden, M.H.: Development of Quasi-Seamless Hydrographic Separation Models Based On Satellite Altimetry and Coastal Tide Gauges in Malaysia. Doctor of Philosophy (Geomatic Engineering). Faculty of Built Environment and Surveying, Universiti Teknologi Malaysia, Malaysia (2022)
21. Chen, M., Murali, K., Khoo, B.C., Lou, J., Kumar, K.: Circulation modelling in the Strait of Singapore. J. Coastal Res. **21**(5), 960–972 (2005)

Satellite based Image Processing & Analysis

Can Sentinel-1 SAR Deliver Accurate High-Resolution DEMs in Vegetated Humid Tropics?

Musa Muhammad Chindo[1,3] and Mazlan Hashim[1,2(✉)]

[1] Faculty of Built Environment & Surveying, Universiti Teknologi Malaysia, 81310 Johor Bahru, Johor, Malaysia
mazlanhashim@utm.my
[2] Geoscience and Digital Earth Centre (INSTeG), Research Institute for Sustainable Environment, Universiti Teknologi Malaysia, 81310 Johor Bahru, Johor, Malaysia
[3] Department of Survey & Geoinformatics, Federal Polytechnic Nasarawa, Nasarawa 962101, Nigeria

Abstract. The continuous access to high-resolution all-weather, day-and-night large-scale synthetic aperture radar (SAR) data from the Sentinel-1 SAR, a critical satellite mission under the European Union's Copernicus program, has revolutionised satellite-based environmental monitoring applications, including three-dimensional (3D) mapping of the earth surface over the last decade. Consequently, synthetic aperture radar interferometry (InSAR) has emerged as a valuable technique for Earth's surface Digital Elevation Model (DEM) generation using the Sentinel-1 SAR, especially in the arid and semi-arid regions. However, in the vegetated humid tropics (VHT), where InSAR holds the promise of mitigating the persistent cloud cover challenges associated with especially optical satellite-based DEM generation in the humid tropics, InSAR-DEM generation using Sentinel-1 SAR has remained challenging due to coherence constraints and the difficulties of acquiring image pairs with optimal baseline characteristics. This study assesses the impact of perpendicular and temporal baselines on Sentinel-1 InSAR coherence in VHT. The result revealed that while the optimal perpendicular baseline for achieving optimal coherence is between 201 m and 202 m, the optimal perpendicular baseline for generating InSAR-DEM with optimal precision is between 176 m and 188 m. The study also affirms the smaller the temporal baseline the better for improved coherence and precision of InSAR-DEM, with a critical temporal baseline of 36 days revealed. The findings shall provide a guide to the geoscientific community on InSAR-DEM generation and other InSAR applications with Sentinel-1 in the VHT especially on perpendicular and temporal baselines consideration.

Keywords: InSAR · perpendicular baseline · temporal baseline · SNAP

A. Abdul-Rahman et al. (Eds.): GeoWeek 2024, LNGC, pp. 205–218, 2025.
https://doi.org/10.1007/978-3-031-86654-8_16

1 Introduction

Despite being merely Earth's topographic model, the Digital Elevation Model (DEM) [1], a three-dimensional (3D) topographic map containing height information [2], owing to its simple structure and ease of implementation on computers, has emerged as the most commonly used method of representing Earth's surface topography over the decades [3]. DEM is an important source of geographic information [4], recognised by the United Nations (UN) Office for Outer Space Affairs as a crucial geospatial dataset used for mapping and modeling natural hazards and risks that are influenced by topography [5]. Traditionally, elevation data for DEM generation were primarily obtained from field surveys, topographic maps and aerial photographs [6, 7]. However, with technological advancements, the traditional topographic mapping methods typified by their labor-intensiveness, high cost, and variable map quality have been replaced with new satellite remote sensing-based techniques [8, 9]. These techniques facilitate the rapid collection of diverse Earth data [10].

Satellite-based DEM generation techniques facilitate the creation of DEMs over extensive, contiguous regions, including remote and inaccessible locations. Indeed, satellite-based DEMs of medium spatial resolution from various sources are readily accessible online at no cost [9, 11–14]. The Advanced Spaceborne Thermal Emission and Reflection Global Digital Elevation Model (ASTER GDEM), derived from the photogrammetric processing of remotely sensed optical images [9], is among the most widely utilised freely available satellite-based DEMs [15, 16]. However, because of the limitations of the optical remote sensing sensors, ASTER GDEM exhibits considerable noise issues in persistently cloud-covered environments such as the humid tropics [17] and is generally characterised by larger height errors in highly forested areas [11, 16, 18]. The widely acknowledged limitations of the optical sensors for satellite-based monitoring justify the development and utilisation of synthetic aperture radar (SAR) as an alternative means of Earth's surface monitoring, including modelling of the Earth's surface topography through SAR interferometry (InSAR) technique [19].

InSAR is a widely accepted SAR processing technique within the geoscientific community [20–22], which uses the phase difference between SAR image pairs to produce Earth's surface 3D information, such as digital elevation model (DEM) generation [19, 23]. The technique, which was introduced for topographic mapping five decades ago by [20], and further explored by [21] and [22], only became popular with the availability and open access to Sentinel-1 SAR in 2014 [9, 24], courtesy of the European Space Agency's (ESA) policy of free data distribution from 2014 [25]. Furthermore, the availability of the Sentinel Application Platform (SNAP) at no cost, which significantly streamlined the processing workflow, has encouraged the effective utilization of the freely available Sentinel-1 SAR for the generation of relatively higher resolution (10 m) satellite-based DEM applying the InSAR techniques [9, 26–34]. InSAR and Sentinel-1 offer significant advantages for satellite-based DEM generation, including wide-area coverage, all-time and weather data acquisition capability, and high temporal resolution [35].

However, in the vegetated humid tropics (VHT), InSAR-DEM generation using Sentinel-1 has remained underexplored. The VHT, a unique and ecologically diverse region typically found around the equator, is characterised by high temperatures and abundant rainfall, which create a humid climate (Wohl et al., 2012) that supports dense vegetation. The VHT provides numerous ecosystem services such as carbon sequestration, water retention, and habitat for a wide array of flora & fauna. While the rich biodiversity of the VHT is crucial for environmental balance & holds immense value for great scientific interest, the vegetation's canopy, whose elements are large for the C-band's smaller wavelength of 5.6 cm to penetrate [36], presents the challenge of volume decorrelation. Furthermore, the Sentinel-1 has a narrow orbital tube design, originally intended to favour differential InSAR [9].

While the narrow orbital tube design of the Sentinel-1 helped in reducing the temporal baseline achievable with Sentinel-1 due to its frequent revisits [37, 38], it limits the separation between the orbital paths of successive passes of the Sentinel-1 satellites, which determines the perpendicular baseline lengths achievable between Sentinel-1 image pairs for InSAR processing. While Sentinel-1 image pairs with at least 100 m, or an optimum range of 150 m to 400 m perpendicular baselines have been recommended for InSAR-DEM generation, for precise description of topographic changes by observed fringes, using image pairs with short temporal baseline has been suggested as the most efficient means of overcome decorrelation [39]. This introduces another challenge of acquiring suitable image pairs in terms of both perpendicular and temporal baselines since the perpendicular baselines for most Sentinel-1 image pairs with 6 or 12 days temporal baselines often fall below 25 m, which is far lower than even the lower limit of the recommended range of 150 m to 400 m [9, 40] Therefore, this study aimed to assess the impact of perpendicular and temporal baselines on Sentinel-1 InSAR products such as coherence and InSAR-DEM in VHT. The result shall serve as a guide to the geoscientific community on the use of the Sentinel-1 SAR for InSAR-DEM generation in the VHT and has the potential to advance Sentinel-1 InSAR application in VHT in general.

2 Material and Method

2.1 Study Area

This study was conducted in part of Johor, covering an approximate land area of 4939.216 km2. Johor is Peninsular Malaysia's southernmost state and third-largest in terms of land area [41] and Malaysia's third most-populous after Selangor and Sabah [42]. Owing to Its Geographic Location Around Latitude 1° 37′ 30″ N and longitude 103° 52′ 50″ E (see Fig. 1), the area experiences a humid climate with all-year high temperatures (between 21 to 32 °C) and high humidity, influenced by two distinct monsoon seasons. The Southwest Monsoon, occurring from late May to September, brings relatively drier weather with occasional showers. In contrast, the Northeast Monsoon, from November to March, is characterized by heavier rainfall and more frequent thunderstorms [43].

Because of the humid climate and the significant agricultural development and urbanization in Johor [41, 42], the study area is mainly composed of built-up, natural forests and plantations (oil palm and rubber) land covers [19]. The area's terrain is gently undulating, with elevations ranging from sea level up to 500 m [44]. Frequent successive days of extreme rainstorms during the Northeast monsoon season often lead to serious flooding, particularly in the Johor River Basin (JRB). The December 2006/2007 flood

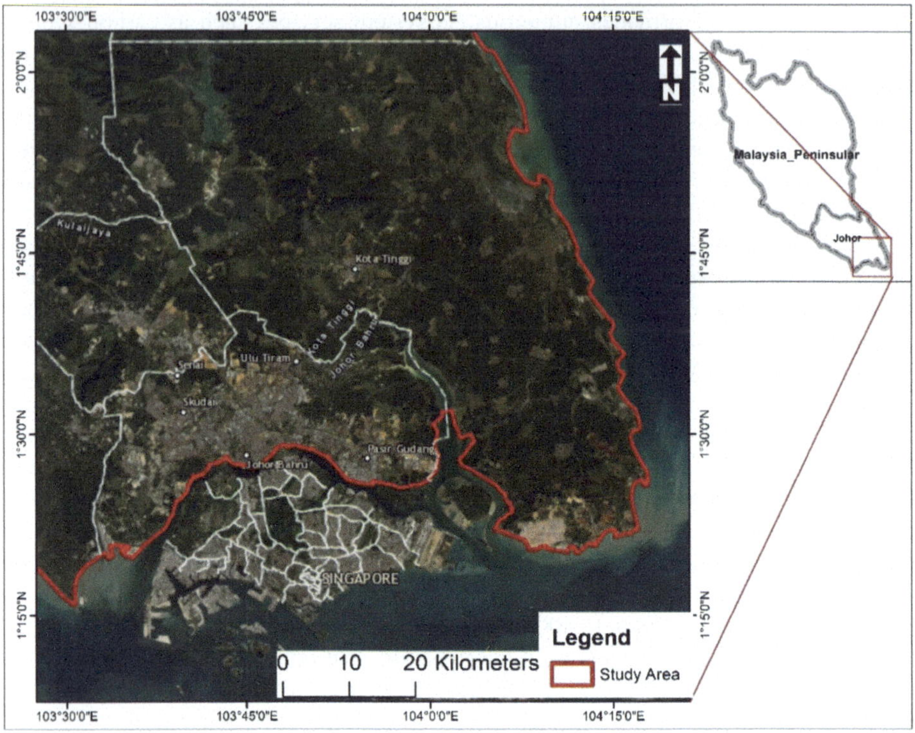

Fig. 1. The study area.

incident was the most disastrous flood ever recorded in Johor, resulting in total losses amounting to 1.5 billion, making it the costliest flooding event in Malaysia [43].

2.2 Datasets

Three primary datasets comprising Sentinel-1 SAR satellite images, Copernicus open-source (OS) DEM, and benchmarks (BMs) were utilised to achieve the objectives of this study. Consequently, six Sentinel-1 single look complex (SLC) image pairs, summarised in Table 1, were used for the study. These pairs, with perpendicular baselines ranging from 150 m to 400 m, were freely downloaded from the Alaska Satellite Facility (ASF) (https://search.asf.alaska.edu/#/). Alongside these images, the Copernicus OS-DEM was employed for InSAR processing to generate the respective InSAR products, including coherence maps, interferograms, and InSAR-DEMs. The InSAR products were then evaluated using the acquired benchmarks (BMs) as ground control points (GCPs). This assessment aimed to determine the impact of perpendicular and temporal baselines on coherence and InSAR-DEMs, ultimately identifying the optimal baseline characteristics for achieving the best coherence and InSAR-DEM quality in the vegetated humid tropics (VHT).

Table 1. Summary of Sentinel-1 SAR datasets used.

	Primary		Secondary		Perp. baseline (m)	Temp. baseline (days)
	Scene ID	Burst	Scene ID	Burst		
1	S1A_IW_SLC__1SSV_20150626T112501_20150626T112528_006543_008B41_C971	IW3/3-6	S1A_IW_SLC__1SDV_20150708T112451_20150708T112519_006718_008FEA_710C	IW3/6-9	228	12
2	S1A_IW_SLC__1SDV_20150121T112448_20150121T112515_004268_005312_67FD	IW3/6-9	S1A_IW_SLC__1SSV_20150226T112455_20150226T112522_004793_005F51_B066	IW3/3-6	176	36
3	S1A_IW_SLC__1SSV_20150109T112456_20150109T112523_004093_004F43_716F	IW3/3-6	S1A_IW_SLC__1SDV_20150214T112447_20150214T112514_004618_005B11_1A69	IW3/6-9	155	36
4	S1A_IW_SLC__1SSV_20150109T112456_20150109T112523_004093_004F43_716F	IW3/3-6	S1A_IW_SLC__1SDV_20150708T112451_20150708T112519_006718_008FEA_710C	IW3/6-9	201	180
5	S1A_IW_SLC__1SDV_20150614T112451_20150614T112519_006368_008645_14A0	IW3/6-9	S1A_IW_SLC__1SSV_20160128T112456_20160128T112523_009693_00E252_6B53	IW3/5-8	188	228
6	S1A_IW_SLC__1SSV_20150602T112500_20150602T112527_006193_008131_656A	IW3/3-6	S1A_IW_SLC__1SDV_20150708T112451_20150708T112519_006718_008FEA_710C	IW3/6-9	202	36

2.3 Data Processing and Results Evaluation

To process the datasets and generate the Sentinel-1 InSAR products, SNAP, a robust open-source tool developed by ESA for the efficient handling of its Sentinel satellite data, was used. The processing largely involves preprocessing the Sentinel-1 SLC image pairs and InSAR processing to generate DEMs, which were used to assess the impact of perpendicular and temporal baseline on achievable coherence and generated InSAR-DEM precision. To evaluate the influence of perpendicular and temporal baselines on coherence, the coherence images of all image pairs were first terrain-corrected and then vectorised to extract coherence values for all pixels. The mean coherence values for each image pair were subsequently calculated. Finally, the image pairs were sorted in ascending order of their perpendicular and temporal baselines and plotted against their corresponding mean coherence values to analyze the effect of the perpendicular and temporal baselines on coherence using trend lines.

Furthermore, to evaluate the influence of perpendicular and temporal baselines on InSAR-DEM precision, the accuracy of all the InSAR-DEMs relative to the acquired BMs control point was first determined using root mean square error (RMSE) metrics (see Eq. 1). Finally, the image pairs were sorted in ascending order of their perpendicular and temporal baselines and plotted against their corresponding RMSE values to analyse the effect of the perpendicular and temporal baselines on InSAR-DEMs precision using trend lines.

$$\mathrm{RMSE} = \sqrt{\frac{\sum_{i=1}^{n} \left(Z_{InSAR-DEM.i} - Z_{BM.i}\right)^2}{n}} \tag{1}$$

where ZInSAR-DEM is the elevation of InSAR-DEM at BM point ZBM which is the ground truth elevation.

3 Result and Discussion

The InSAR technique was employed to produce six Digital Elevation Models (DEMs) for a section of Johor, a region characterized by dense vegetation and a humid tropical climate. This was accomplished using Sentinel-1 SLC SAR image pairs labelled 1, 2, 3, 4, 5, and 6, each with varying perpendicular and temporal baselines. This section presents an evaluation of the impact of perpendicular and temporal baselines on the achievable coherences of Sentinel-1 InSAR and the precision of the generated InSAR-DEMs in the VHT.

3.1 Impact of Perpendicular and Temporal Baselines on Sentinel-1 InSAR Coherence in VHT

Figure 2 depicts the terrain-corrected coherence images for the six image pairs, labelled 1 through 6, as detailed in Table 1. These images serve as the foundation for assessing the impact of perpendicular and temporal baselines on the coherence achieved by Sentinel-1 InSAR in VHT.

Figures 3 and 4 illustrate the plots of mean coherence achieved by the six image pairs against perpendicular and temporal baselines, respectively. These plots reveal the relationships between coherence and the respective baselines (perpendicular and temporal).

3.2 Impact of Perpendicular and Temporal Baselines on Sentinel-1 InSAR-DEM in VHT

Figure 5 displays the InSAR-DEMs for the six image pairs, labelled 1 through 6, as detailed in Table 1. These InSAR-DEMs were fundamental in evaluating the impact

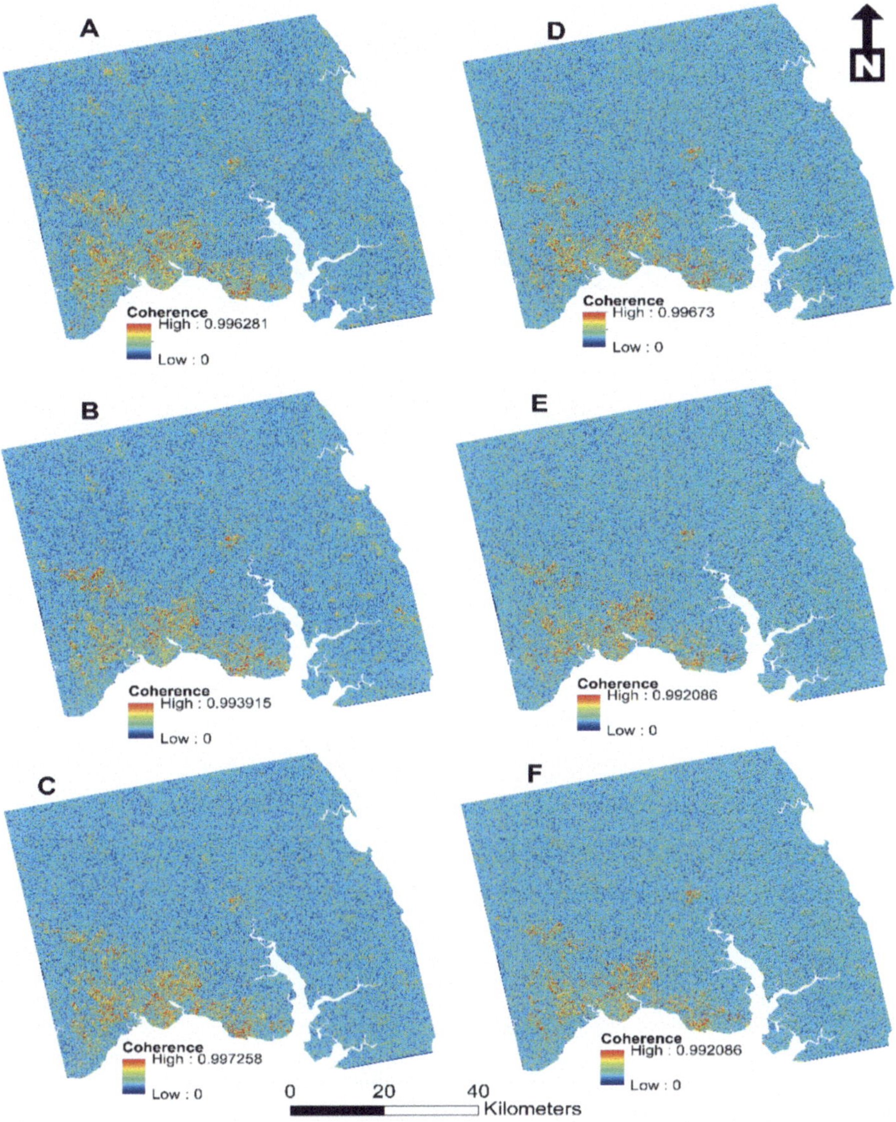

Fig. 2. Coherence image of the six image pairs 1 to 6

of perpendicular and temporal baselines on the precision of Sentinel-1 InSAR-DEMs I VHT.

Figures 5 and 6 illustrate the precision achieved by the six image pairs' InSAR-DEMs against perpendicular and temporal baselines, respectively. These plots reveal the relationships between the accuracy of Sentinel-1 InSAR-derived DEMs and the respective baselines (perpendicular and temporal) (Fig. 7).

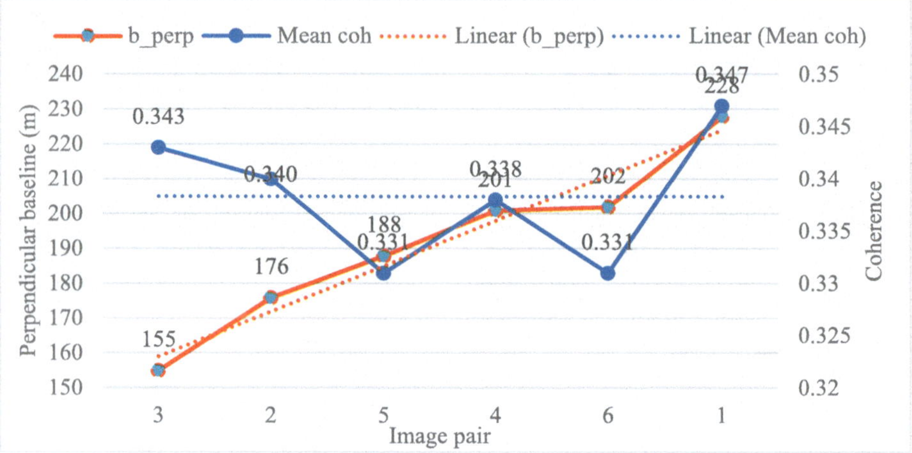

Fig. 3. Impact of perpendicular baseline on coherence.

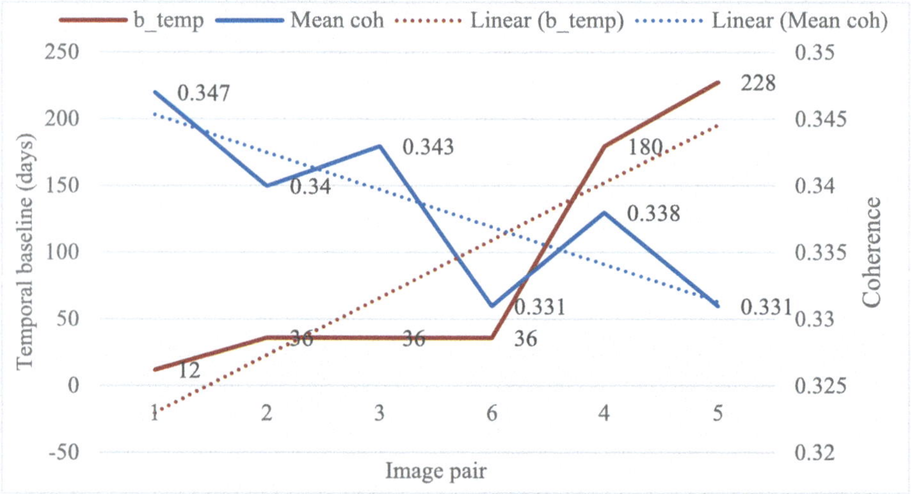

Fig. 4. Impact of temporal baseline on coherence.

Sentinel-1 SAR has significantly improved the ability to monitor and map the Earth's surface, particularly in persistently cloud-covered regions like the VHT. However, the interaction between SAR signals and dense vegetation makes both perpendicular and temporal baselines critical for generating accurate InSAR-DEMs with Sentinel-1 in such environments. Using image pairs with perpendicular baselines within the range of 150 m and 400 m and a short temporal baseline can enhance accuracy. However, due to the difficulty of finding image pairs with large perpendicular and small temporal baselines. The importance of baseline configuration for improved InSAR-DEM generation with Sentinel-1 SAR has been overlooked. Thus, the low accuracy reported for Sentinel-1

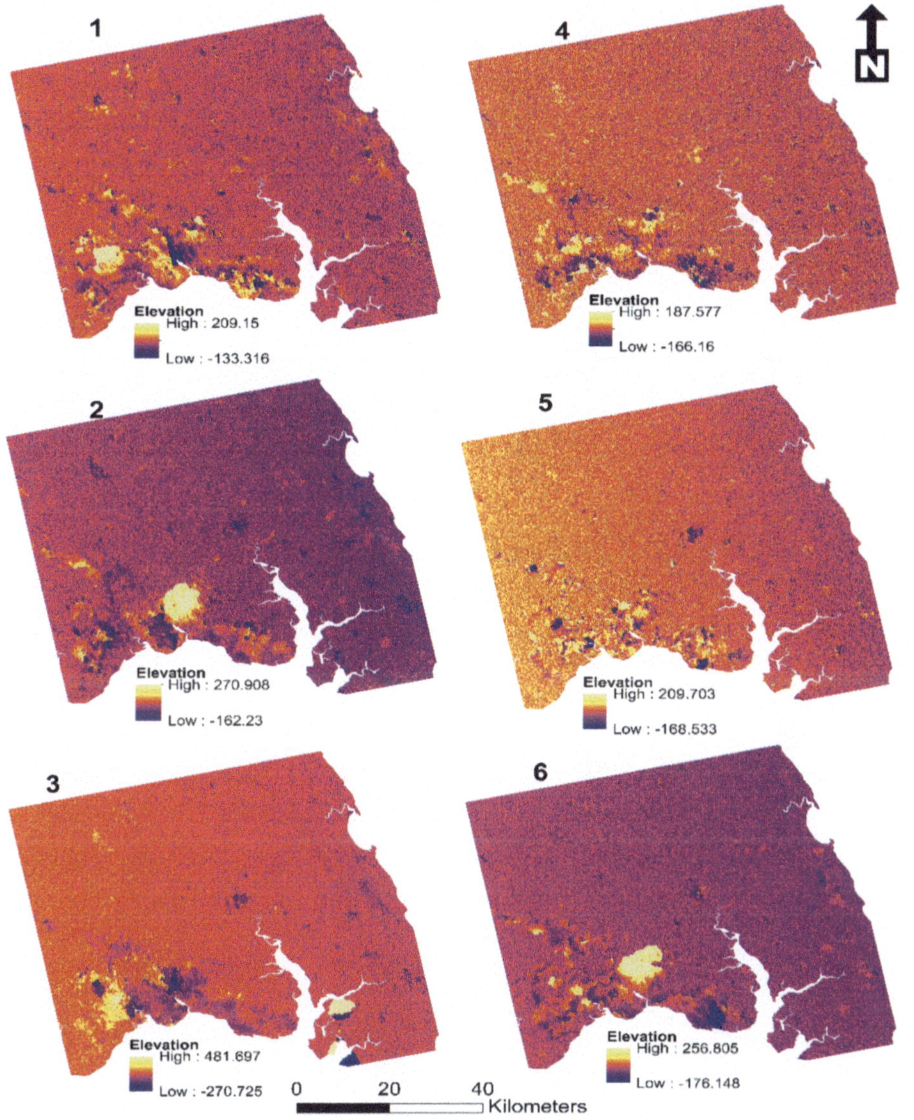

Fig. 5. InSAR-DEMs of the six image pairs 1 to 6.

InSAR-DEM generated in the VHT in even areas with relatively good coherence [32, 45].

Therefore, this study assessed the impact of both perpendicular and temporal baselines on InSAR products, specifically coherence and InSAR-DEM. The assessment of the impact of perpendicular baseline on coherence, as shown in Fig. 3 reveals no direct relationship exists between variations in the perpendicular baseline and coherence. However, the intersection of mean coherence and perpendicular baseline trendlines between

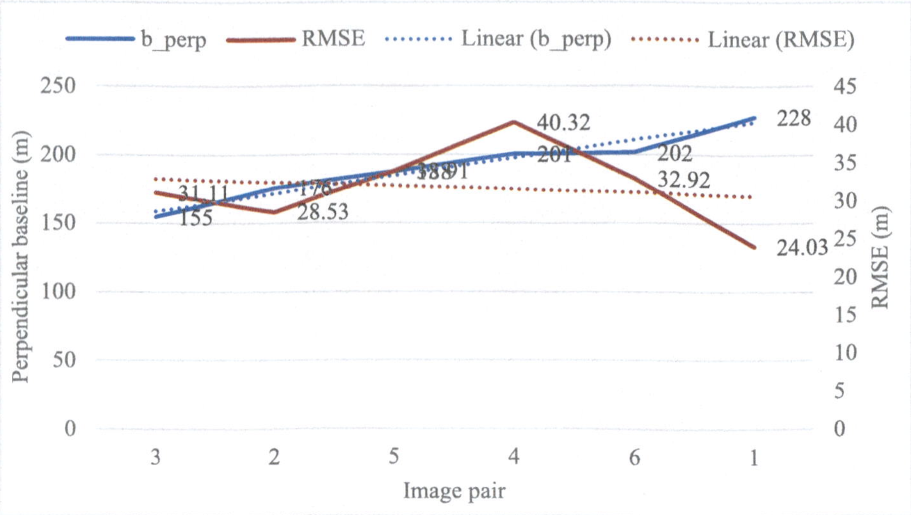

Fig. 6. Impact of perpendicular baseline on InSAR-DEM accuracy.

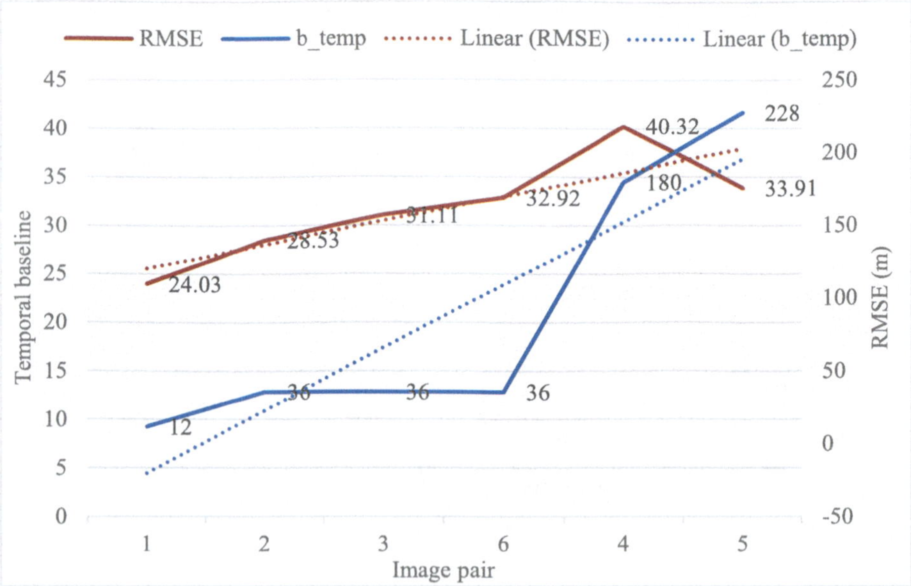

Fig. 7. Impact of temporal baseline on InSAR-DEM accuracy.

image pairs 4 and 6 with perpendicular baselines of 201 and 202 m indicates the optimal perpendicular baseline for achieving optimal InSAR coherence with Sentinel-1 in VHT is between 201 m and 202 m. Conversely, the assessment of the impact of temporal baseline on coherence, as depicted in Fig. 4 indicate an inverse relationship between temporal baseline and coherence. The smaller the temporal baseline, the better the coherence achievable with Sentinel-1 in the VHT. Although the least possible temporal baseline with Sentinel-1 for InSAR-DEM generation, which is considered the optimal temporal baseline, is 12 days, the intersection of mean coherence and temporal baseline trendlines at image pair 6 with temporal baselines of 32 days indicates the critical temporal baseline for achieving optimal InSAR coherence with Sentinel-1 in VHT is 36 days.

Furthermore, assessment of the impact of perpendicular and temporal baseline on the precision of Sentinel-1 InSA-DEM reveals almost similar pattern with coherence. While the assessment of the impact of perpendicular baseline on InSAR-DEM precision, as shown in Fig. 5 reveals a slightly inverse relationship between the perpendicular baseline and InSAR-DEM precision, the intersection of the InSAR-DEM precision and perpendicular baseline between image pairs 2 and 5 with perpendicular baselines of 176 m and 188 m indicates the optimal perpendicular baseline for generating InSAR-DEM with the best possible precision with Sentinel-1 in VHT is between 176 m and 188 m. However, the inverse relationship between temporal baseline and InSAR-DEM precision is more distinct, indicating the least possible obtainable temporal baseline between image pairs is the best for generating InSAR-DEM with the best possible precision in VHT.

4 Conclusion

The study evaluated the influence of perpendicular and temporal baselines on Sentinel-1 InSAR products, specifically coherence and InSAR-DEMs in VHT. This assessment provided valuable insights into optimising these baselines to enhance the accuracy and reliability of InSAR measurements in such challenging environments. While the study reveals the optimal perpendicular baseline for achieving optimal InSAR coherence with Sentinel-1 in VHT is between 201 m and 202 m, the optimal perpendicular baseline for generating InSAR-DEM with the best possible precision with Sentinel-1 in VHT is between 176 m and 188 m. The study also reveals the least possible obtainable temporal baseline between image pairs as the best for generating InSAR-DEM with the best possible precision in VHT, however, with a critical temporal baseline of 36 days.

Acknowledgement. The authors acknowledge Universiti Teknologi Malaysia for the Junior Visiting Researcher grant (ref Q.J130000.21A6.00P75).

References

1. Miller, C.L.: The theory and application of the digital terrain model (1958)
2. Moon, S.H., Choi, H.L.: Super-resolution based on deep learning technique for constructing digital elevation model. In: AIAA Space and Astronautics Forum and Exposition, SPACE 2016, American Institute of Aeronautics and Astronautics Inc, AIAA (2016). https://doi.org/10.2514/6.2016-5608
3. Wilson, J.P., Gallant, J.C.: Terrain analysis: principles and applications. Wiley (2000)

4. Guo'an, T., Shanshan, G., Fayuan, L., Jieyu, Z.: Review of Digital Elevation Model (DEM) Based Research on China Loess Plateau. J. Mt. Sci. **2**, 265–270 (2005). Accessed 19 Feb 2022. [Online]. Available: http://www.imde.ac.cn/journal
5. United Nations: Data application of the month: digital elevation models. Accessed 17 Feb 2022. [Online]. Available: https://www.un-spider.org/links-and-resources/data-sources/daotm-elevation
6. Yue, L., Shen, H., Yuan, Q., Zhang, L.: Fusion of multi-scale DEMs using a regularized super-resolution method. Int. J. Geogr. Inf. Sci. **29**(12), 2095–2120 (Dec. 2015). https://doi.org/10.1080/13658816.2015.1063639
7. Deilami, K., Hashim, M.: Very high resolution optical satellites for DEM generation: a review. Eur. J. Sci. Res. **49**(4), 542–554 (2011). Accessed 19 Feb 2022. [Online]. Available: http://www.eurojournals.com/ejsr.htm
8. Zandbergen, P.: Applications of shuttle radar topography mission elevation data. Geogr. Compass **2**(5), 1404–1431 (Sep. 2008). https://doi.org/10.1111/j.1749-8198.2008.00154.x
9. Braun, A.: Retrieval of digital elevation models from Sentinel-1 radar data - Open applications, techniques, and limitations. De Gruyter Open Ltd (2021, Jan. 1). https://doi.org/10.1515/geo-2020-0246
10. Guo, H., Goodchild, M.F., Annoni, A.: Manual of digital earth (2020)
11. Hirt, C., Filmer, M.S., Featherstone, W.E.: Comparison and validation of the recent freely available ASTER-GDEM ver1, SRTM ver4.1 and GEODATA DEM-9s ver3 digital elevation models over Australia. Aust. J. Earth Sci. **57**(3), 337–347 (Apr. 2010). https://doi.org/10.1080/08120091003677553
12. Yang, L., Meng, X., Zhang, X.: SRTM DEM and its application advances. Taylor and Francis Ltd (2011). https://doi.org/10.1080/01431161003786016
13. Bolch, T., Kamp, U., Olsenholler, J., Bolch, T., Kamp, U.: Using ASTER and SRTM DEMs for studying geomorphology and glaciation in high mountain areas (2005). https://doi.org/10.5167/uzh-137249
14. Wang, X., Holland, D.M., Gudmundsson, G.H.: Accurate coastal DEM generation by merging ASTER GDEM and ICESat/GLAS data over Mertz Glacier, Antarctica. Remote Sens. Environ. **206**, 218–230 (Mar. 2018). https://doi.org/10.1016/j.rse.2017.12.041
15. Lin, X., et al.: A DEM super-resolution reconstruction network combining internal and external learning. Remote Sens. (Basel) **14**(9) (May 2022). https://doi.org/10.3390/rs14092181
16. Uuemaa, E., Ahi, S., Montibeller, B., Muru, M., Kmoch, A.: Vertical accuracy of freely available global digital elevation models (Aster, aw3d30, merit, tandem-x, srtm, and nasadem). Remote Sens. (Basel) **12**(21), 1–23 (Nov. 2020). https://doi.org/10.3390/rs12213482
17. Earthdata: NASADEM: creating a new NASA digital elevation model and associated products. Accessed 31 Oct 2023. [Online]. Available: https://www.earthdata.nasa.gov/esds/competitive-programs/measures/nasadem
18. Yamazaki, D., et al.: A high-accuracy map of global terrain elevations. Geophys. Res. Lett. **44**(11), 5844–5853 (Jun. 2017). https://doi.org/10.1002/2017GL072874
19. Chindo, M.M., Hashim, M., Rasib, A.W.: Revising digital elevation models generation from Sentinel-1 synthetic aperture radar and interferometric technique in densely-vegetated humid tropical environment (2022). https://doi.org/10.1080/10106049.2022.2136263, https://doi.org/10.1080/10106049.2022.2136263
20. Graham, L.C.: Synthetic interferometer radar for topographic mapping (1974)
21. Zebker, H.A., Goldstein, R.M.: Topographic mapping from interferometric synthetic aperture radar observations. J. Geophys. Res. **91**, 4993–4999 (1986). https://doi.org/10.1029/jb091ib05p04993

22. Li, F.K., Goldstein, R.M.: Studies of multibaseline spaceborne interferometric synthetic aperture radars. IEEE Trans. Geosci. Remote Sens. **28**(1), 88–97 (1990). https://doi.org/10.1109/36.45749
23. Suryanti, D.I., Wahyudiono, A.: Interferometric synthetic aperture radar (InSAR) for mapping forest area in the tropics (2017). [Online]. Available: https://www.researchgate.net/publication/316862309
24. Potin, P., et al.: Sentinel-1 mission status. In: Proceedings of EUSAR 2016: 11th European Conference on Synthetic Aperture Radar, pp. 1–6 (2016)
25. Amitrano, D., et al.: Sentinel-1 for monitoring reservoirs: a performance analysis. Remote Sens. (Basel) **6**(11), 10676–10693 (2014). https://doi.org/10.3390/rs61110676
26. Ghannadi, M.A., Enayati, H., Khesali, E.: Generating digital elevation model of the earth using Sentinel-1 images and interferometry. Sci. Res. Q. Geograph. Data (SEPEHR) **27**(108), 109–121 (Feb. 2019). https://doi.org/10.22131/SEPEHR.2019.34623
27. Ahmadabadi, A., Karam, A., Saffari, A., Yazdanpanah, M.: Estimation of surface and elevation displacement of Ardestan sand dunes using radar interferometry and spectral indices. پژوهشهای ژئومورفولوژی کمّی, **8**(4), 1–17 (Apr. 2020). https://doi.org/10.22034/GMPJ.2020.106407
28. Ghannadi, M.A., Alebooye, S., Izadi, M., Moradi, A.: A method for Sentinel-1 DEM outlier removal using 2-D Kalman filter. Geocarto Int. (2020). https://doi.org/10.1080/10106049.2020.1815866
29. Sefercik, U.G., Buyuksalih, G., Atalay, C., Jacobsen, K.: Validation of Sentinel-1A and AW3D30 DSMs for the Metropolitan Area of Istanbul, Turkey. PFG – J. Photogrammetry, Remote Sens. Geoinform. Sci. **86**(3–4), 141–155 (Oct. 2018). https://doi.org/10.1007/s41064-018-0054-3
30. Soni, C., Chaudhary, A., Sharma, U., Sharma, C.: Satellite radar interferometry for DEM generation using Sentinel-1A imagery. In: Advances in intelligent systems and computing, pp. 26–33. Springer Science and Business Media Deutschland GmbH (2021). https://doi.org/10.1007/978-981-15-6067-5_4
31. Nikolakopoulos, K., Kyriou, A., Charalampopoulou, V.: DSM generation from Sentinel and COSMO-SkyMed data using interferometry and radargrammetry: a case study from Mykonos, Greece, **9535,** 143–151 (Jun. 2015). https://doi.org/10.1117/12.2193977
32. Sunu, H.A., Yuwono, D., Suprayogi, A.: DSM accuracy analysis of Semarang city using InSAR method using Sentinel-1 imagery. Original title: Analisis ketelitian DSM kota Semarang dengan metode INSAR menggunakan citra sentinel-1 (2019)
33. Mohammadi, A., Bin Ahmad, B., Shahabi, H.: Extracting Digital Elevation Model (DEM) from SENTINEL-1 satellite imagery: case study a part of Cameron Highlands, Pahang, Malaysia (2018). [Online]. Available: http://iraj
34. Devaraj, S., Yarrakula, K.: Evaluation of Sentinel 1–derived and open-access digital elevation model products in mountainous areas of Western Ghats, India. Arab. J. Geosci. **13**(20), 1–14 (Oct. 2020). https://doi.org/10.1007/S12517-020-06108-W/TABLES/4
35. Druce, D., et al.: An optical and SAR based fusion approach for mapping surface water dynamics over mainland China. Remote Sens. (Basel) **13**(9) (May 2021). https://doi.org/10.3390/rs13091663
36. Woodhouse, L.H.: Introduction to microwave remote sensing. CRC Press (2006)
37. Geudtner, D., Torres, R., Snoeij, P., Davidson, M.: Sentinel-1 system overview (2012)
38. Miranda, N.: Sentinel-1 mission overview (2013)
39. Chindo, M.M., Hashim, M., Rasib, A.W.: Challenges of Insar Dem derivation with sentinel-1 SAR in densely vegetated humid tropical environment. Int. Arch. Photogrammetry, Remote Sens. Spatial Inform. Scie. **XLVIII-4/W6-2022**, 93–98 (Feb. 2023). https://doi.org/10.5194/isprs-archives-XLVIII-4-W6-2022-93-2023

40. Prats-Iraola, P., et al.: Role of the orbital tube in interferometric spaceborne SAR missions. IEEE Geosci. Remote Sens. Lett. **12**(7), 1486–1490 (Jul. 2015). https://doi.org/10.1109/LGRS.2015.2409885
41. Pakiam, G. K., Mui, H., Terrace, K.: Trends in Southeast Asia agriculture in Johor: what's left? (2018). [Online]. Available: http://bookshop.iseas.edu.sg
42. Hutchinson, F.E.: Situating Johor. In: Johor: Abode for Development?, ch. One, pp. 5–25. ISEAS Publishing, Singapore (2020)
43. Banjir, R., et al.: Flood risk index assessment in Johor River Basin (2015)
44. Obaid, H.A., Shahid, S.: Soil erosion susceptibility of Johor River basin. Water Environ. J. **31**(3), 367–374 (Aug. 2017). https://doi.org/10.1111/wej.12252
45. Zylshal, Z., et al.: Space-borne interferometric SAR for digital elevation model in the tropical region: an initial result of Sentinel-1 RADAR data over the Indonesian archipelago. In: Proceedings - 2021 7th Asia-Pacific Conference on Synthetic Aperture Radar, APSAR 2021, Institute of Electrical and Electronics Engineers Inc. (2021). https://doi.org/10.1109/APSAR52370.2021.9688483

Damage Assessment with YOLOv9 Instance Segmentation: An Analysis of the Marrakech Earthquake Case Study

Hakima Zair(✉), Zineb El Faraj, Marouane Ouazani Chahidi, Safia Loulad, Samir Zouhri, and Hassan Rhinane

Department of Geology, Faculty of Sciences Aïn Chock, Hassan II University, Casablanca, Morocco
zairhakima1@gmail.com

Abstract. In the present study, we address the urgent need for an accurate assessment of building damage in the aftermath of the recent earthquake in Marrakech, Morocco. Assessing the damage is necessary to determine the impact on buildings since they bear the brunt of most natural disasters. The aim of this study is to enhance the efficiency of detecting house damage by utilizing UAV images and advanced deep-learning algorithms. The visual surface characteristics were used to categorize the impact of the post-earthquake UAV imagery into multiple classes using the state-of-the-art SOTA model, YOLOv9. The methodology employed involved preparing an annotated dataset of 337 images and dividing it into training (70%), validation (20%), and test sets (10%). The model was set up with pre-trained weights, fine-tuned with the SGD optimizer, and augmented for dataset diversity. This approach effectively detected the damage within buildings, with evaluation metrics showing a precision of 0.913, recall rate of 0.7604, mean average precision at 50%(mAP50) of 0.834, and a mean average precision from 50% to 95%(mAP50-95) of 0.543. The experiment outcomes have demonstrated the ability of YOLOv9 in detecting structural damage in buildings.

Keywords: YOLOv9 · Damage detection · Earthquake

1 Introduction

Natural disasters, particularly earthquakes, have far-reaching and frequently catastrophic effects on human lives and infrastructure. Earthquakes can cause broad destruction, resulting in considerable loss of life, population displacement, and extensive damage to structures and infrastructure. The immediate aftermath of such incidents necessitates prompt assessment to guide emergency action, deploy resources effectively, and commence recovery procedures. Traditional damage assessment methods, such as on-site inspections and manual surveys, while detailed, are time-consuming and often impractical in the chaotic aftermath of a disaster [1]. The magnitude and immediacy of the necessary action in these situations underscore the significance of a prompt and precise evaluation of damage. Planning for long-term rehabilitation and reconstruction as well

A. Abdul-Rahman et al. (Eds.): GeoWeek 2024, LNGC, pp. 219–227, 2025.
https://doi.org/10.1007/978-3-031-86654-8_17

as for urgent relief operations depend heavily on an accurate evaluation of the damage. Precise evaluation aids in determining the areas most impacted, setting priorities for the distribution of resources, and organizing activities for restoration. The techniques for assessing damage after a disaster have seen a major change in the last few years due to technological advancements, especially in the areas of artificial intelligence, geographic information systems (GIS), and remote sensing. Unmanned aerial vehicles (UAVs) and satellite imaging are examples of remote sensing technologies that enable quick assessment even in inaccessible locations by enabling the capture of detailed photographs of impacted areas without the need for physical presence [2]. These technologies have demonstrated significant potential in automating images data analysis and speeding up the damage assessment process when paired with deep learning algorithms. Deep learning is a branch of machine learning that simulates the neural networks found in the human brain and has completely changed tasks related to picture recognition and categorization. One kind of deep learning model that has proven particularly useful for interpreting visual data is Convolutional Neural Networks (CNNs). The YOLO (You Only Look Once) method is one of these that has become a potent tool for object recognition in real time. Because of its rapid and accurate item detection and classification capabilities, YOLO is a good fit for disaster response applications where accurate and timely information is essential [3]. Over time, YOLO models have undergone significant evolution, with each iteration gaining efficiency, speed, and accuracy over the previous one. YOLOv9, the most recent version, is a prime option for quick damage assessment in disaster-affected areas since it uses cutting-edge algorithms to improve its performance in identifying objects in difficult surroundings [4]. Rapid and accurate analysis of vast amounts of data can greatly boost emergency response activities' effectiveness, preventing additional harm and potentially saving lives.

Following the recent earthquake that affected the Marrakech region, it was imperative to determine the exact extent of property and infrastructure damage, with an emphasis on the rural parts of the Al Haouz province. Conventional damage assessment methods were restricted by the earthquake's epicenter's location in an area with a mix of modern and traditional building structures. To ensure that rescue efforts happen as soon as possible in these circumstances, it is critical to determine the impact levels fast and precisely. Traditional damage assessment methods, which usually involve human inspections and ground-based surveys, are labor- and time-intensive. As seen in the province of Al Haouz's rural areas, these methods might not work in areas with a lot of damage or difficult terrain [2]. YOLO (You Only Look Once), one of the deep learning models, offers a viable substitute in this situation. YOLO is a real-time, one-stage object identification technique that is perfect for applications that need quick analysis and decision-making. Due to the algorithm's ability to forecast numerous bounding boxes and class probabilities at once, the detection procedure can be completed much more quickly [3].

YOLOv9, the most recent version, improves on the strengths of its previous by providing improved network design and training methods that result in increased performance. Because of these enhancements, YOLOv9 is especially well-suited for identifying and categorizing damage in intricate and varied surroundings, like those present in Marrakech's earthquake-affected districts. Our study suggests using YOLOv9 to pinpoint locations that have been severely affected in an effort to improve damage detection procedures' efficacy and efficiency. We aim to provide quick and precise structural damage assessments by utilizing YOLOv9, which will help disaster response teams make decisions quickly and intelligently. Particularly in rural locations with varied and difficult conditions, the incorporation of YOLOv9 into the damage assessment process has the potential to revolutionize the way post-disaster evaluations are carried out.

Despite significant advancements in the application of deep learning models for damage assessment, a notable gap exists in their deployment in rural settings. Most studies to date have focused on urban environments where building structures are more uniform, and extensive datasets are available for training AI models [5]. Rural areas, such as those in Al Haouz province, present unique challenges: building materials and structures are highly variable. This research aims to bridge this gap by exploring the effectiveness of one-stage algorithm models, specifically YOLOv9, in rural settings. The goal is to enhance the accuracy of detecting structural damage within buildings that exhibit a high degree of variability in construction materials and design. By addressing these specific issues, our study seeks to expand the applicability of deep learning models to more diverse environments, ultimately improving disaster response and management in rural areas [2]; [1].

1.1 Materials and Methods

The material and methodology section introduces first the study area and then describes the general identification framework (Fig. 2). The identification framework includes data collection, labelling, preprocessing, model training, evaluation and finally, prediction.

1.2 Study Area

In this study our target area is small village located near to the epicentre of the 6.8 magnitude (Mw) Al Haouz earthquake, which occurred on September 8, 2023, at a depth of 19 km, in Marrakech, Morocco (USGS). The epicenter of this seismic event was located at 31.055° N, 8.389° W, and it had a major impact on the surrounding areas. As Morocco threatened to high seismic activity because of its proximity to the African plate and the Eurasian plate [6]. The earthquake was caused by a faulting system located 75 km southwest of Marrakech in the High Atlas Mountains [7]. One of the villages that was badly affected is Douar Tassaout (Fig. 1), which is 22 km away from the epicenter.

The geological features of the Al Haouz region are complicated and have a considerable impact on seismic activity. The High Atlas Mountains, which include the faulting system that caused the earthquake, are the result of a collision between the African and Eurasian plates. This geological setting results in a highly stressed region susceptible to earthquakes [8]. This area is mainly composed of geological formations from largely Mesozoic and Cenozoic sedimentary rocks such as limestone, dolomite, and marl [9]. The interaction of various rock types under stress from tectonic movements can amplify the impact of seismic occurrences.

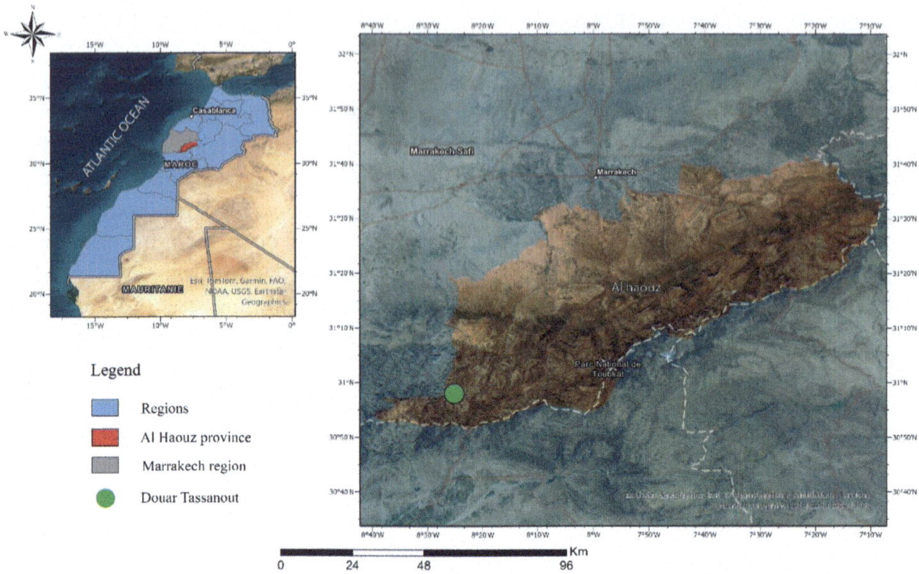

Fig. 1. Geographic Context of the Study Area.

1.3 Data

We completed a series of preparatory steps, including the collection of high-resolution image data captured by the WingtraOne GEN II drone. The flight mission covered approximately 207 ha, and 334 images were acquired shortly after the earthquake and then we processed to generate detailed elevation models and orthophotos, ensuring precise georeferencing.

In our investigation, we examined the disaster's impact on buildings by categorizing damage into several intensity categories. We classified buildings based on a visual evaluation of their roof surfaces, determining whether they were unaffected, badly damaged, or permanently damaged. To improve our analysis and reduce misclassification, we also included detectable objects as categories. Consequently, we defined the following damage levels (Table 1).

2 Methodology

2.1 Data Collection and Labelling

The methodology includes a complete data collection and analysis architecture that employs advanced image processing and deep learning techniques. Initially, 337 high-resolution photographs covering 207 hectares were obtained using an aerial mission, resulting in orthophotos and elevation models. These data types are critical because they provide extensive visual and topographic observations, which contribute in the precise identification of damage. The second phase was data labelling, which involved establishing damage categories and developing annotations using Roboflow, a robust annotation

Table 1. Surface characteristics of the main categories.

Categories	Roof surface characteristic
Demolished (219 instances/annotations)	This class describes structures that have been totally demolished and are no longer functional. Identifying demolished properties is critical for determining the degree of the disaster and planning future reconstruction
Vegetation (161 instances/annotations)	This category refers to places covered in vegetation rather than constructions. This class helps distinguish between damaged buildings and regions with natural cover, lowering the possibility of mistaking vegetation for damaged structures
Survivors' Tents (146 instances/annotations)	This class includes temporary shelters established by survivors in the aftermath of the event. Identifying survivors' tents is important for distinguishing between improvised lodgings and damaged or undamaged structures, which helps to better understand the urgent requirements of affected communities
Damage (121 instances/annotations)	This class refers to structures that are partially damaged but still standing. By categorizing these structures independently, we can better estimate the extent of the damage and prioritize repair efforts
Undamaged (91 instances/annotations)	This class identifies buildings that have not been damaged by the disaster. Recognizing intact structures counts for distinguishing them from damaged ones and determining their overall impact on the built environment

tool. This approach was crucial in creating exact masks for the areas of interest, increasing the dataset's quality and utility.

2.2 Data Preprocessing and Model Training

For data preprocessing, the dataset was separated into three sets: training (70%), validation (20%), and testing (10%), guaranteeing a balanced distribution for model training and evaluation. Data augmentation techniques were used to boost dataset diversity, which improved the model's predictability. Furthermore, the photos were separated into 7,050 tiles and optimized for training the YOLOv9 model [10]. The YOLOv9 model, noted for its efficiency and accuracy in object detection but also has shown its effectiveness in instance segmentation tasks, it was started with pre-trained COCO weights. This technique, as noted in recent publications, such as Chien et al.'s work on employing YOLOv9 for fracture identification in pediatric wrist trauma X-ray pictures, proves the model's robustness and adaptability across multiple domains [10].

The model's performance was thoroughly assessed using precision, recall, mAP50 (mean Average Precision at IoU threshold 0.5), and mAP50-95 (mean Average Precision at IoU thresholds from 0.5 to 0.95) metrics. These measures demonstrate the model's ability to accurately detect and categorize damage, making it a useful tool for real-world applications. This method framework exemplifies a complex approach to damage detection by combining high-quality data, precise labeling, strong pre-processing, and advanced deep learning algorithms (Fig. 2).

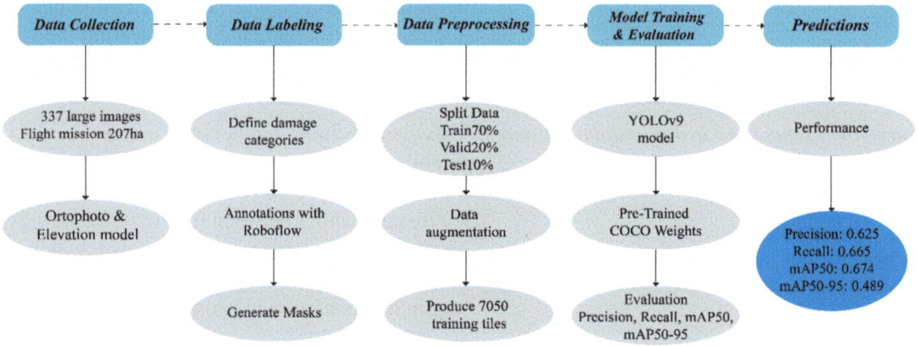

Fig. 2. Framework of automatic identification of damage in Tassaount village.

3 Results

In our case study, the Yolov9 model has demonstrated a remarquable performance in instance segmentation tasks, reaching a mean Precision about 0.625, mean Recall of 0.665, mAP50 of 0.674, and mAP50-95 of 0.489. The predictions did successfully classify and recognized most of the classes, the obtained evaluation metrics reflect the effectiveness of the YOLOv9 model in instance segmentation, specifically the demolished buildings.

To further verify the recognition ability of YOLOv9, we presented the segmentation performance of YOLOv9 on the various categories of Tassaount data in Fig. 4. In addition to the Survivors-tent class that showed a significant high value for all metrics, we can find that Demolished and Undamaged had comparatively higher values of Precision and Recall than the other categories.

Here are some samples of the testing dataset's aerial photos and the correspondingly generated predicted segmentation results for a more informative demonstration (Fig. 3).

The YOLOv9 model has proven to be an effective tool for instance segmentation, particularly when recognizing damage in high-resolution UAV pictures. Its high-performance metrics and capacity to handle several categories make it an invaluable tool for automatic damage assessment in disaster-affected areas. Moving forward, increasing the dataset to better represent underrepresented classes and fine-tuning the model can improve its applicability and performance in difficult situations (Fig. 4).

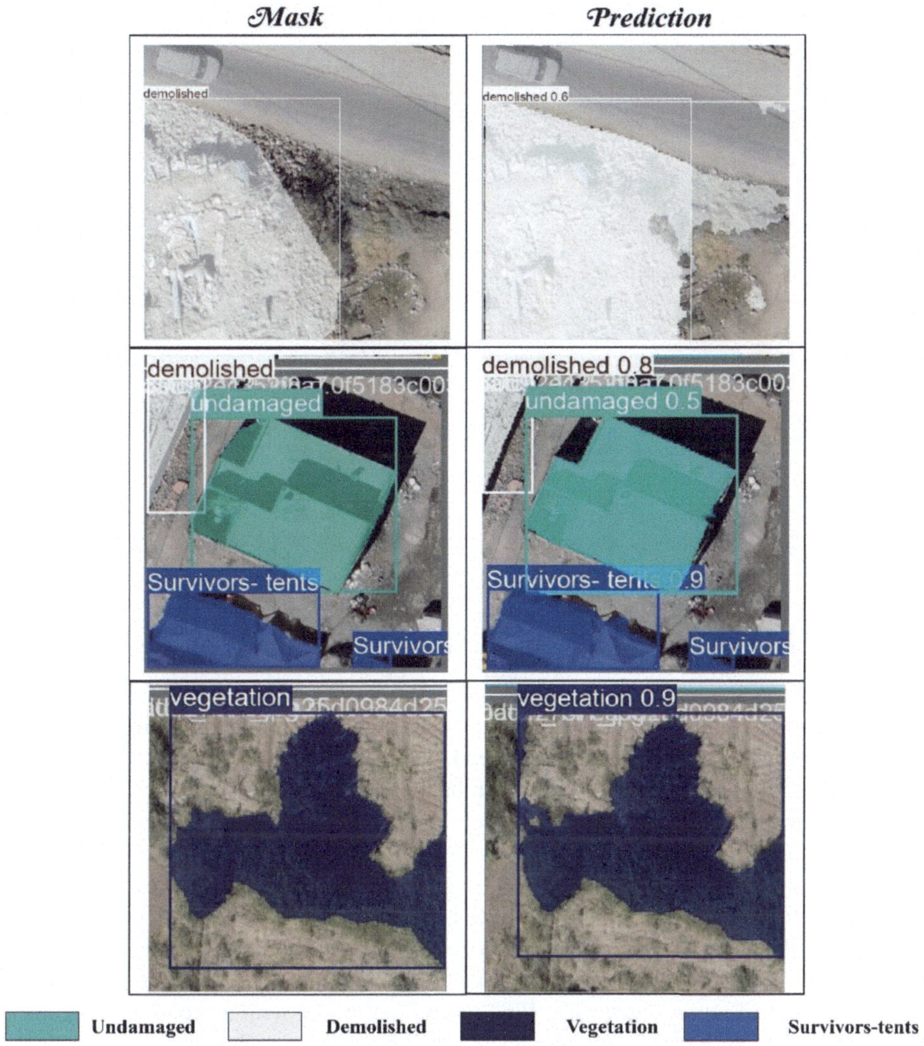

Fig. 3. Visualisation examples of predicted results.

4 Discussion

Moreover, the model has successfully identified the main damage categories with an unexpected value, still faced challenges in misclassifying some categories. We reasoned that some categories that were well classified could be because of the well-defined shape of the objects such as survivors-tents and also due to the large amount of data representing this class, which what helped to be easily predicted by the model, at the other hand we explain the low performance of the model in detecting other classes such as the demolished is due to the unshaped and unfamiliar category for the pretrained model,

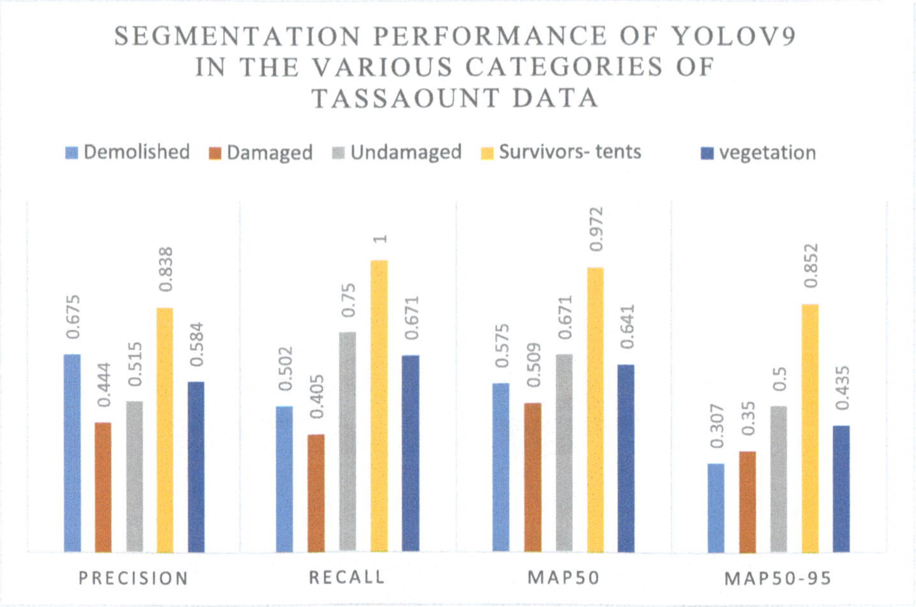

Fig. 4. Segmentation performance of YOLOv9 in the various categories.

thus for the misrecognition of damage category, we think it was due to the similar surface appearance between damage and undamaged category.

This is relying mainly on the class imbalance between the number of instances generated during the labelling phase, which makes it difficult to learn and effectively categorize underrepresented classes.

Addressing class imbalance is the key to increase the model performance. Effective solutions include oversampling the minority class and undersampling the majority class can improve the accuracy of its performance. Furthermore, combining transfer learning with specific to a domain pretraining can improve the model's capability to identify uncommon and irregular categories. Fine-tuning the model with a broad and balanced dataset that includes a variety of samples for each category can also help in boosting its suitability and accuracy.

5 Conclusion

This research proposes a segmentation method for autonomously identifying impacted areas in the Al Haouz region using high-resolution UAV pictures and the YOLOv9 model. The experimental results showed that YOLOv9 has good feature extraction and classification capabilities for predicting various levels of impact. The model succeeds at detecting well-defined and frequently represented categories, but it struggles with misclassification due to irregular forms and class imbalance. The successful recognition of categories like survivors' tents emphasizes the significance of precise object definitions and adequate data representation. In contrast, challenges in recognizing categories such as demolished structures and bias between damaged and undamaged areas demonstrate the model's limitations when dealing with less recognized and similar-looking classes.

To improve the model's performance, class imbalance must be addressed using approaches such as oversampling and undersampling. Furthermore, using domain-specific pretraining and fine-tuning on a balanced dataset can increase the model's strength and accuracy. Implementing these strategies will result in more precise and reliable damage detection, making the model a valuable tool for real-world disaster management and recovery efforts.

In future research, the Tassaount segmentation dataset can be used to detect disaster-affected areas in other locations. We also propose expanding the dataset to better represent underrepresented classes, which would improve performance in difficult cases.

References

1. Dong, L., Shan, J., Tang, Z.: A comprehensive review of earthquake-induced building damage detection with remote sensing techniques. ISPRS J. Photogrammetry Remote Sens. **84**, 85–99 (2013). Link
2. Nguyen, H.Q., Tran, V.H., Pham, V.T.:. Damage assessment of buildings following natural disasters using UAV images and deep learning techniques. Sensors **20**(18), 5241 (2020). Link
3. Redmon, J., Divvala, S., Girshick, R., Farhadi, A.: You only look once: unified, real-time object detection. In: Proceedings of the IEEE Conference on Computer Vision and Pattern Recognition (CVPR), pp. 779–788 (2016). Link
4. Bochkovskiy, A., Wang, C., Liao, H.M.: YOLOv4: Optimal speed and accuracy of object detection. arXiv preprint arXiv:2004.10934 (2020). Link
5. Howard, A.G., et al.: MobileNets: efficient convolutional neural networks for mobile vision applications. arXiv preprint arXiv:1704.04861 (2017). Link
6. Nayak, K., López-Urías, C., Romero-Andrade, R., Sharma, G., Guzman-Acevedo, G.M., Trejo-Soto, M.E.: Ionospheric Total Electron Content (TEC) Anomalies as earthquake precursors: unveiling the geophysical connection leading to the 2023 Moroccan 6.8 MW earthquake. Geosciences (Basel) **13**(11), 319 (2023)
7. Akilan, A., Azeez, K.K.A., Schuh, H., Yuvraaj, N.: Large-scale present-day plate boundary deformations in the eastern hemisphere determined from VLBI data: implications for plate tectonics and Indian Ocean growth. Pure Appl. Geophys. **172**(10), 2643–2655 (2014)
8. Beauchamp, W., Allmendinger, R.W., Barazangi, M., Demnati, A., Alji, M.E., Dahmani, M.: Inversion tectonics and the evolution of the High Atlas Mountains, Morocco, based on a geological-geophysical transect. Tectonics **18**(2), 163–184 (1999). https://doi.org/10.1029/1998tc900015
9. El Harfi, A., Lang, J., Salomon, J., Chellai, E.: Cenozoic sedimentary dynamics of the Ouarzazate foreland basin (Central High Atlas Mountains, Morocco). Int. J. Earth Sci. **90**(2), 393–411 (2000). https://doi.org/10.1007/s005310000115
10. Chien, C., Ju, R., Chou, K., Chiang, J.: YOLOV9 for fracture detection in paediatric wrist Trauma x-ray images. arXiv:2403.11249 (2024)

Satellite Based Assessment on the Water Yield Deficit at Sub-Watershed Level in Peninsular Malaysia During 2005 Drought

Mohd. Rizaludin Mahmud[1,2](✉), Mazlan Hashim[1,2], Mohd. Nadzri Md. Reba[1,2], Nurul Hazrina Idris[1,2], Suzanna Noor Azmy[1,2], Noordyana Hassan[1,2], and Nurul Nadiah Yahya[1,2]

[1] Geoinformation, Faculty of Built Environment and Surveying, Universiti Teknologi Malaysia, 81310 Johor Bahru, Johor, Malaysia
rizaludin@utm.my

[2] Geoscience and Digital Earth Centre, Faculty of Built Environment and Surveying, Universiti Teknologi Malaysia, 81310 Johor Bahru, Johor, Malaysia

Abstract. This paper assessed the 2005 drought impact on water yield deficit at sub-watershed level of Peninsular Malaysia using satellite-based algorithm. Year 2005 has been identified as one of the warmest years by the National Oceanic and Atmospheric Administration and Malaysia has been severely affected by large synoptic drought encompassing all the states in Peninsular Malaysia. The satellite-based run off model developed by Hashim et al. (2016) and Mahmud (2012) has been adapted with appropriate data from 2000 and 2007. The use of simplistic original water balance equation was able to simulate the run-off with good agreement with in-situ (NSME = 0.3; Bias = 37%). Comparable run-off curve was achieved with actual streamflow ($r = 0.771$, $P = 0.001$). The sub-watershed scale information on severely affected catchment that unable to recover from the rainfall deficit for 12 months. Almost 78% out of the total catchments in Peninsular Malaysia were experiencing run-off deficit. Two drought waves. Were identified, one at early year, and another one in September and October. The drought recovery sign showed towards November and December. Nonetheless, some forested catchments in the west and south Peninsular Malaysia still experiencing run-off deficit. The findings of this study are useful for assessing the drought impacts to catchment productivity that incorporate spatial based dynamics.

Keywords: Run-off · Rainfall · El-Nino · Evapotranspiration

1 Introduction

There is an increasing concern about the regional and global temperature anomalies have affected rainfall intensity and variations over Malaysia for instance in year 2005 [2]. According to the Malaysian meteorological reports too, the increasing global sea-surface temperature, especially at the equatorial belt is believed to be the major cause of this El-Nino event [1]. The decreasing trend of rainfall has leads to shortage in river flow and

A. Abdul-Rahman et al. (Eds.): GeoWeek 2024, LNGC, pp. 228–242, 2025.
https://doi.org/10.1007/978-3-031-86654-8_18

reduction of dam storage. The daily river flow readings were recorded as only 128 m^3/s and 183 m^3/s which were significantly lower than the average (261 m^3/s and 196 m^3/s in normal condition at Kelantan River and Hulu Perak respectively) [4]. This large-scale synoptic drought event of 2005 has impacted the water resources in the entire Peninsular Malaysia. Nonetheless, there spatial pattern regarding water yield deficit during the significance drought events were remained unknown in term of spatial catchment dynamics; including where are the most affected catchments and also the recovery pattern. This is because the recovery of water yield deficit is the key information that signified the offset of a drought event. To examine such spatial catchment dynamics were constrained by using ground observations due to station limitation, or hydrological data situation conflict [3]. Such discrete assessment was unable to accurately represent spatial drought assessment at larger coverage areas due to their dependent on spatial interpolation [5]. Consequently, satellite observation is perceived as a useful supporting tool for monitoring drought due to its spatial, temporal and economic capacity. Downscaling of the satellite derived hydrometeorology variables has been demonstrated with satisfactory results at Malaysian sites. For instance, the improvement on coarse resolution of satellite precipitation was critical for sub-watershed level uses by [6] and evapotranspiration localization by [7]. Hence, the subsequent use of those variables for operational run-off estimation and spatial mapping has been developed with reasonable performance [9].

Prior to the earlier statement, this study take initiative to utilize the satellite-based run-off method in analyzing the drought impact to the catchment productivity. The chosen operation grid is 4 km with monthly temporal resolution. The run-off was simulated using the modified of [10] water balance with parameterized and downscaled rainfall and evapotranspiration products from Tropical Rainfall Measuring Mission (TRMM) satellites and Moderate Resolution Imaging Spectroradiometer (MODIS). The drought assessment is carried out in 2005, a period of rainfall deficit in the tropics due to El-Nino phenomena. It was also considered among the warmest year in the century [8]. The prospective spatial based catchment drought analysis is useful in providing references for future alerts, adaptation, and sustainable water resources planning.

2 Materials and Method

2.1 Study Site

The study was undertaken in Peninsular Malaysia (Fig. 1). The region experienced wet and humid tropical climate (26°–34 °C) with heavy annual rainfall (~2500 mm). The surface water of Peninsular Malaysia is produced by nearly 120 major river basins, primarily a forested watersheds fueled by run-off generated from heavy rainfall. Due to the high average temperature, the rate of evapotranspiration (ET) is high and consistent, ranging from 120–160 mm each month. At present, there were estimated of 4.9 million hectares of those watershed with elevation ranged from lowland (~300 m) to highlands (~2500 m), remotely located, and difficult access. Figure 1 showed the study area.

2.2 The Satellite-Based Drought Assessment

The concept behind the development of spatial-based drought assessment was to use run-off information of the scaled down multi-temporal rainfall and evapotranspiration

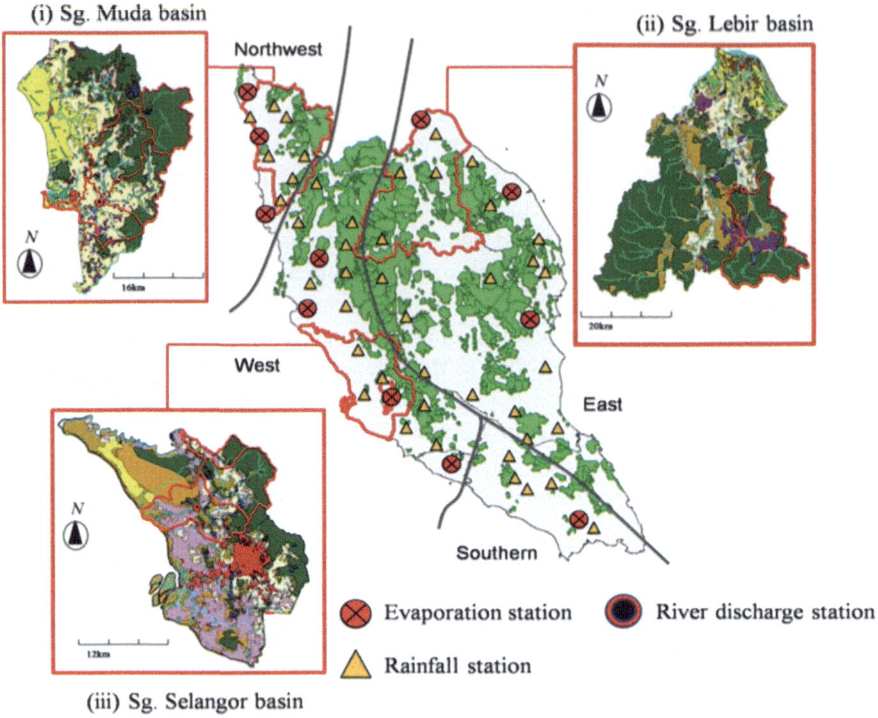

Fig. 1. Study area

datasets that were acquired from the operational satellite during normal and drought conditions at sub-watershed. Two spatial analyses on drought were conducted: (1) run-off deficit calculation (Eq. (1)), (2) qualitative and quantitative assessment on temporal run-off deficit in 2005.

The run-off deficit is calculated as follows:

$$\text{Water yield deficit} = (\text{Nr}) - (\text{Dr})/(\text{Nr}) \times 100\% \tag{1}$$

where Nr denotes the normal run-off value, and Dr denotes the drought season run-off value. (+) positive RD value means run-off decrement and indicating the impact of drought, and vice versa for negative RD value.

In this study, the monthly run-off maps were produced during eight consecutive years; 2000 to 2007. The monthly average run-off years from 2000 to 2007 (excluding 2005) were assumed as the normal run-off condition (pre-drought), while 2005 was used as drought sample year. The calculation of the run-off deficit was done on per pixel basis at each sub-watershed level to produce continuous and regular run-off maps. The area of each forested sub-watershed was acquired from the Forestry Department of Peninsular Malaysia. In total, there were 196 sub-watersheds in all 10 states in Peninsular Malaysia. Watershed that is smaller than 1200 hectares were not considered due to the limitation of the map's spatial resolution.

2.3 A Fully Satellite-Based Run-Off Model (SRO)

To obtain the run-off from satellite-based data, the algorithm developed by [9] were adapted in this study with minor improvision especially on the soil moisture components. The two main inputs of the equation: rainfall and evapotranspiration were acquired from the latest Tropical Rainfall Measuring Mission (TRMM) data product which also known as TMPA 3B43 version and MODIS 16 satellite product respectively. There are nine basic steps in the run-off estimation procedures using water balance approach (Eq. 2), covering from determination of water balance components until the computation of the estimated run-off which modified from [11] by [12].

$$N = P - Etp \pm SM \tag{2}$$

where N is the net water excess, P is the rainfall from TRMM, Etp is the evapotranspiration from MODIS 16, and SM is the soil moisture.

The soil moisture value was estimated using the soil moisture depletion equation (Eq. (3)). The available water capacity was fixed to 250 mm based on two main assumptions: (1) the average rooting depth of the vegetation is about 1.5 m, and (2) the soils in Peninsular Malaysia are predominantly silt loams, clay loams and clays, as reported by the [4]. The soil moisture retention constants of the soil moisture depletion equation were considered as 249.5 and −0.0040 respectively. Soil moisture (SM) is calculated as follows:

$$S_M = a.e^{b.A_{PWL}} \tag{3}$$

where a and b are constants, and A_PWL is accumulated potential water losses.

2.4 Satellite Rainfall & Evapotranspiration Data

2.4.1 Rainfall Retrieval and Downscaling from TRMM Multi Satellites Precipitation Analysis Data (TMPA)

Rainfall was retrieved from the rain rate information given in the TMPA 3B43 version 7 data product in this study. The data can be directly downloaded from the Internet Link: http://daac.gsfc.nasa.gov/data/datapool/TRMM. The rainfall data was radiometrically and spatially downscaled using the Geographical Difference Analysis (GDA) approach and site specific co-efficient (SISCO) technique, as introduced by [14] and [13] respectively.

2.4.2 Evapotranspiration (E_T) Retrieval from MODIS 16 and Downscaling

Evapotranspiration data product from the Moderate Resolution Imaging Spectrometer (MODIS) 16 was used. The algorithm, used to estimate the E_T from the satellite data was adapted from the procedures developed by [15] and detail of its downscaling can be found in [9]. The satellited estimated ET then were multiplied by a scale factor to convert the pixel number to standardized unit (mm/month). The data were rescaled to 4 km grids, similar to the rainfall. The downscaling approach was implemented using the linear regression based on monsoon characteristics as demonstrated by [7] and its corresponding co-efficient. Locations for validation were shown in Fig. 1.

2.5 Accuracy Assessment

To evaluate the downscaled rainfall and ET, three indicators were calculated namely correlation, bias, and root mean square error (RMSE). Comparisons are made between raw and downscaled datasets version in the selected datasets (Fig. 1). Meanwhile, to determine the accuracy of satellite-based derived run-off, two analyses were conducted; (i) correlation between the satellite run-off against the actual river discharge, (ii) Nash Sutcliffe model efficiency index (NSME) (Eq. 4) and bias (Eq. 5) between the satellite run-off and in-situ simulated run-off. Due to differences in measurement unit between satellite run-off and river discharge, the second analysis were used to determine the relative efficiency of satellite-based run-off model. The in-situ simulated run-off were calculated using similar Thornthwaite and Mather water balance equation with inputs from discrete rain gauge and evapotranspiration estimation from ground temperature data. Procedures used by [12] was followed. Three sub-watersheds, (i) Sg. Muda (north), (ii) Sg. Lebir (east) and (iii) Sg. Selangor (west & southwest) were selected to represent each local climatic region (Fig. 1).

$$\text{bias} = \sum_{i=1}^{N} \frac{S_i}{G_i} \times 100 \tag{4}$$

$$E_{RMS} = \sqrt{\left[\frac{1}{N}\sum_{i=1}^{N}(S_i - G_i)^2\right]} \tag{5}$$

where G_i is observed ground rainfall, S_i is satellite-based precipitation, $\overline{O_i}$ is average ground rainfall and $\overline{S_i}$ is average satellite-based rainfall and N is number of samples.

3 Results

3.1 Model and Parameter Validation

3.1.1 Downscaled Rainfall and Evapotranspiration

The calibrated satellite derived rainfall and evapotranspiration have showed improved performance at both quantitative (bias, RMSE) and temporally qualitative scale (correlation) (Table 1). In comparison between the two parameterized inputs, the overall performance of downscaled ET (bias = 43%, RMSE = 24, r = 0.64) was higher than rainfall (bias = 43%, RMSE = 70, r = 0.610). The downscaled rainfall has higher ability to depict the rainfall variation during Northeast (NEM) and Inter-Monsoon 2 (IM2) than other monsoon seasons, but experienced trade-off by producing higher quantitative errors. Both NEM and IM2 seasons were associated with wetter seasons. There was less seasonal effect on the performance of ET except with moderate underestimation. The results showed that the scaled down satellite rainfall and ET were capable to represent the hydrological conditions at sub-watershed level.

Table 1. Accuracy between the raw and downscaled satellite variables; (a) TRMM Rainfall and (b) MODIS 16 Evapotranspiration.

(a) TRMM Rainfall						
Season	Bias (%)		RMSE (mm)		Correlation, r	
	Raw	Downscaled	Raw	Downscaled	Raw	Downscaled
Northeast Monsoon	+70	+50	128	90	0.701	0.806
Inter-Monsoon 1	+30	+22	86	55	0.425	0.488
Southwest Monsoon	+40	+29	83	62	0.412	0.474
Inter-Monsoon 2	+30	+18	95	71	0.656	0.754
(b) MODIS 16 – Evapotranspiration Product						
Season	Bias (%)		RMSE (mm)		Correlation, r	
	Raw	Downscaled	Raw	Downscaled	Raw	Downscaled
Northeast Monsoon	+61	−45	31	19	0.665	0.764
Inter-Monsoon 1	+63	−42	41	22	0.651	0.745
Southwest Monsoon	+57	−43	36	25	0.603	0.693
Inter-Monsoon 2	+65	−42	35	28	0.640	0.736

Table 2. Validation on satellite derived run-off against the actual river discharge and in-situ simulated run-off.

Streamflow station	Watershed area (ha)	River discharge (m^3/year)	Satellite run-off & river discharge		Satellite run-off & in-situ simulated run-off	
			Correlation, r	P-value	NSME	Bias (%)
Sg. Lebir	326100	1433	0.844	0.001	0.82	−27
Sg. Muda	268600	968	0.764	0.001	0.71	+41
Sg. Selangor	192900	638	0.719	0.001	0.66	+43

3.1.2 Satellite Derived Run-Off

The satellite-based estimated run-off (SRO) showed good correlation and significance with the actual river discharge at all selected sub-watersheds (Table 2). 60 months period of time-series analysis showing good agreement between SRO, IRO and actual river flow discharge (Fig. 2). Thus, this result indicated that the SER can depict the actual run-off and streamflow condition on the site. However, the strength of SRO to predict observed in-situ simulated run off (IRO) were varied from good (Sg. Lebir; NSME = 0.62) to moderate (Sg. Selangor; NSME = 0.01). There was tendency that the satellite run-off model performance is proportional with the size of the sub-watershed, ranged from a large to the small one.

3.2 2005 Drought Analysis aT Sub-Watershed Level

At monthly scale, two major drought period are identified (Fig. 2). We identified that the variation of monthly run-off deficit was strongly associated with declination in rainfall. The most intense drought impact was occurred from January to March. Almost 98% of the forest watersheds were experiencing a run-off deficit (average deficit, 62%). Then, the run-off deficit decreased from April to June (average deficit, 18%). The second wave of weaker drought began in July to October (average deficit, 30%). The drought is fully recovered for most catchment in November to December (average deficit, 0%). Interestingly, we notified that despite the drought recovery for most areas in November and December (Fig. 3), certain part in the west coast and southern part were still experiencing a run-off deficit. Plausible factors contributed to this phenomenon can be the rainfall deficit occurrence, size and elevation of the catchments. This gave us evidence that during drought period, smaller, lower elevation and low rainfall intensity watersheds were more severely affected.

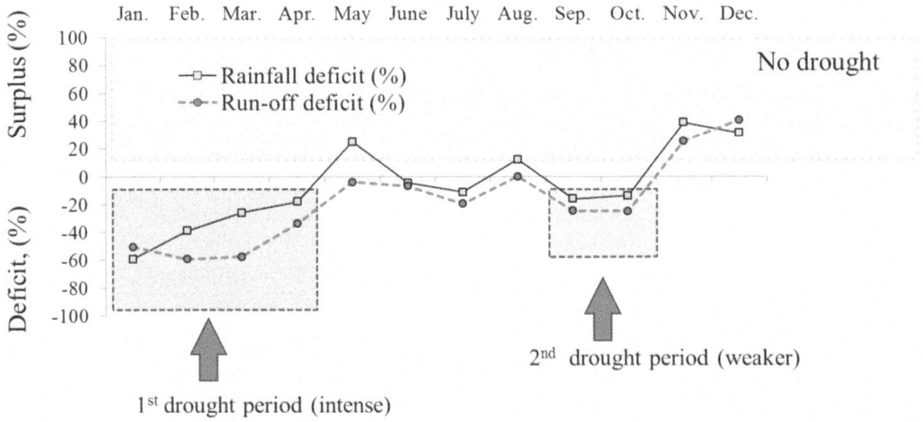

Fig. 2. Peninsular scale of monthly forest run-off deficit in 2005 drought

On annual scale, 164 out of 199 forested sub-watersheds were experiencing run-off deficit in 2005. Total affected watershed was close to 3.14 million hectares (78%).

The average deficit was 24%. Subsequently, the average rate of run-off in 2005 has decreased 16.4% from 53m/ha/year (normal) to 44m/ha/year. From these 164 affected sub-watersheds, 71 of them faced severity of 24% deficit and higher. The most severely affected sub-watershed was identified in Jemaluang (Johor) with a loss rate of 824mm/year run-off. The drought intensity was varied via west and east regions. The drought has affected most of the watershed in the western part. Seven states in three respective regions were involved; northwest (Perlis, Kedah), west coast (Perak, Selangor, N. Sembilan, Malacca, 24–39%) and southern (Johor, 41% deficit). Meanwhile, the east coast region experienced moderate effects (7–16% deficit). There was a good relationship between the rate of run-off deficit and 2005 run-off readings in the watershed located at the western part (r = 0.640), suggesting the severe impact of the drought in this region.

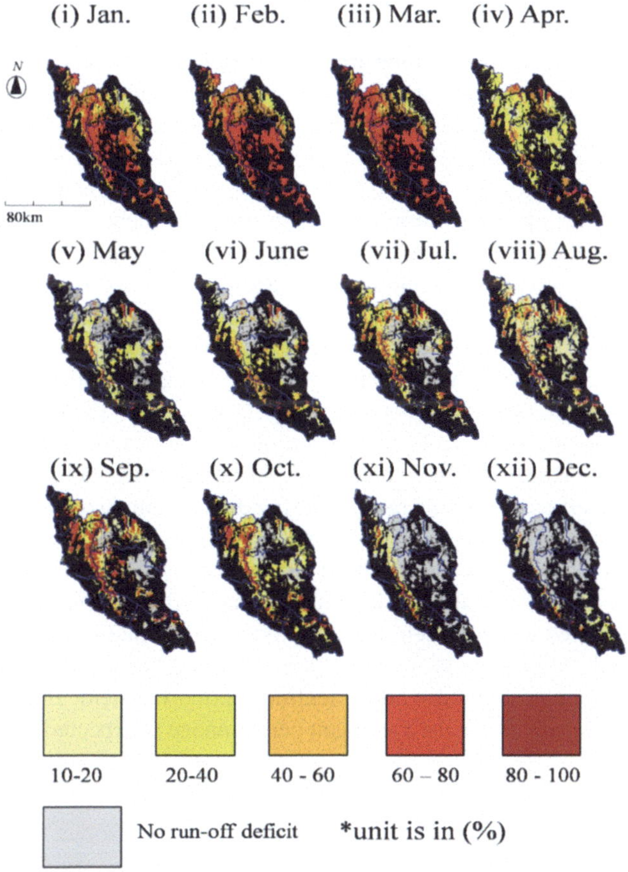

Fig. 3. Monthly forest run-off deficit in Peninsular Malaysia in 2005. *Black areas are non-forested and non-catchment areas.

4 Discussion

Based on the findings, there is a temporal lag between the water yield deficit pattern at Malaysian scale with the global El-Nino index and anomaly temporally. Although both the El-Nino index had offset by mid-2005 and followed by La-Nina, the water yield deficit only showed significant reduction in November and December. Instead, it seems like that there is a second drought peak in September. This indicated that the localized hydrometeorology in Peninsular Malaysia were also influenced by other regional factors. Hence, this requires further investigation.

The findings showed that the 2005 drought to forested catchment in Peninsular Malaysia with effective spatio-temporal visualization. Such valuable information serves good opportunity to increase our understanding about the contemporary gap of knowledge on the humid tropical forest-atmospheric dynamics at sub-watershed scale due the scarcity or drought at large-spatial scale including mountainous area (Wohl et al. 2012). Hence, that watershed part is critical to accurately predicting, analyzing and assessing water yield because ground observation are scarce. In addition, the critical and vulnerable forested sub-watershed areas affected by the drought can be identified at high precision. The establishment of systematic and effective drought management systems such as Famine Early Warning System (FEWS-NET) and Monitoring Agricultural Resources (MARS) in African continents that suits condition of local climate and hydrometeorology characteristics can be materialized. With more frequent extreme events ahead, managing climate change and increased climate variability requires these fundamental technical capacities and creative coping strategies to be devised (van Dijk et al. 2013; Wilhite et al. 2014).

Despite reliable performance of the SRO to evaluate drought, two majors concerns should be considered. First is the predictive power with regards to the decreasing watershed size. This suggest that as the watershed is getting smaller, the limitation of spatial resolution and geographical errors had constituted to the inefficiency of estimation. Attention to this weakness would be pivotal for the smaller size sub-watershed application. Second is the effect of satellite rainfall during different monsoon seasons. The large errors during the heavy rainy seasons and low temporal predictive ability in dry seasons might influence the quality of the results. This leads us to two plausible explanations. First, if the errors were systematics, this tells us that further downscaling was needed. This factor is believed to have been implicitly occurred in many satellite data embedded studies to hydrology (e.g., Wolf et al. 2005) but often being rarely emphasized and discussed due to lack of literature and scientific reports on them. Interestingly, if the errors were unsystematic, it means a stagnant performance is expected even after further downscaling.

5 Conclusion

This paper presented the scaled down algorithm of satellite rainfall & evapotranspiration to assess drought at sub-watershed level in humid tropical forest. The use of simplistic original water balance equation was able to simulate the run-off with good agreement with in-situ (NSME = 0.3; Bias = 37%). Comparable run-off curve was achieved with actual streamflow ($r = 0.771$, $P = 0.001$). The proposed algorithm was capable to identify two temporal drought periods in Peninsular Malaysia that occurred at local catchment scale. Sub-catchment scale information on severely affected catchment that unable to recover from the rainfall deficit for 12 months was also identified. The algorithm performance is influenced by two major factors; (i) seasonal rainfall accuracy, and (ii) watershed size. This study provides a basic guideline on how the drought phenomena can be monitored and assessed via fully satellite measurements is challenging in dynamic humid tropical forested watershed areas. To get the best information from in situ set up with many stations, high spatial density coverage are needed, which would be logistically expensive. Fully satellite-based drought monitoring at sub-watershed can supplement in situ data by providing consistent observations at high spatial density with intensive coverage and cost effective.

Acknowledgment. The authors would like to express his gratitude to all data contributors especially from the Dept. of Irrigation and Drainage. Special thanks also go to all that directly or indirectly assisted in the completion of this study. This research is funded by the Kursi Sarawak Research Grant (R.J130000.7352.1R064) entitled Developing an Online Big Geospatial Data Dissemination, Analysis and Decision for Water Yield Monitoring and Forecasting in Sustaining Raw Water Security.

Appendix 1: Annual Forest Water Yield Assessment in Peninsular Malaysia at Sub-Watershed Level Using Satellite (Selected)

Watershed	Region	Sub-watershed name	Area (ha)	Normal annual run-off (mm/year)	2005 annual run-off (mm/year)	Deficit (%)
Perlis	Northwest	Bukit Bintang	2787	258	125	52
		KurungBatang	705	549	374	32
		Bukit Wang Yu	2812	261	180	31
Kedah	Northwest	Gunung Jerai	8558	247	91	63
		Sungkop	1567	623	383	39
		Bukit Relau	1150	929	626	33
		Gunung Bongsu	8888	540	401	26
		KohMoi	9417	969	721	26
		Bukit Enggang	6828	712	535	25
Perak	West	BatangHijau - Larut	50716	978	566	42
		Behrang	7808	621	366	41
		Bukit Slim	40453	742	443	40
		Bukit Tapah	63260	697	462	34
		Bubu - Kuala Kangsar	18021	534	361	32
		Bukit Kinta	134955	807	551	32
		BatangHijau - Hulu Perak	60399	772	533	31
		Bujang Melaka - Kinta Manjung	4425	750	532	29
		Kledang - Sayong 2	20296	1290	917	29
		Kledang - Sayong	11104	1290	917	29
		Piah	74826	1174	857	27
		Korbu	81797	953	698	27

(continued)

(continued)

Watershed	Region	Sub-watershed name	Area (ha)	Normal annual run-off (mm/year)	2005 annual run-off (mm/year)	Deficit (%)
		Bukit Larut Matang	7242	1625	1200	26
Selangor	West	Batang Kali	8482	619	272	56
		Bukit Kutu	6511	652	296	55
		Semangkok	1523	505	254	50
		Serendah	4080	790	416	47
		Hulu Selangor	15188	750	416	45
		Sg. Lalang	19271	1094	653	40
		Hulu Gombak	19983	1153	700	39
		Bukit Gading	18956	876	545	38
		Hulu Langat	11699	1479	1010	32
		Bukit Cherakah	1924	1646	1204	27
Perlis	Northwest	Bukit Bintang	2787	258	125	52
		KurungBatang	705	549	374	32
		Bukit Wang Yu	2812	261	180	31
Kedah	Northwest	Gunung Jerai	8558	247	91	63
		Sungkop	1567	623	383	39
		Bukit Relau	1150	929	626	33
		Gunung Bongsu	8888	540	401	26
		KohMoi	9417	969	721	26
		Bukit Enggang	6828	712	535	25

(continued)

(*continued*)

Watershed	Region	Sub-watershed name	Area (ha)	Normal annual run-off (mm/year)	2005 annual run-off (mm/year)	Deficit (%)
Perak	West	BatangHijau - Larut	50716	978	566	42
		Behrang	7808	621	366	41
		Bukit Slim	40453	742	443	40
		Bukit Tapah	63260	697	462	34
		Bubu - Kuala Kangsar	18021	534	361	32
		Bukit Kinta	134955	807	551	32
		BatangHijau - Hulu Perak	60399	772	533	31
		Bujang Melaka - Kinta Manjung	4425	750	532	29
		Kledang - Sayong 2	20296	1290	917	29
		Kledang - Sayong	11104	1290	917	29
		Piah	74826	1174	857	27
		Korbu	81797	953	698	27
		Bukit Larut Matang	7242	1625	1200	26
Selangor	West	Batang Kali	8482	619	272	56
		Bukit Kutu	6511	652	296	55
		Semangkok	1523	505	254	50
		Serendah	4080	790	416	47
		Hulu Selangor	15188	750	416	45
		Sg. Lalang	19271	1094	653	40
		Hulu Gombak	19983	1153	700	39

(*continued*)

(*continued*)

Watershed	Region	Sub-watershed name	Area (ha)	Normal annual run-off (mm/year)	2005 annual run-off (mm/year)	Deficit (%)
		Bukit Gading	18956	876	545	38
		Hulu Langat	11699	1479	1010	32
		Bukit Cherakah	1924	1646	1204	27

References

1. Malaysian Meteorological Dept.: Climate change report in Malaysia. Malaysian Meteorological Department (2005)
2. Malaysian Meteorological Dept.: El-Nino impact on climate change in Malaysia. Malaysian Meteorological Department (2009)
3. Mahmud, M.R., Mohd Yusof, A.A., Mohd Reba, M.N., Hashim, M.: Mapping the daily rainfall over an ungauged tropical micro-watershed: a downscaling algorithm using GPM data. Water **2020**(12), 1661 (2020). https://doi.org/10.3390/w12061661
4. DID (Dept. of Irrigation and Drainage): Drought report in Peninsular Malaysia. Hydrology and Water Resources Division, Department of Irrigation and Drainage, Malaysia, Kuala Lumpur (2005)
5. Du, L., Tian, Q., Yu, T., Meng, Q., Jancso, T., Udvardy, P., et al.: A comprehensive drought monitoring method integrating MODIS and TRMM data. Int. J. Appl. Earth Obs. Geoinf. **23**, 245–253 (2013)
6. Mahmud, M.R.:. Downscaling the satellite precipitation data in humid tropical region. Doctoral Dissertation, Tokyo Metropolitan University, Japan (2015)
7. Nadzri, M.I.: Water yield from TRMM and MODIS in Peninsular Malaysia. Master thesis, Universiti Teknologi Malaysia (2014)
8. NOAA National Climatic Data Center: 2015 is Earth's warmest year by widest margin on record (2016). Accessed 20 Jan. 2016
9. Hashim, M., et al.: Satellite-based run-off model for monitoring drought in Peninsular Malaysia. Remote Sens. **8**, 633 (2016). https://doi.org/10.3390/rs8080633
10. Thornthwaite, C.W., Mather, J.R.: The water balance, p. 104. Drexel Institute of Technology, Laboratory of Climatology, Centerton, New Jersey (1955)
11. Strangeways, I.: Precipitation: theory, measurement, and distribution. Cambridge University Press, London (2007)
12. Mahmud, M.R.: Run-off modeling and mapping using rainfall and evapotranspiration estimates from remote sensing satellite data in Peninsular Malaysia; Master Thesis. UniversitiTeknologi Malaysia, Johor Bharu, Malaysia (2012)
13. Mahmud, M.R., Numata, S., Hosaka, T., Matsuyama, H., Hashim, M.: Assessment of effective seasonal downscaling of TRMM precipitation data in Peninsular Malaysia. Remote Sens. **7**, 4092–4111 (2015)
14. Cheema, M.J.M., Bastiaanssen, W.G.M.: Local calibration of remotely sensed rainfall from the TRMM satellite for different periods and spatial scales in the Indus Basin. Int. J. Remote Sens. (2011). https://doi.org/10.1080/01431161.2011.617397. Wohl, E., et al.: The hydrology of the humid tropics. Nat. Climate Change **12** (2012)
15. Mu, Q.Z., Zhao, M.S., Running, S.W.: Improvements to a MODIS global terrestrial evapotranspiration algorithm. Remote Sens. Environ. **115**, 1781–1800 (2011)
16. Wolf, D.B., et al.: Ground validation for the Tropical Rainfall Measuring Mission (TRMM). J. Atmos. Ocean. Tech. **22**, 365–380 (2005)

GIS & Land Administration

An Interactive Spatial Web-Based System for Exploring Sri Aman Tourism, Sarawak

Serene Anak Joseph Goyang[1], Abdul Rauf Abdul Rasam[1,2(✉)], Muhammad Abdul Hakim Muhamad[1], Rosmadi Ghazali[1], and Mayuree Nasa Khan[3]

[1] School of Geomatics Science and Natural Resources, College of Built Environment, Shah Alam, Malaysia
rauf@uitm.edu.my

[2] Malaysia Institute of Transport (MITRANS), Universiti Teknologi MARA, 40450 Shah Alam, Selangor, Malaysia

[3] Faculty of Tourism and Hotel Management, Mahasarakham University, Kantharawichai, Thailand

Abstract. Tourism is a significant source of income and employment for locals, particularly in rural areas like Sri Aman in Sarawak. However, the region may not be documented as a tourist attraction due to outdated promotion, lack of infrastructure, and lack of interactive geospatial web-based maps. This study developed an interactive web-based tourist system using a Geographical Information System (GIS). The system entailed user requirement analysis (URA), database system design, data collection, web-based development, and system assessment. The system underwent testing by distributing questionnaires to 24 respondents, with over 95% finding it more interactive than the previous version. The interactive GIS tourism web-based system provides easy access to the Earth's surface, infrastructure, and information.

Keywords: GIS · Spatial Web-based · Tourism

1 Introduction

Tourism significantly contributes to global social, cultural, environmental, heritage, and economic sectors. Before the pandemic, it accounted for 1 in 4 global new jobs [1]. In 2007, Malaysia's Ministry of Tourism introduced the 'Malaysia Truly Asia' theme to attract international tourists, making it Southeast Asia's second most visited country in 2018. [2]. Former Sarawak Chief Minister Tan Sri Adenan Satem announced that Sri Aman, known for its 'Benak' tidal bore and annual Tidal Bore Festival, is no longer a demoted town due to a rural transformation program. The Sarawak government plans to continue development to improve the community's standard of living and boost tourism in the area, with the new web-based system playing a key role in this transformation. GIS is a powerful tool that integrates hardware, software, and data collection to manage, analyze, and capture spatial data. It allows users to quickly organize and update data through maps, tables, and charts [3]. GIS applications have various uses beyond the

A. Abdul-Rahman et al. (Eds.): GeoWeek 2024, LNGC, pp. 245–254, 2025.
https://doi.org/10.1007/978-3-031-86654-8_19

tourism industry [4], such as displaying the Earth's surface characteristics and accessing infrastructure and activity-based information [5–8]. Advanced GIS technologies, including 3D models, virtual reality, augmented reality, and crowdsourcing platforms, can also be utilized in the study [9–11]. Web-based GIS is one of the techniques used to promote tourism in most countries [12–15].

This study aims to design a geospatial-based system for exploring Sri Aman Tourism in Sarawak, Malaysia. A web-based GIS is a popular technique for promoting tourism and has gained interest in the industry [16–18]. The system provides information about attractive places in the Sri Aman District without additional fees. The study aims to enhance the sharing and visualization of information in the tourism industry with three main objectives: i. to study the user requirement of the proposed system, ii. to design the proposed system, and iii. to test the proposed system.

2 Methodology

2.1 Study Area: Sri Aman, Sarawak

The area for this study is Sri Aman, located in Sarawak, Malaysia. The Sri Aman Division, formerly known as Simanggang, is located in the south of Sarawak with an area of 5,499.6 sq km, and Bandar Simanggang is located 193 km from Kuching City. The Sri Aman division at Sungai Batang Lupar is famous for the tidal bore (Benak) phenomenon, which can reach up to two to three meters at high tide. This phenomenon is celebrated yearly at the Benak Festival and is an official annual event in the Sarawak State Tourism Calendar (Fig. 1).

Fig. 1. Sri Aman, Sarawak, Malaysia (Google Map, 2023)

2.2 Research Framework

The Software Development Life Cycle (SDLC) approach in the study is a methodology divided into four stages: requirements and analysis, system design, system development, and system evaluation. In the first phase, User Requirement Analysis (URA) was conducted to determine users' responses to the information needed to provide in an interactive web-based system. In addition, a well-organized response from the 43 respondents from local people and tourists was required to construct a structural database that could allow analysis and storage. This methodology was conducted using a questionnaire divided into three sections: respondent background, the current information system, and the proposed system. Tourists and locals answered the questionnaires via Google Forms, which were published on social media sites such as WhatsApp, Facebook, and others. This was done to learn more about Sri Aman's tourist hotspots and to ask for feedback on the web-based system that would be created. The outcome of URA also allowed for an understanding of necessary functions and database design (Fig. 2).

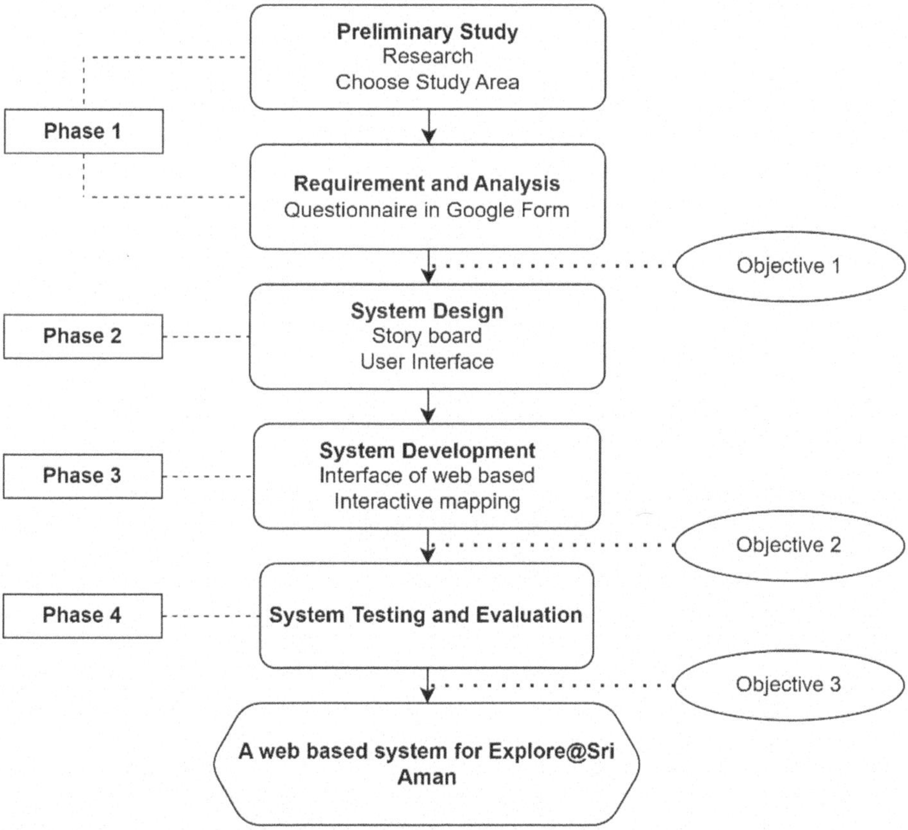

Fig. 2. Flowchart of the study methodology

System design for Sri Aman was essential in developing the web-based system as it involved the interaction between the system and the user using ArcGIS Online, Google Earth, OpenStreetMap (OSM), and HTML. On the other hand, concept design or storyboards were crucial in developing the web-based system before the user could efficiently operate it. The user can easily access the data and perform functions within the system [13]. Additionally, in designing the web-based system, many essential elements were highlighted: text structure, the presentation of map information, color combination, and interactive visualization of the website as required in cartographic and GIS information system guidelines.

Then, the web-based system was developed with six sub-topics of the system menus, where the user could find more details about the attractions in Sri Aman through the web-based system, as shown in Fig. 3. The interface of the web-based system is designed to be simple to conduct and learn even by the first-time user. In this phase, the web-based system's development will be created using ArcGIS Online, which is compatible to integrate with the interactive mapping. Some elements are highlighted in this stage, such as color usage, graphics, map display, and text structure.

The system testing was performed to ensure the system's capabilities and that it can meet the user's needs. The system was tested after the completion of the web-based system development process. Moreover, this test was necessary as the final phase of the research to ensure the system functions well and can be operated properly without any errors or inabilities. In addition, maintaining continuity and effectiveness is essential for developing interactive mapping that integrates seamlessly with the GIS function and the information system within Sri Aman's web-based system. The completed questionnaire had two parts. The first part consists of a respondent's profile, and the second part is about the satisfaction of web-based development. All questions were designed for quick answers. Twenty-four respondents tested the web-based system for preliminary results. The questionnaire determined the level of satisfaction with the user's perspective on the functionality of the web-based system.

3 Result and Discussion

3.1 User Requirement and Design of Interactive GIS Web-Based System of Tourism Sri Aman (Explore@Sri Aman)

A general user requirement study of the 43 respondents reveals that the existing websites used by respondents are four websites where you can find information about tourism in Sri Aman: the Sri Aman Division Administration, Google, Tripadvisor, and Sarawak Tourism Board (Fig. 4). Most respondents stated that they frequently use Sarawak Tourism to get tourism information. Sarawak Tourism usually provides some details about tourism in Sri Aman, but some information has not been updated.

Therefore, users assess respondents on existing websites to gather tourism information. This questionnaire has five scales: bad, poor, medium, good, and excellent. The evaluation focuses on six aspects: user-friendliness, complete information, a map, an easy-to-understand and exciting map, an interesting multimedia element, and interactivity. Most respondents said all of the existing websites above are user-friendly, but they lack map information. The websites are also not equipped with a complete web-based

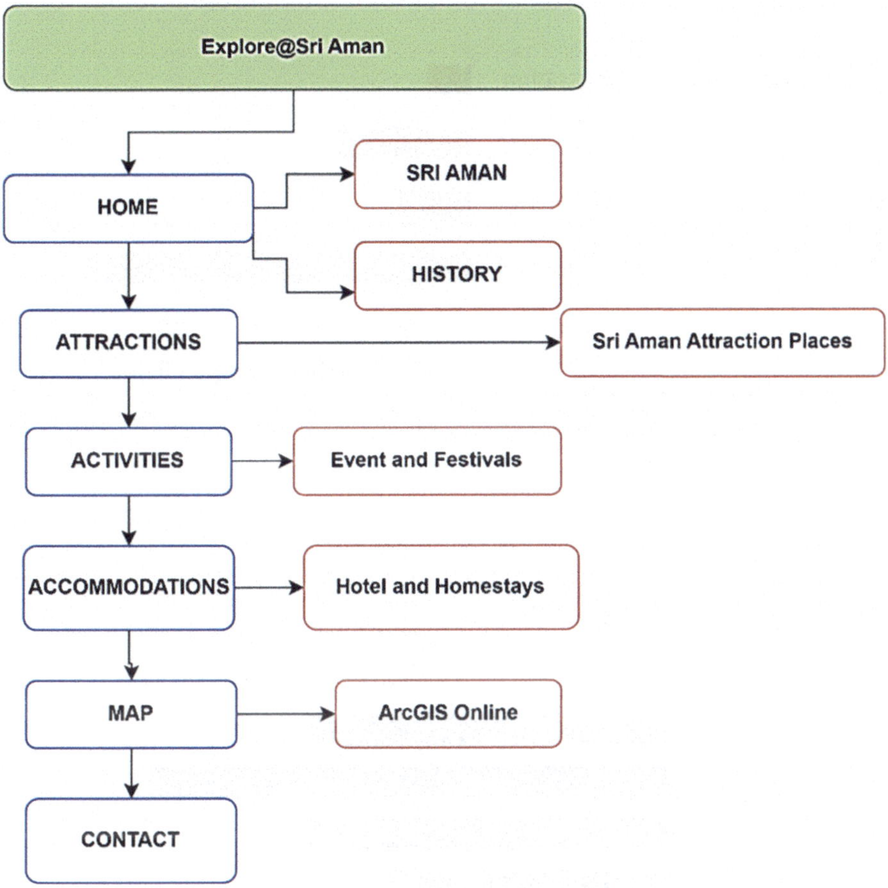

Fig. 3. The concept of the web-based system of Explore@Sri Aman

reference, as there is no interactive map for the tourist to refer to. The Sri Aman infrastructure was not updated and cannot be seen on Google Earth. Additionally, the website only displays a public map image.

Figure 5 shows the user suggestions for developing a new or proposed system. The aspects of suggestions are a reliable source of information, creating an interface that is reliable, interactive, easy to access, and fast. Most respondents suggested that all aspects should be considered when developing the Explore@Sri Aman, with more than 50% indicating this requirement for each aspect. Most respondents, accounting for 100%, deemed the easy-to-access aspect of the proposed system essential. In comparison, 72.1.4% out of 43 respondents identified a reliable source of information as the second most crucial aspect. 69.8% also identified interactivity and speed as important system development aspects. 55.8% of the respondents also chose to create a reliable interface as an aspect that should be included in the proposed system. The last questions are about the elements needed in the proposed system. The results indicate that most respondents strongly support the basic interface design, a user-friendly system, and a completed

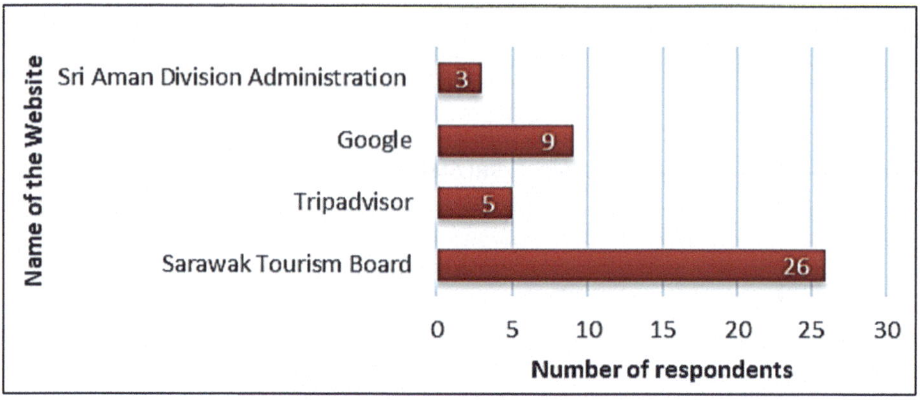

Fig. 4. Existing tourism websites used by the respondents

database element as essential components of the proposed system. The respondents also agreed that the other element, the GIS and mapping system development with spatial analysis, should be included in the proposed system. All these elements are suggested to be included in the proposed system.

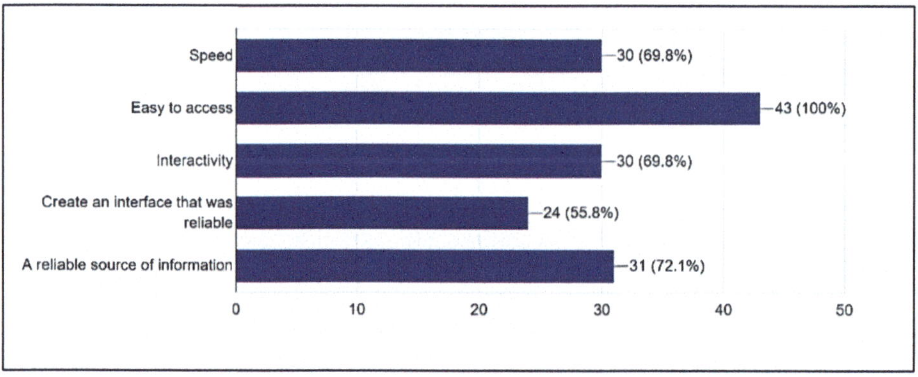

Fig. 5. Criteria of user suggestion on the proposed system

The SDLC steps were followed to develop a GIS web-based system for Explore@Sri Aman based on the results of user requirements. The system interface is divided into six sections: About Sri Aman, Map, Events, Accommodations, Map Tour, and Food Hunting. The home page is creative and unique, focusing on nature, mountainous landforms, scenery, leisure, and tourism (Fig. 6). The dynamic map page provides information about tourist attractions, hotels, and foodies and videos about available and existing activities in Sri Aman. The Home Page of every web-based system is crucial for attracting users through its creativity and uniqueness. Explore@Sri Aman was designed with the concept of nature, mountainous landforms, scenery, leisure, and tourism in terms of a color theme. The design of the Explore@Sri Aman system also considered the user's requirements, incorporating seven interface pages, including the home page and About Sri Aman. Tourist attractions inspired the Home Page for the Explore@Sri Aman system on the Sarawak web. Explore@Sri Aman is tailored for tourists from other places who wish to explore Sri Aman tourism. The background image of the Home Page features a video introducing the Sri Aman episode from TVS Entertainment of Borneo from Above, symbolizing Sri Aman town.

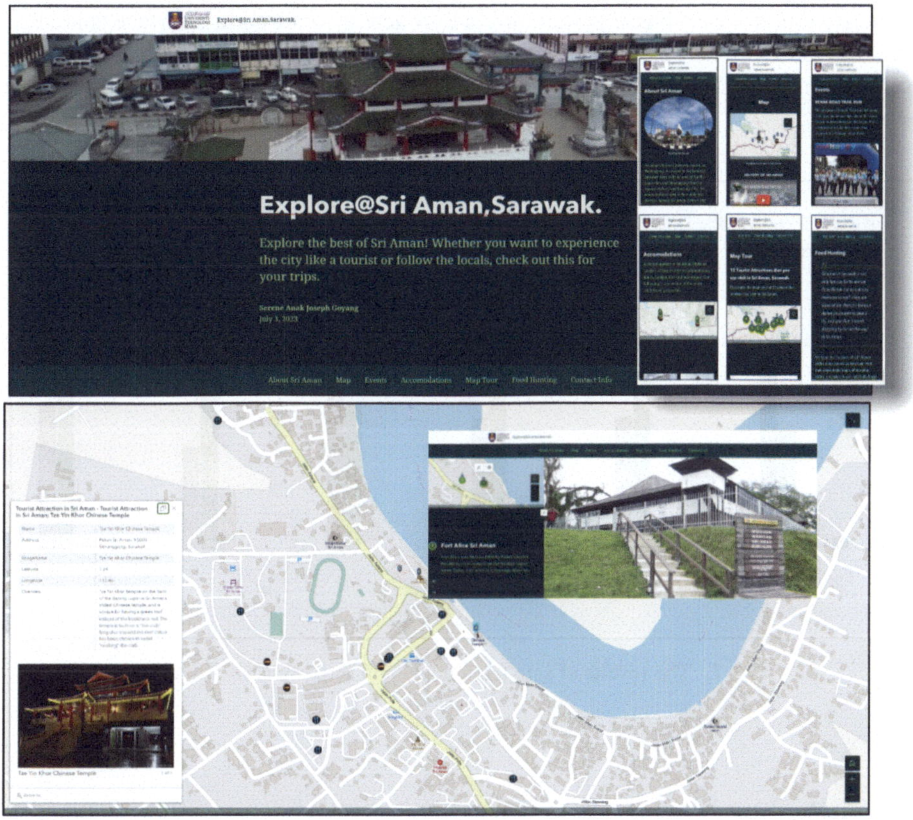

Fig. 6. Main Home Page and Map Page of Explore@Sri Aman

The Map Page created digital maps showing tourist attractions, hotels, and eateries. Users can zoom in and out of the map and click on points of interest to access location information, as depicted in the figure below. Users can smoothly navigate the map by moving the cursor without using zoom-in and zoom-out buttons. Furthermore, the map page features videos showcasing attractions and activities in Sri Aman, such as the tidal bore, Mount Hosanna, and more.

3.2 System Testing and Evaluation of Explore@Sri Aman

A web-based tourism system in Sri Aman was evaluated using Google Forms on a five-point Likert scale. The six items of the evaluation consisted of a primary interface of a sound system, a user-friendly system, a completed database, GIS information, and mapping, spatial analysis, and additional information (Fig. 7). 24 respondents completed the questionnaires and provided feedback on the system's content. Most respondents (95%) were delighted with the system, with some choosing neutral or delighted responses. The interactive tourism web-based system offers easy access to the earth's surface characteristics, infrastructure accessibility, and activity-based information, making it a valuable tool for tourism in Sri Aman. However, some suggested using more vibrant colors for the website.

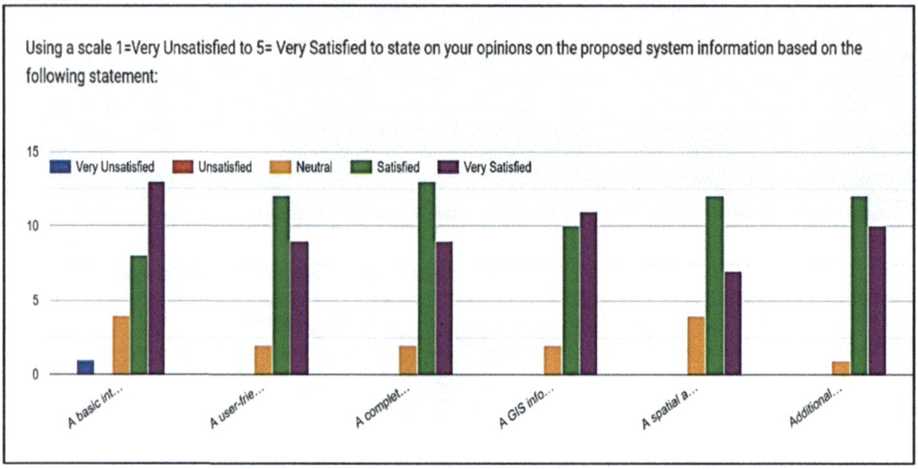

Fig. 7. The results of satisfaction level on Explore@Sri Aman

4 Conclusion

Explore@Sri Aman is a spatial web-based system developed by the SDLC in a GIS environment, offering an interactive map system with various cartographic features and functions such as symbology, map elements, a completed database, mapping, spatial analysis, and additional geoinformation. This user-friendly system helps tourists get a first impression of the place based on descriptions and history. Over 95% of respondents were also satisfied with the web-based content, allowing people worldwide to stay connected with attractive locations. This has a financial benefit for local villages in the socioeconomic sector. Further studies will explore the advanced GIS and Artificial Intelligence (AI)-based tourism website, which offers interactive maps and analysis, location-based services, route planning, attraction information, user reviews, and travel itineraries.

Acknowledgement. The authors thank UiTM Shah Alam for financially sponsoring this research through Geran Kolaborasi Entiti Penyelidikan (KEPU—600-RMC/KEPU 5/3 (015/2021)). The authors have no conflicts of interest to declare in the study.

References

1. World Travel and Tourism Council. Economic Impact Research (2022). https://wttc.org/research/economic-impact
2. Tourism Malaysia (2019). Last accessed 25 Oct 2023, https://www.tourism.gov.my/media/view/tourism-contributes-rm84-1-billion-to-malaysia-economy-with-25-8-million-tourists-in-2018
3. Mohan, L., Dhawan, S.C., Krishna, A.: Web-based tourism information system using geographical information system (GIS) - A Case Study Journal of Hospitality Application & Research, Birla Institute of Technology Mesra, Ranchi (2008)
4. Azizur Rahman, MD.: Application of GIS in ecotourism development: a case study in Sundarbans, Bangladesh (2010)
5. Grigore Vasile, D.C., Stefan, B., MĂDUȚA, M.F., Alexandru, I., Wendt, J.A.: The tourist map is a scientific tool supporting exploring protected areas in Bihor County, Romania (2016)
6. Jovanovic, V., Njegus, A.: The application of GIS and its components in tourism. Yugoslav J. Oper. Res. **18**(2), 261–272 (2008)
7. Masron, T., Mohd Med, B., Marzuki, A.: GIS-based tourism decision support system for Langkawi Island, Kedah, Malaysia. Theoretical and Empirical Research (2015)
8. Mustapha, N.I., Rasam, A.R., Saraf, N.M., Idris, R., Wakijan, A.: Cycling route mapping via cartography and GIS techniques. IOP Conf. Ser. Earth Environ. Sci. **1240**(1), 012008 (2023). https://doi.org/10.1088/1755-1315/1240/1/012008
9. Fedeline, R., Abdul Rauf, A.R., Rosmadi, G., Saiful, A.J.: Geovisualising 3D cultural heritage building for urban tourism via mobile augmented reality application. J. Inf. Knowl. Manag. (JIKM) **13**(2) (2023)
10. Koga, Y., Yamamoto, K.: Tourism support system to utilize virtual reality space reflecting dynamic information in real time. ISPRS Int. J. Geo-Inf. **12**, 301 (2023). https://doi.org/10.3390/ijgi12080301.(2023)
11. Sha'aban, S.N., Abdul Rasam, A.R., Asmat, A.: A 3D GIS mapping for Landscape Geovisualisation and park management: a case study in Kuala Keniam Station, Taman Negara Malaysia. IOP Conf. Ser. Earth Environ. Sci. **685**(1), 012010 (2021). https://doi.org/10.1088/1755-1315/685/1/012010
12. Hassannia, R., Barenji, A.V., Li, Z., Alipour, H.: Web-based recommendation system for smart tourism: Multiagent technology. Sustainability (Switzerland) **11**(2) (2019). https://doi.org/10.3390/su11020323
13. Kamarudin, M.K., Mohd, F.A., Mohamed, A.H., Rahim, A., Shaari, M.S.N., Abd Samad, M.A.: GIS tourism web based of Kampung Seberang Ramai, Perlis. J. Tourism, Hospitality Environ. Manag. **7**(29), 123–138 (2022). https://doi.org/10.35631/jthem.7290
14. Mohd Hasmizi, N.S., Abdul Rasam, A.R., Mohamed Saraf, N.: E-mosque tourism: discovering mosques via digital mapping in Kuala Lumpur, Malaysia. In: Charting the sustainable future of ASEAN in science and technology, pp. 507–517 (2020). https://doi.org/10.1007/978-981-15-3434-8_44
15. Sherrylye, S.J., Abdul Rauf, A.R., Rosmadi, G., Saiful, A.J., Farrah Diana, S.: Development of a web mapping system for tourism attractive places in Bintulu, Sarawak. J. Inf. Knowl. Manag. (JIKM) **13**(2) (2023)

16. Markos, A.: Web GIS for tourism development of Bahir Dar town and its surroundings, Ethiopia, in partial fulfilment of the requirement for the degree of Master in Geo-information science (2012)
17. Othman, H., Abdul Rasam, A.R., Jaini, N.: Geopark management and GIS: Geospatial Tree Information Inventory System. In: Charting the sustainable future of ASEAN in science and technology, pp. 553–567 (2020). https://doi.org/10.1007/978-981-15-3434-8_48
18. Halim, M.A.: Discovering new tourist attractions through social media data: a case study in Sabah Malaysia. 2018 IEEE 8th International Conference on System Engineering and Technology (ICSET) [Preprint] (2018). https://doi.org/10.1109/icsengt.2018.8606373

Unveiling Legitimacy and Analyzing Sarawak's Maritime Boundary Through Historical Archives

Ching Yen Sim[1](✉), Kelvin Kang Wee Tang[2], and Mohd Razali Mahmud[3]

[1] Faculty of Built Environment and Surveying, Universiti Teknologi Malaysia, 81310 Johor Bahru, Malaysia
simchingyen@graduate.utm.my
[2] Geomatic Innovation Research Group (GnG), Universiti Teknologi Malaysia, 81310 Johor Bahru, Malaysia
tkkelvin@utm.my
[3] GeoCoastal Research Unit (Gi2RG), Universiti Teknologi Malaysia, 81310 Johor Bahru, Malaysia

Abstract. There is a contention between the Federal Government and the Sarawak Government over the legitimacy of the Sarawak (Alteration of Boundaries) Order in Council (OIC) 1954 and its relation to the Territorial Sea Act 2012 (TSA 2012). The Sarawak OIC 1954, enacted under the Colonial Boundaries Act 1895, extended Sarawak's boundaries to cover the "continental shelf" and allowing resource exploration. The historical and legal implications of the OIC for Sarawak's maritime boundary and its alignment with international law, particularly the 1958 Geneva Convention on the Continental Shelf and the United Nations Convention on the Law of the Sea (UNCLOS) 1982 remain unclear. This paper aims to provide an initial understanding of the Sarawak OIC 1954 by examining its history and archives to clarify the complexities of this important document. It offers insights based on archived UK documents from 1945 to 1974, exploring the historical legitimacy and motivations behind the OICs and their role in delineating maritime boundaries adjacent to Sarawak.

Keywords: Order in council · Continental Shelf · Sarawak maritime boundary

1 Introduction

Sarawak, situated on the northwest coast of Borneo, was a crucial British colony before becoming one of Malaysia's federation states in 1963. The legitimacy of Sarawak's maritime boundaries, particularly regarding its jurisdiction over resources (oil & gas) on the continental shelf, remains a contentious issue in Malaysian politics [1]. The polemic centers on the historical legal foundations of Sarawak's territorial claims, specifically those derived from the document of Sarawak (Alteration of Boundaries) Order in Council 1954 (Sarawak OIC 1954) [2]. This order extended the state's borders to include the adjacent continental shelf by stating:

A. Abdul-Rahman et al. (Eds.): GeoWeek 2024, LNGC, pp. 255–271, 2025.
https://doi.org/10.1007/978-3-031-86654-8_20

...The boundaries of the Colony of Sarawak are hereby extended to include the area of the continental shelf being the seabed and its subsoil which lies beneath the high seas contiguous to the territorial waters of Sarawak...

Sarawak contends that this maritime territory was established prior to its incorporation into the Federation of Malaysia in 1963 [1] under the principle of *Utis Possidetis Juris*, but remains unclear and yet to be clarified. Furthermore, the polemic has also led to interpretation variances between the Federal Government and the Sarawak Government regarding territorial divisions and the legitimacy of the Sarawak OIC 1954 in relation to current legislation, specifically the Territorial Sea Act 2012.

The objective of this paper is to reviews archive from The National Archives of United Kingdom (TNA) to clarify the historical legitimacy of the maritime boundaries set by the OIC 1954 and Sarawak (Definition of Boundaries) OIC 1958 No. 1518 [3], assessing their consistency with the Convention on the Continental 1958 (Geneva 1958) [4] and United Nations Conventions on the Law of the Sea (UNCLOS) 1982 [5].

2 Background

Continental shelf is the submerged edge of continent, extending from the coast to a drop-off point (shelf break), and then descending toward the deep ocean floor as shown in Fig. 1 [6].

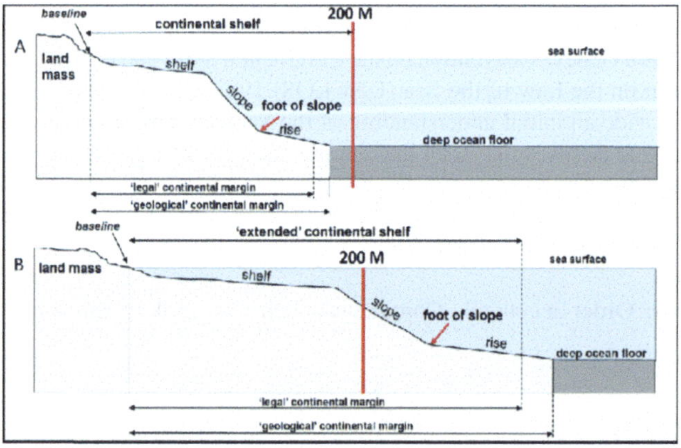

Fig. 1. Legal concept of a continental shelf [7]

In 1947, the Truman Proclamation set the limit of continental shelf to a water depth 100 fathoms (shelf break), which about 200 m [7]. It evolved under international law particularly Article 76 (UNCLOS 1982), where the continental shelf defined as:

the continental shelf of a coastal state comprises the seabed and subsoil of the submarine areas that extend beyond its territorial sea throughout the natural prolongation of its land territory to the outer edge of the continental margin, or to a distance of 200 nautical miles from the baselines from which the breadth of the territorial sea is measured where the outer edge of the continental margin does not extend up to that distance [5]. *If the continental shelf extends beyond 200 M to the outer limits of its continental shelf, Article 76 allows the coastal state to submit scientific and technical justifications to the entitlement of an extended continental shelf and the delineation of its outer limits to the Commission on the Limits of the Continental Shelf (CLCS).*

The concept of "sovereignty" over the continental shelf was explicitly rejected by the Geneva 1958, where the first Convention was formulated. The continental shelf, by the terms of the Convention does not form part of the coastal state's territory, the Convention merely allows "sovereign rights" to that state for particular purposes, viz exploring and exploiting its natural resources. The concern was that states might claim exclusive rights of waters above the shelf, thus encroaching on the freedom of the high seas. With this proviso, the distinction between "sovereignty" and "sovereign rights" becomes subtle. The latter implies that property rights to the shelf's natural resources, as defined in the Convention, are vested in the coastal state, and these resources cannot be transferred to others except through an express grant [8].

As we can observe, the content in paragraph 3 of the Sarawak OIC 1954 serves the same purpose which stated that.

Nothing in this Order shall be deemed to affect the character as high seas of any waters above the said area of the continental shelf.

However, an archival research method must be conducted which involving the periodic review of the timeline of early history of Colonial's administration to present days (as illustrated in Fig. 2). The question also arises regarding the term "the boundaries" used in this Order in Council and which specific boundary it refers to.

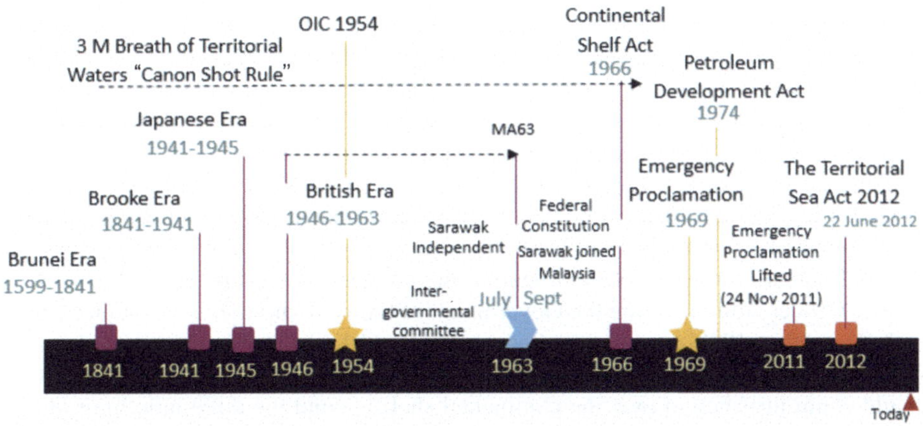

Fig. 2. Timeline of early history of Colonial's administration to present days

The Sarawak OIC 1954 was issued on 24 June 1954, at Buckingham Palace under the authority of the Queen, based on the Colonial Boundaries Act of 1895 [2, 9]. This Act allowed the British Crown to define and alter the boundaries of British colonies, including their territorial waters, through Orders in Council. Sarawak further enact the regulation such as the Land Code 1958 and the Oil Mining Ordinance 1958. In 1969, federal waters of Malaysia extended to 12 M (nautical mile) under the Emergency Ordinance No. 7 (Amendment No. 11) 1969, declared under Article 150 of the Malaysian Constitution [10]. The ordinance froze the states' territorial waters, including Sarawak's, and was enforced through the Continental Shelf Act 1966 [11], the Petroleum Mining Act 1966 [12], and the National Land Code 1965 [12].

On November 24, 2011, the Federal Government lifted the emergency under Article 150(7) of the Federal Constitution, which expired on May 24, 2012. Before this expiration, the Territorial Sea Act 2012 (Act 750) was enacted, retaining similar provisions [13]. This raised questions about whether the lifting of the emergency annulled the ordinances and related laws, restoring Sarawak's right over its continental shelf or whether Sarawak's "right" or "territory" restored by the Federal Constitution (Articles 1 and 2) [14] and historical agreements such as the Malaysia Agreement 1963 [15] and the Inter-Governmental Committee Reports [16].

3 Archived Documents

A substantial amount of highly relevant archives is discussed in this paper. These archives can be categorized into different eras and correspondence subject, as summarized in the Table 1.

4 Findings

4.1 De Factor of Oil Exploration

On March 9, 1909, Sir Charles Johnson Brooke granted exclusive oil rights over Sarawak to The Anglo-Saxon Petroleum Co. Ltd., with a lease lasting until 1989 upon finding oil. This agreement for Sarawak's oil industry was modified in 1921 and 1923. In 1923, Sarawak Oilfield Limited, a subsidiary of Anglo-Saxon, signed a key agreement with Charles Vyner Brooke, focusing on onshore exploration. The lease also extended to offshore areas, anticipating future claims over these resources [17].

After the cession of Sarawak in 1947, British officials, Mr. Creech Jones [17], suggested fresh agreements/lease for oil exploitation in British territories, aiming to increase royalty rates and control beyond territorial waters. By 1948, the Model Colonial Oil Mining Regulation was established, reinforcing this expansionist policy.

The Officer Administering the Government of Sarawak through the Colonial Office on 7 September 1948 proposed to Sarawak Oilfield Limited that the oil exploration license would include the territorial waters of Sarawak. If the government of Sarawak should claim jurisdiction over the continental shelf beyond the three-mile limit of territorial waters, the company's rights would include the area of the seabed concerned in respect of the winning of oil. Their rights would not extend to the control of shipping, fishing, or any other similar matters [18], this has marked the early idea to control the seabed beyond territorial sea in this region.

Table 1. Summary of archives found at TNA

Record year	Correspondence subject	Party involved
1945–1953	Oil exploration, mining lease at Sarawak, Brunei, North Borneo and etc., acquiring control over continental shelf include British Dominion around the world	Colonial Office (CO), Foreign and Commonwealth Offices (FCO), Oil and Gas Company (Sarawak Oilfield Limited, Shell and etc.), Ministry of Fuel and Power, UK British High Commissioner, United Kingdom to the United Nations (New York), South West Pacific Department and etc
1954–1956	Drafting and proclamation of OIC 1954 & Boundary Definition at Brunei Bay, seismic survey by O&G company	
1956–1958	Boundary Definition of OIC 1958, Convention on Continental Shelf (Geneva 1958), First Conference UNCLOS	
1958–1960	UK legislation and ratification, territorial waters policy, Brunei continental shelf, UK Malaya Agreement, Sarawak Territorial Waters and Law of the sea	
1960–1963	Geneva Conventions on Continental Shelf, 1958 UK Continental Shelf Bill	
1963–1974	Malaysia internal affair Sarawak, discussion the validity of Sarawak OIC 1954. Dispute between Federal Government and Sarawak Over Continental Shelf	

Table 2. Known Limits of territorial waters on the Far East Station (part of table) [27]

Coastal state	Limits of territorial waters (Nautical Mile = M)
United Kingdom	3 M
Malaya, Brunei, North Borneo and Sarawak	3 M
Cambodia	5 M
Ceylon	6 M
China	12 M

4.2 Policy of Colonial: Alteration of Boundaries

When discussing about draft of the oil lease, Foreign Office on 8 January 1951 (Ref. WN 1533/1), said it is customary for a country, before granting oil concessions, to pass a petroleum act containing regulations and conditions like those in the model oil mining lease drawn up by the Colonial Office. They also suggested operation covers extra-territorial waters [19]. Letter from the Colonial Office, 17 November 195 (Ref. Leg 115/04, GEN. &0/77/1) acknowledged that four Orders in Council have been made under the Colonial Boundaries Act 1895 to extend the boundaries of certain British colonies to include the continental shelf through order in council such as [18] the Bahamas (Alteration of Boundaries) 1948, the Jamaica (Alteration of Boundaries)1950, the British Honduras (Alteration of Boundaries) 1950 and the Falkland Islands (Continental Shelf) 1950. The primary purpose of the first three Orders in Council was to grant the colonial government control over natural resources that might be found in the adjacent continental shelf. While the Falkland Islands set the limit at the 100-fathom line (approximately 200 m) due to political reasons. A further note by the Legal Division on 17 November 1953 explained that the expression "boundaries of a colony" means the outward limits or frontier of the colony adjacent to the high seas or some other territory that is not part of the colony. Thus, an extension of the boundaries of a colony involves the annexation of territory to the colony. However, legislation enacted for a colony is normally limited in effect to the colony itself, it does not have extra-territorial effect.

During the conversation between Gordon Smith (Colonial Office) and the officer in the Department of External Affairs. High Commissioner's (6 November 1953, Ref. GEN. 70/77/1). The officer asked what exactly "boundaries" meant in those orders and the purpose of the orders, questioning whether they were following the suit of American republics claiming sovereignty over the continental shelf. Gordon Smith clarified that the extension of the boundaries of certain British colonies to include the continental shelf, aimed to clarify and assert territorial rights over the seabed adjacent to these colonies [20].

Refer to further inquiry in two savingram each (on 7 March 1953 from the Secretary of State for the Colonies to the Officer Administering the Government of North Borneo, Sarawak, and Brunei, the Commissioner General of Southeast Asia and on 6 January 1953, the Governor of Sarawak's confidential savingram No. 4). The edge of the continental shelf (100-fathom line) was discussed regarding the delimitation of territorial waters and the continental shelf at Brunei, Sarawak—North Borneo, and Sarawak—Brunei, and Her Majesty's government has consistently refused to recognize any areas beyond the 3-mile limit as territorial waters, in principle, due to the extensive maritime interests of the United Kingdom [21]. If the 100-fathom line was the implicit limit of and apply to the Sarawak OIC 1954, the map of Sarawak's "boundary" would resemble as Fig. 3. This map utilizes depth data from Admiralty Charts 2660A and 2660B (data year-1923), illustrating the 100-fathom limit, which marks the continental shelf break, off the coast of Sabah (formerly known as North Borneo). The boundary extends across the sea adjacent to Brunei Bay and continues towards the coast of Vietnam. Notably, the majority of the sea area in front of Sarawak is characterized by a shallow continental shelf.

4.3 Continental Shelf Delimitation at Brunei Bay

Relevant information about the Sarawak OIC 1954 is found in the file related to continental shelf delimitation at Brunei Bay. British intend to divide the Continental Shelf among

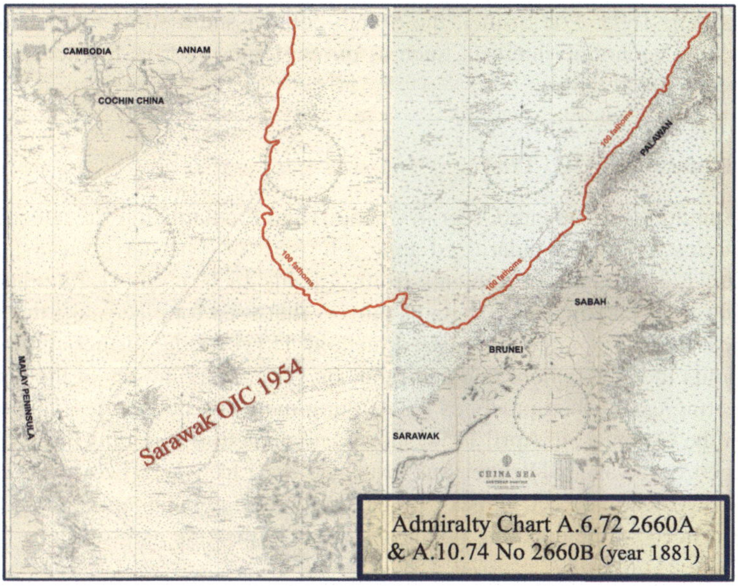

Fig. 3. Show the position of Sarawak OIC 1954 related to 100 Fathom Limit

neighboring states since absent of foreign power's interest [22]. However, British government refuses to recognize territorial waters beyond a 3-mile limit (extracted from letter on 6 March 1953 Ref. PMD.115/321/01). They particularly rejected the Philippines' claim over waters in 1932 convention as it was intended to divide land areas between the Philippines and North Borneo. This reflects the UK's strict adherence to the 3-mile territorial waters principle, influencing its stance on maritime boundary issues. This principle is critical for understanding the UK's perspective on Sarawak's maritime claims and the importance of historical agreements in defining boundaries. In the meantime, Shell is eager to conduct a geophysical survey, including areas beyond Brunei's territorial waters, to explore potential oil reserves as the Seria Oilfield production declines (letter of Colonial Office to E.H. Peck, Foreign Office on 16 November 1953 ref. PMD.115/271/02). This motivated Britain to establish the legal framework before conducting exploration activities, such as geophysical surveys.

In the letter Colonial Office (R.H. Burt) to E.H. Peck on 16 November 1953 ref. PMD.115/271/02 said, it is standard practice to issue proclamations claiming jurisdiction over the continental shelf and then negotiate the exact boundaries with neighboring states. On 25 November 1953, the Foreign Office agreed that a proclamation could be drafted, provided that North Borneo, Brunei, and Sarawak agreed on boundaries. This emphasizes the need for consensus among local governments before making formal proclamations, aligning with the collaborative approach in boundary demarcation. In the savingram from Secretary of State for the Colonies to Government Officers, 17 December 1953, Shell's interest in conducting a geophysical survey beyond British Borneo's territorial waters, a formal declaration of sovereignty over the continental shelf is recommended. For North Borneo and Sarawak, this requires an Order in Council under the Colonial

Boundaries Act, 1895. For Brunei, the Sultan could issue a similar proclamation. It showcases how economic interests, such as oil exploration, drive the need for formal declarations of sovereignty.

4.4 Proclamation of OIC 1954

The draft Statutory Instrument (Order in Council) and Proclamation for extending colonial boundaries follow a standardized approach used for other territories, such as the Persian Gulf [23]. Notably, comments from officials such as Sir G. Fitzmaurice and B.O. White reveal that the legal language mirrors that used for places like the Bahamas and Jamaica, indicating a template-based method. However, he further comments that the draft omits the 100-fathom line and relies on general definitions of the continental shelf, reflecting established practices for defining maritime boundaries. The Secretary of State for the Colonies, in April 1954, showed no objection to issuing these Orders for North Borneo, Sarawak, and Brunei. Subsequently, letter from E.H. Peck to R.H. Burt, 20 April 1954 revealed that Peck agrees with the draft wording but notes that interpretation will be influenced by the covering public statement requesting its provision.

4.5 Public Statement

The "Savingram from the Secretary of State for the Colonies" dated 5 June 1954, details the extension of jurisdiction over the continental shelf and includes final drafts of the OIC along with a public statement as follow:

> *For a number of years oil has been mined in Brunei and Sarawak and the potentialities er North Borneo are now being investigated. Drilling has recently been carried out in the territorial waters of Brunei, and it is desirable that the resources of the seabed beyond the territorial waters of the three territories should also be explored so far as possible. It is becoming increasingly practicable to exploit such submerged resources, and it is desirable in the interests of protection, conservation and orderly development that the exploitation of them should be properly controlled. The right of a littoral state to claim sovereignty over the seabed and subsoil adjacent to its coasts in order to control the exploitation of the natural resources therein has become established recently in international practice. Accordingly, the boundaries of North Borneo, Sarawak and Brunei have been extended under the provisions of the North Borneo (Alteration of Boundaries) Order in Council, 1954, the Sarawak (Alteration of Boundaries) Order in Council, 1954 and Brunei Proclamation particulars to be inserted to permit the Governments of these territories to exercise jurisdiction over the exploitation of the natural resources of the continental shelf adjacent to their coasts. The status as high seas of the waters above the continental shelf outside territorial water limits is not affected. In cases in which the continental shelf of any of the three territories mentioned extends to or is adjacent to the continental shelf of another State, it is intended that the boundary of the shelf shall be determined on equitable principles by agreement with the State in question.*

Sir G. Fitzmaurice on April 8, 1954 informed that the whole area is naturally the 100-fathom line [24], this proved British have already aware about the geomorphology of the seabed at that area. On 15 June 1954, Sarawak accepted the draft proclamation and agreed on the delimitation line with Brunei and North Borneo at Brunei Bay. The document stated that two additional Orders in Council are required to precisely define the boundaries [25]. Britain does not define the limits when annexing continental shelves unless in special circumstances and complicated situation in Brunei Bay which requires a specific definition of the continental shelf boundaries. In the meantime, the British Malayan Petroleum Company has urged for legal permissions to exploit Sarawak's continental shelf resources because they has invested heavily in preliminary exploration off Sarawak and Brunei [26].

4.6 Geneva 1958 and Territorial Sea Breath

Territorial Waters Law of The Sea International Conferences from 1958 to 1960 aimed to codify the Law of the Sea. Agreements reached include conventions on the territorial sea, high seas, fishing, and the continental shelf [27]. Britain (UK) signed these conventions in 1958, but they only ratify it in 1964. During the vote of UNCLOS II, Various coastal states have specified the nautical miles each state claims. Notably, the United Kingdom, Malaya, Brunei, North Borneo, Sarawak, Japan, and the Maldives all claim a 3 M limit, while other states like China, Indonesia, and the USSR claim up to 12 M.

4.7 Validity of OIC 1954

A circular from Secretary of State for the Colonies, Colonial Office (Mr. Alan Lennox-Boyd) dated 5 November 1959 (C.O. Ref: IRD 313/269/01 Circular 1176/59) to the Officer Administering the Government of Sarawak No. 1137 [27] said:

> *...Order in Council was made under the prerogative annexing part of the seabed to Her Majesty's Dominions and attaching it to a colony "for administrative purposes". Whatever may have been their effect in international law, these Orders were regarded for the purposes of municipal law as annexing the continental shelf and establishing full sovereignty over it so as to enable the colonies concerned to make laws with respect to it in the same way as they would legislate for the sea-bed beneath their territorial waters. It is clear, however, that it would be inconsistent with the Convention, which the United Kingdom expects to ratify, to maintain that the continental shelf is part of Her Majesty's dominions and that Her Majesty exercises full sovereignty over it, since Article 2(1) of the Convention provides that the coastal State shall exercise only limited sovereign rights over the continental shelf. For the same reason it would be inconsistent with the Convention to "annex" the adjacent shelf to a Protectorate or Trust Territory so as to make it an integral part of the territory. Further "annexation" by Order in Council of the continental shelves of Colonial territories must therefore be ruled out...He also willing to assist in the preparation of detailed arguments If the validity of such legislation is challenged...It is not proposed to alter or revoke existing Orders*

relating to continental shelves, and if asked by other States how they are reconciled with the Convention, it would be said that the Orders were made before the nature of the coastal States' rights under international law in respect of the Continental Shelf had been clarified by the Convention; that the purpose of the Orders made under the Colonial Boundaries Act, 1895, was to extend the jurisdiction of the Governments of the Colonies concerned so as to give them control over the seabed and subsoil contiguous to their coasts with a view to the exploitation of its natural resources; and that any rights now exercised under these Orders, or under the Submarine Areas of the Gulf of Paria (Annexation) Order, 1942, would be the rights recognised by the Convention... [28]

In the international second Conference on The Law of the Sea held at Geneva (UNCLOS II) in 1960 (Ref: IRD 313/017, Circular 1311/60, 30 December 1960), the United Kingdom emphasized that as the Conference unsuccessful to agree on a specific breadth of the territorial sea, Her Majesty's government would continue to regard the 3 M as the only breadth recognized under international law.

At Brunei Bay, the relevant boundaries in this area are set out in the North Borneo (Definition of Boundaries) Order in Council, 1958 (S.I. 1958 No. 1517, 1958 II, page 1706) and the Sarawak (Definition of Boundaries) Order in Council 1958 (S.I. 1958 No. 1518, 1958 II, page 1708). The outer limit of the continental shelf is at present normally taken in practice to be the 100-fathom contour since this is shown on Admiralty charts and approximates to the 200 m depth referred to in Article 1 of the Convention [29].

Mr. Gordon Smith informed Mr. Titchener on 8 September 1960 that while local legislation by Colonial governments for the specified purposes would likely be valid, there remained some uncertainty, suggesting that UK legislation might be used to clarify colonial legislative powers [30].

Regarding boundary questions, any UK legislation concerning Colonial continental shelves would be enabling, with the power to define boundaries either conferred on H.M. in Council or left to local legislatures, if local legislatures can regulate their shelves, they can also define their boundaries, but for shelves annexed under the Colonial Boundaries Act, 1895, only H.M. in Council can define these boundaries, the replacement of these existing Orders in Council. This is important so as to remove the inconsistency that exists between these Orders and Article 2 (1) of the Convention. U.K. act ought to confer express authority on colonial legislatures to make laws with respect to the exploration and exploitation of the continental shelf.

4.8 Status of OIC 1954

When UK draft the Bill of Continental Shelf, a letter of Colonial Office on 13 July 1961 to Ministry of Power (Ref. IRD 313/01) [30] revealed that they have different opinion about the OIC 1954. They think that these Orders appear to be inconsistent with Article 2 of the Convention, but they "have hitherto proposed to leave them in force and if their inconsistency with the Convention is raised to explain that they were made before the Convention was drawn up and that the rights now exercised under them would be only those recognized by the Convention." But again this is not perhaps an entirely satisfactory course and they wish to consider whether the Bill should provide for the revocation or amendment of those existing Orders in order to remove the inconsistency with the Convention and to put the Colonies concerned in the same position as other Colonial territories. They then decided to seek the views of the Foreign Office, and the territories concerned before reaching a firm decision.

On 19 November 1962, the Ministry of Power discusses the necessity of U.K. legislation to implement the Convention on the Continental Shelf 1958 (Ref. PD 1146/30/4). They mentioned that they will maintain their tentative view on extending legislation to colonial territories by Order in Council, pending the content of the U.K. Act. The U.K. Law Officers had previously provided an opinion on whether Colonial Legislatures could legislate for their continental shelves. Their opinion was twofold: Public international law: "yes" and Municipal law: "probably yes". On the question of the inconsistency with the Convention of certain existing Orders in Council which annex certain shelves or parts of shelves to adjacent Colonies, they stated:

> *...we agree with the Foreign Office that we should take the opportunity of making provision by Act of Parliament to renounce the claim to full sovereignty over the areas in question. We would propose that the Act should revoke the existing Orders, which in all cases but one were made under the Colonial Boundaries Act 1895...*

In so far as they annex submarine areas outside territorial waters to Her Majesty's Dominions but should leave any provisions as to boundaries in force as if they had been made under the provision of the Act empowering the Queen to define boundaries. He then acknowledges two of the territories have already become independent, and it is possible that others will become so by the time the Act is passed, in any such cases it should be done about the order in question will be a matter for the territory concerned to decide. He further stated that:

> *As regards the others, we may wish to consult them about revocation of the orders, but this should not cause difficulty since they are already aware from paragraph 8 of our Circular 1176/59 of 5th November 1959, that only those rights recognised by the Convention should now be exercised under the Orders.*

The Geneva Convention on the Continental Shelf was adopted on 29 April 1958 and entered into force on 10 September 1964. The UK ratified the Convention on 11 May 1964. This timeline highlights the formal adoption and implementation of international legal standards regarding continental shelves, setting the stage for subsequent national legislation and international negotiations.

The UK Bill addressing the high seas, including the continental shelf, is under discussion. The bill would need to align with the Geneva Convention and address the legislative framework for continental shelves, reflecting the broader context of implementing international maritime law domestically. These rights are not based on occupation or proclamation. This reinforces the idea that rights to the Continental Shelf are granted by the Convention rather than through occupation, aligning with the broader principles of the treaty.

Departure from Res Nullius: The Convention marks a departure from the concept of the sea bed as "res nullius" (unclaimed). It uses "sovereign rights" instead of "sovereignty" to clarify that while coastal states have specific rights, the Continental Shelf is not part of their territory. This distinction is critical for understanding the legal status of the Continental Shelf and how it differs from traditional claims of territorial sovereignty. The Convention stipulates that exploration and exploitation must avoid unjustifiable interference with navigation, fishing, and scientific research. It also requires notice for construction of installations and prohibits interference with international navigation routes. These regulations aim to balance the rights of coastal states with the rights of other states and international interests, reflecting the Convention's goal of harmonizing various maritime activities.

While the Convention addresses exploration and exploitation, it does not cover other rights such as defense. These issues remain governed by customary international law. This underscores the Convention's focus on resource rights while leaving other aspects of maritime law to traditional practices and agreements. The Convention is expected to apply to all territories for which the UK is responsible, excluding certain exceptions like the Persian Gulf Sheikhdoms. If the Convention does not confer full sovereignty, UK Orders in Council extending to the Continental Shelf should be revoked, though this primarily affects territories like the Cayman Islands and Turks and Caicos Islands. This provision highlights the practical implications of the Convention for former and current British territories, addressing how existing legal instruments must adapt to new international standards.

In 1969, Mr. Ramani's Legal Adviser to the Malaysian Ministry of Foreign Affairs enquiry concerning background information about the seabed boundaries of Sarawak and Sabah. United Kingdom Mission to the United Nation 20 August 1969, New York, FWM 3/305/2 J.L Simpson Deputy Legal Adviser Foreign and Commonwealth Office replied that

> *...in 1954 the limitation of sovereign rights for the purposes of exploring and exploiting the continental shelf had not yet become accepted doctrine and the Orders might have been based on earlier extensions such as the Truman proclamation and the Persian Gulf extensions which were not based on that limitation...* [31]

In the file of [32] titled as "Dispute Between Federal Government and Sarawak Over Continental Shelf", A letter from South West Pacific Department (Commonwealth Office) Registry No. 52, 2 April 1970 (FWM 4/1) D.F.B. Le Breton To R.C. Clift British High Commissioner Kuala Lumpur stated:

> *...It is clear that there are two differing opinions among lawyers about relative strengths of Federal and State legal claims. There appear to be a number of precedents in other countries with Federal constitutions which suggest that the Malaysian Federal Court would be likely to take a policy decision favouring the Federal rather than the State Government. The continental shelf, by the terms of the Convention does not form part of the coastal state's territory, the Convention*

merely allows "sovereign rights" to that state for particular purposes, viz exploring and exploiting its natural resources. The concept of "sovereignty" shelf was explicitly rejected by the Geneva Conference of 1958 at which the Convention was formulated. Since it was considered that, so drafted, many states would claim exclusive fishing rights in the waters above the shelf, thereby seriously encroaching on the freedom of the high seas. Is this particular danger is eliminated in the proclamation by its second section which provides that annexation of the shelf shall not affect the character as high seas of the water above it. With such a proviso, the distinction between "sovereignty" and "sovereign rights" becomes a fine one, since the latter carries the clear implication of that rights of property in the shelf's natural resources, as defined in the convention, are vested in the coastal state, and that the property in such resources can not pass to other persons except by way of express grant. Once the provisions of the Convention have been enacted in the law of Malaysia, none of Malaysia's existing laws should be seen to be inconsistent with the terminology or concepts used in the convention... [33].

Regarding the Ramani inquiry, M.G. De Winton, deputy legal adviser, wrote on 19 March 1970 to the South West Pacific Department, stating there was no issue with providing Ramani a copy of the public statement for the Orders in Council [34], but sharing the sensitive 1959 confidential circular despatch (Circular 1176/59) was not advisable due to strained relations between Sarawak and the Federal Government [32]. On 25 March 1970, M.G. De Winton further explained said that Federal Court in Malaysia likely to take policy decision favoring the Federal Government like all the Federal Courts (high court of Australia and Supreme court of Canada) and the Tideline cases U.S V. Louisiana (339 US), U.S v. California (332 US), U.S v. Texas (339 US). He then advice if the Federal Court follows the O'Connell's reasoning it might hold that the nine additional miles to which the Federal Government has laid claim as territorial waters are within exclusive Federal Jurisdiction.

5 Conclusion

The Sarawak OIC 1954, which granted Britain the rights to explore the subsoil and seabed of the continental shelf, was part of a broader practice applied to Britain's dominions. The term "boundary" in the Sarawak OIC 1954 implied administrative control over the continental shelf, including the authority to issue exploration licenses and mining leases to oil and gas companies. The colonial administration of Sarawak exercised this right from 1954 until the formation of Malaysia in 1963.

British legal practice typically defined the continental shelf up to a depth of 100 fathoms (about 200 m) since the Truman Proclamation. This depth when the shelf break served as a standard measure for defining the limits of exploration rights and has been codified into international law (Geneva 1958 & UNCLOS 1982).

During the ratification process of the 1958 Geneva Convention, the British government, particularly the Foreign Office, proposed the possible revocation of colonial orders that were inconsistent with the Convention. This suggestion was based on Her Majesty's original intention to assert full sovereignty over the area using the OIC (such as Sarawak

OIC 1954). However, they proposed maintaining specific boundaries already defined (such as the boundary between Sarawak and Brunei at Brunei Bay under OIC 1958).

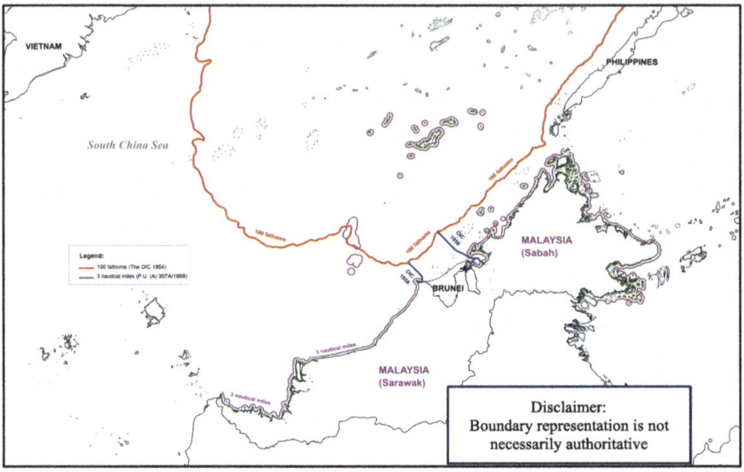

Fig. 4. Illustration of boundary refer to the OIC 1954 and 1969 emergency ordinance

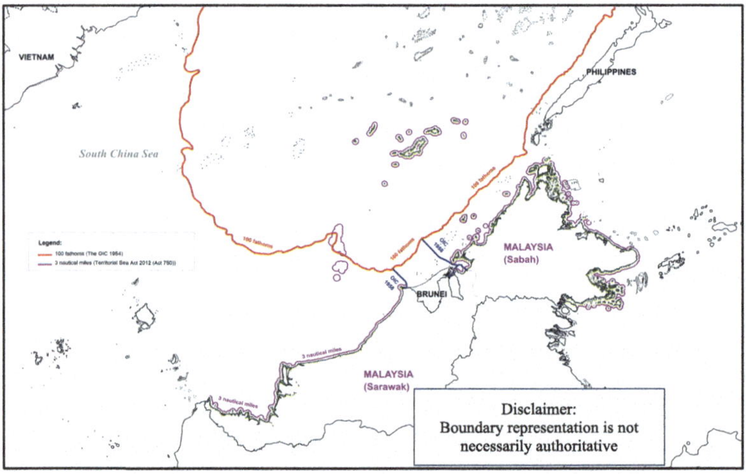

Fig. 5. Illustration of boundary refer the OIC 1954 and Territorial Sea Act 2012

The question of whether the Sarawak OIC 1954 was revoked or left unchanged when the UK ratified the Geneva Convention in 1958 hinges on legal interpretation. Britain did remind its dominions to exercise the order only the rights recognized by the Convention. Archival findings presented in this paper also indicate that before Malaysia Day on 16 September 1963, Sarawak's territorial waters were defined as extending 3 M from its coast since Britain consistently adhered to the 3 M limit when OIC 1954 was issued and up to the Malaysia Day. The extension to 12 M was subsequently implemented by

Federal Proclamation in 1969 through an Emergency Ordinance. So it can be concluded that the territorial waters of Sarawak as defined in the past, remained unaffected by the 1969 emergency ordinance or the Territorial Waters Act 2012 as showed in Figs. 4 and 5.

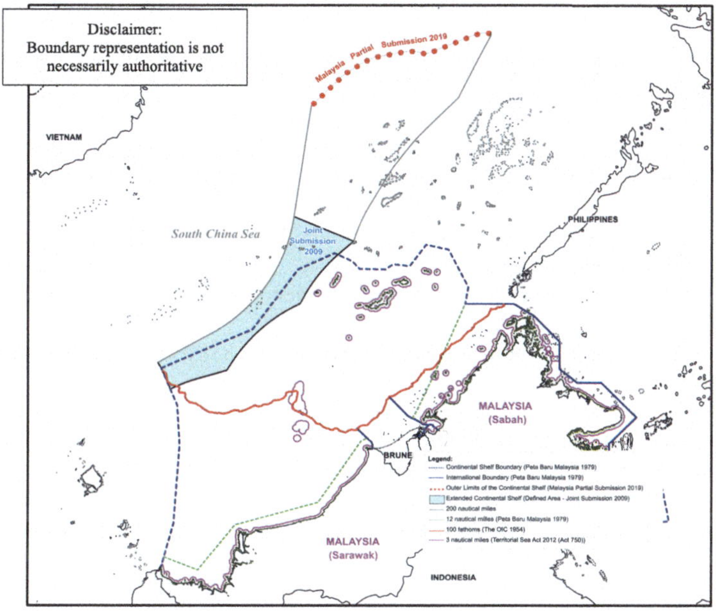

Fig. 6. Overall of current of Malaysia's boundary without extended continental shelf

The historical record of Sarawak's activities between 1954 and 1969 has been overtaken by numerous events and legislative changes, including delimitation of continental shelf between Malaysia-Indonesia in 1969 [35], the proclamation of the Malaysia's continental shelf boundary as depicted in the *Peta Baru Malaysia* 1979 [36] and extended continental shelf claim [37, 38], shown in Fig. 6. Future studies and legal interpretations will be essential to clarify the legal status of Sarawak's exploration rights at the area of continental shelf, considering the federal constitution, international regulations, and the archival findings presented in this paper. Despite internal challenges, Malaysia and its state of Sarawak can complement each other and continue to assert their rights over the continental shelf and the South China Sea. Historical records and evidence demonstrate Malaysia's long-standing effective governance and strong legal and administrative frameworks in the South China Sea. Malaysia has consistently upheld its entitlement to an extended continental shelf beyond 200 M and its outer limits of the continental shelf in the South China Sea which is still awaiting recommendations from the CLCS.

References

1. Tang, A.R., et al.: Sarawak's Claim on the State's Jurisdiction over Oil and Gas on the Continental Shelf: A Legal Historical Perspective. KnE Social Sciences (2023)

2. Sarawak, K. (ed.): Sarawak Government Gazette The Sarawak Government Gazette Extraordinary Part II, The Sarawak (Alteration of Boundaries) Order in Council 1954 (1954)
3. Britain: The Sarawak (Definition of Boundaries) Order in Council 1958, Authority, Editor. London, UK (1958)
4. United Nations: Convention on the Continental Shelf 1958, U. Nation, Editor. Geneva (1958)
5. UNCLOS: United Nations Convention on the Law of the Sea U. Nation, Editor. United Nation: Montego Bay (1982)
6. National Geographic Continental Shelf (2024). https://education.nationalgeographic.org/resource/continental-shelf/
7. Heidar, T.H.: Legal Aspect of Continental Shelf Limits (2017)
8. Schrijver, N.: Sovereignty over Natural Resources: Balancing Rights and Duties. Cambridge Studies in International and Comparative Law. Cambridge University Press, Cambridge (1997)
9. Legislation.gov.uk: Colonial Boundaries Act 1895. 1895 [cited 2024 24 April]. https://www.legislation.gov.uk/ukpga/Vict/58-59/34
10. Malaysia Government: Emergency (Essential Powers) Ordinance, No. 7, 1969, as amended in 1969. United Nation (1969)
11. Malaysia Government: Continental Shelf Act1966 (Act 83). Percetakan National Malaysia Berhad (1966)
12. Malaysia Government: National Land Code 1965, in Interpretation, State Land. Percetakan National Malaysia Berhad, Kuala Lumpur (1965)
13. Malaysia Government: Act 750, Territorial Sea Act 2012. Percetakan National Malaysia Berhad (2012)
14. Malaysia Government: Federal Constitution. 2020, Attorney General's Chambers of Malaysia. https://lom.agc.gov.my/
15. Malaysia Government: United Kingdom of Great Britain and Northern Ireland and Federation of Malaysia, North Borneo, Sarawak and Singapore_Agreement relating to Malaysia (with annexes, including the constitutions of the state of Sabah, Sarawak and Singapore. United Nation, London (1963)
16. Daily Express: Why Act 750 can't simply be enforced: Control over Sabah and Sarawak territorial waters. [cited 2024 6 May] (2023). https://www.dailyexpress.com.my/read/5372/why
17. The National Archives—FCO 141/12379: Sarawak: Oil Mining Lease between Sarawak Government and Sarawak Oilfields Ltd. Kew, UK (1947–1952)
18. The National Archives—FCO 141/12382: Sarawak: Oil Mining Lease between Sarawak Government and Sarawak Oilfields Ltd. Kew, UK (1946–1950)
19. The National Archives—FCO 371/89398: Oil Mining Rights in Sarawak. Code WN file 1533. Kew, UK (1950)
20. The National Archives—CO 936/220: Territorial Waters Boundaries of Continental Shelf boundaries. Kew, UK (1953)
21. The National Archives—CO 1029/12: Labuan Exploration. Kew, UK (1953)
22. The National Archives—FO 371/105746: Territorial Waters. Kew, UK (1953)
23. The National Archives—FO 371/110747: Continental Shelf and Territorial Waters in Borneo, Brunei and Labuan. Kew, UK (1954)
24. The National Archives—CO 1029_129: Appropriation of Oil Resources of Continetal Shelf Outside Territorial Waters Borneo territories. Kew, UK (1954)
25. The National Archives—CO 1029_131: Oil Licenses and Leases over Continental Shelf of Brunei and North Borneo. Kew, UK (1954–1955)
26. The National Archives—FCO 141/12469: Brunei Oil Lease Continental Shelf. Kew, UK (1954–1955)

27. The National Archives—FCO 141/12751: Sarawak Territorial Waters Law of The Sea. Kew, UK (1954–1955)
28. Marston, G.: The incorporation of continental shelf rights into United Kingdom Law. Int. Comparative Law Q. **45**(1), 13–51 (1996)
29. The National Archives - CO 1030/1045: The Continental Shelf of Brunei. Kew, UK (1960–1961)
30. The National Archives - CO 936/710: Law of the Sea Geneva Convention 1958. Kew, UK (1960–1961)
31. The National Archives - FCO 24/493: Malaysia: Political Affairs Ext-Bilateral Indonesia: Relation with. Kew, UK (1969)
32. The National Archives - FCO 24/833: Dispute between Federal & Sarawak over Cont Shelf. Kew, UK (1970)
33. The National Archives - FCO 24/232: Brunei: Territorial; Continental Shelf: Definition of Boundaries of. Kew, UK (1970)
34. The National Archives - FCO 24/835: Agreement on continental shelf boundaries between Indonesia and Malaysia. Kew, UK (1970)
35. Malaysia and Indonesia Government: Agreement between the Government of Malaysia and the Government of Indonesia on the Delimitation of the Continental Shelves between the Two Countries, 27 October 1969, D.T. Infobase, Editor. United Nation (1969)
36. Malaysia Government: Peta Baru Malaysia - Sheet 2, Jabatan Ukur dan Pemetaan Malaysia, Editor. Direktorat Pemetaan Negara Malaysia, Kuala Lumpur (1979)
37. Malaysia Government: Joint Submission by Malaysia and the Socialist Republic of Viet Nam to Commission on the Limits of the Continental Shelf pursuant to Article 76, paragraph 8 of the United Nations Convention on the Law of the Sea 1982 in respect of the Southern Part of the South China Sea, Commission on the Limits of the Continental Shelf (CLCS), Editor. Division for Ocean Affairs and the Law of the Sea, Office of Legal Affairs, United Nations, New York (2009)
38. Malaysia Government: Malaysia Partial Submission to The Commission on The Limits of The Continental Shelf Pursuant to Article 76, Paragraph 8 of the United Nations Convention on the Law of the Sea 1982 in the South China Sea, Commission on the Limits of the Continental Shelf (CLCS), Editor. Division for Ocean Affairs and the Law of the Sea, Office of Legal Affairs, United Nations, New York (2019)

Assessing Perceptions of Digitalization of Land Administration and Land Management

Walter Timo de Vries[1](✉), Claudia Lindner[2], and Walter Dachaga[1]

[1] Technical University of Munich, 80333 Munich, Germany
{wt.de-vries,walter.dachaga}@tum.de
[2] Kadaster International, Amsterdam, The Netherlands
Claudia.Lindner@kadaster.nl

1 Introduction

The extent to which digitalisation objectives and processes of land administration and land management are effective depends on multiple factors, such as (inter) organizational conditions, applicable legal and budgetary frameworks of their ecosystems and technological developments and dependencies [1]. Although there exists abundant literature on digital government and the effects of digitalization of public (land) information systems in the broadest sense [2–5], it is however largely unknown how practitioners and affected beneficiaries within these land related (organizational) ecosystems perceive how, how much and why any of these factors influence the effectiveness or efficiency. Given these unknowns in individual, professional and organizational perceptions, it remains furthermore unclear what needs to be changed in the ecosystems to enhance or alter the effectiveness or efficiency. Consequently, it remains unclear which steps need to be taken to make effective use of the digital technologies, and to adapt (inter-)organizational culture, behaviour and choices to truly benefit from the technological advances. In this light, the main objective of this research was to investigate and categorize the different social perspectives, beliefs and values on preferred and non-preferred (technical and non-technical) practices, fears and opportunities related to digital technologies which professionals in land related ecosystems, such that one can improve the strategies on how to effectively move forward with digitalisation processes.

Addressing this question first requires an understanding of how we define digitalization and examples of digital processes and products in land related ecosystems. Next, we describe an effective methodology to assemble and compare perceptions in such ecosystems. The next section addresses this methodology and describes how the research identified and addressed participants who could express their perceptions. After this follows a description of the results obtained so far, and a first interpretation of the results. The conclusion section describes the implications for these interpretations for the formulation of possible strategies for handling and enhancing digitization processes in land related ecosystems.

A. Abdul-Rahman et al. (Eds.): GeoWeek 2024, LNGC, pp. 272–282, 2025.
https://doi.org/10.1007/978-3-031-86654-8_21

2 Defining and Identifying Digitalization in Land Related Ecosystems

Land related ecosystems can be regarded as inter-related and often also inter-organizational processes and dependencies which create and/or derive land related products and services. The latter does not only include the tangible outputs such as titles, registrations, ownership certificates, taxation values or building licenses, but also more complex outcomes of inter-organizational processes of land use planning, land development and land mobilization, amongst others. At the level of individuals, the ecosystems do not only involve the professionals working within the government or state-based organizations responsible for national or sub-national territories, but also (often licensed or recognized) professionals delivering products or services within the land related processes of creating and delivering products and services.

Digitalization in this paper refers to the adaption or redesign of existing business processes to take advantage of digital data and technologies usually for the purposes of increasing operational efficiencies, and may be considered as part of Business Process Re-engineering (BPR) [1]. In other words, it relates to both how and with which types of technologies the processes of data collection and analysis are carried out as well as the format and distribution channels with which products or services are made available and the technologies which make change and re-designing processes visible and possible. Essentially, there are multiple types of ICT and geoICT which may reflect the manifestation of digitalization. The publications [6, 7] identify multiple types of digital technologies used in land management, including (1) Integrative and analytical technologies, (2) Data acquisition and data processing technologies, (3) Smart and artificial intelligence technologies, (4) Visualisation, representation and simulation technologies and (5) Data management technologies. Each of these types exhibit rapid developments and each are gradually entering the fields of land management, land administration and spatial planning. The degree to which these are however successfully adopted and integrated differs however from case to case, and context, contingencies, capacity and willingness to adopt and eventually redesign existing business processes play a crucial role in each. It is therefore important to discover how which technology is adopted why, when and how.

3 Methodology of Data Collection and Analysis

Understanding perceptions in any human or organizational system requires accepting that one must rely on subjectivity, which is dependent on both individual and organizational context, purpose and preference. One of the methodologies to assemble, analyse and compare variations in subjective perceptions is Q Methodology, which is a systematic methodology to measure and compare so-called operant beliefs. It was originally developed by [8] and further operationalized by [9] and [10] amongst others. Operant beliefs in the eyes of [8] are time and context dependent expressions of preferences, choices, favourites in a social context. These beliefs are unique to a specific individual and can thus only be compared based on similarities and differences in the beliefs themselves and not based on the similarities and differences of the individuals. Q methodology follows a strict procedure and logic of defining a data collection and data analysis process. It shares many characteristics of mixed methods, relying on both qualitative and quantitative techniques, inferences and interpretations. The justification of relying on Q methodology is to unveil a hidden structure in a concourse of a specific thematic field, which otherwise remains complex and unstructured. In the field of land management, it has been tested to uncover policy systems [11], operational aspects of land governance [12], spatial policy priorities in city planning [13].

The Q methodology procedure contains 7 steps [14]: (1) Identify concourse and volume of debate—in Q the flow of communication surrounding any topic is referred to as a 'concourse'; concourses are not just restricted to words, but can also consists of expression on social media, blogospheres, pictures, etc.; (2) Define or refine Q sample—this refers to a selection of statements which are either representative or exemplary for positions in a certain concourse; (3) Select P sample—this refers to a selection of participants who are willing and suitable to respond to a Q 'questionnaire' (referred to as Q Sort); (4) Administer Q sort (using the selected statements) and additional data selection—this can take place manually (using cards with individual statement) or using an online interactive software; (5) Run factor analysis (the statistical analysis; could also be a principle component analysis); (6) Interpret and report operant factors (usually the most significant factors and components); (7) Present and discuss findings.

In this study we use the Q method to search and detect operant beliefs and values vis-à-vis digitalisation and digital transformation in the context of land administration and land management. As the Q methodology starts by identifying a concourse related to digitalization, we relied on several literature sources which compiled a summary of several aspects and used a structured approach to derive basic statements. In our case the concourse of statements is a matrix on the one hand based on 6 aspects of digitalization [1] - being products and services, 6 I's, types of models, organisational performance, cyber risks, technological disruptions, and on the other hand based on 5 aspects of land management and land administration [6] - being spatial structuring and design, spatial monitoring and assessment, administration of land and properties, compliance and coercion, participation and mobilisation. Out of the 30 possible statements which this matrix generates we chose 25 statements for the Q sorting process, which was used as the basic Q sample.

The Q sorting was prepared using an online sorting process via the software http://qsortware.net/ and the link https://application.qsortware.net/user/devries2000/. When executing these links, a participant has to complete the sorting of statements in two steps. The **first step** is a more generalized sort to get acquainted with the statements. One has to read each statement and choose first if one generally agrees, do not agree, or considers the statement neutrally (or not having any opinion yet) in view of digital transformations in land management and land administration. The **second step** is to sort more carefully into 5 categories of agreement and disagreement. Here one must sort according to the maximum number of statements in each level of agreement. All statements must be placed in one category. At the extremes of agreement and disagreement you can only place one statement. At the end the 25 statements need to be ranked in a normal distribution-resembling order—as in the following scheme (one card representing one statement; the number of statements indicated in each column; the values −3 up to + 3 representing the scores used for statistical analysis.

Very bad for digital transformations in land administration and land management **1 card** −3	**3 cards** −2	**5 cards** −1	Neutral **7 cards** 0	**5 Cards** +1	**3 cards** +2	Very good for digital transformations in land administration and land management **1 card** +3

The analysis of the data relied on Ken-Q online software, available via https://shawnbanasick.github.io/ken-q-analysis/. The results of the QSoftware could be uploaded Ken-Q as predefined Excel sheets. Once uploaded it is possible to execute different kinds of statistical analyses. Essentially, each Q sort represent a personal single perspective on how to rank each statement against each other statement. The factor analysis then primarily calculates the variety in statement ranking, and not in the variety among participants attributes. The resulting factor loadings then represent clusters of belief systems. The Q survey was distributed in April 2024 via direct mails, via the FIG newsletter, via LinkedIn and among TUM alumni.

4 Results

The initial results (in mid May2024) generated 35 responses, from 23 countries (from 5 continents). Expert/knowledge familiarity levels with digital technology varied from Expert (17), familiar (16), limited knowledge or skills (2). The background was from National government (9), regional local government (6), Academia (11), private sector (3), NGO (1), other (including development agency, volunteer, UN) (5). A first analysis shows that there are 4 or 5 significant factors which explain approximately 61% to 66% of the variation in responses. The degree of correlation between Q participants is relatively small (a maximum correlation 60% between responses of two particular participants), which demonstrates a high degree of different views and beliefs among participants scores.

The resultant factor loadings for the first 5 factors are in the Table 1 (in greyscale the statement with the highest and lowest statement ranking for each factor):

We could read and/or interpret these loadings as follows (Table 2).

The statements with the highest and lowest degree of consensus and disagreement in the factor loadings are as in Table 3 (high consensus if ranking value is low; high disagreement if ranking value is high):

In other words, there is a big disagreement among participants and among value systems on statement - (3) Land titles should always be made available in both a paper and a digital format—and a high consensus for statement - (21) Digital integration of land information leads to loss of historically relevant land information -. One has to be careful how to interpret these statistics however, because for statement (21) the Q score is consistently low (Table 4):

This implies that practically all values systems consider that the loss of historically relevant land information is not relevant (each factor considers the statement the least relevant). For statement (3) the factor loadings vary the most. This means that this may be a distincting issue when comparing different coherent belief systems:

Overall, the first results of the Q survey show a high degree of variation of opinions on how and why digitalisation occurs and in which direction it should develop. There is also a low degree of correlation between participants. Part of the reason may be that the participants of the survey represent a high variation of countries with different levels of digitalisation maturity. This could explain part of the variation. Nevertheless, the correlation between participants from the same country is also low (e.g. Germany: Ger1-Ger2: −6; Australia: Aus1-Aus2: 38; Uganda: Uga1-Uga2: 29—all on a scale of 0 to 100).

Table 1. Factor scores per statement.

Statement number	Factor 1	Factor 2	Factor 3	Factor 4	Factor 5
	Z-score	Z-score	Z-score	Z-score	Z-score
1	−0.1	1.71	1.72	1,73	1,99
2	1.18	−0.29	0.59	0.24	1.39
3	1.09	−2,25	−1.05	0.87	1.27
4	2,12	0.39	0.38	−1.2	0.85
5	−0.88	0.74	0.77	0.37	0.3
6	0.59	0.58	1.36	0.74	1.43
7	0.3	0.41	0.59	0.14	−0.3
8	1.46	0.26	−0.49	1.04	−0.03
9	−0.8	−0.17	−0.57	−0.61	−0.17
10	−1.62	−0.79	1.29	−0.83	0.51
11	0.94	0.62	2,03	1.03	0.04
12	0.52	0.46	−0.66	1.27	−1.06
13	0.89	0.24	−0.21	0.94	−0.93
14	0.5	0.38	−0.59	0.22	−0.46
15	−0.14	0.06	−1.21	−1.52	0.3
16	−0.3	1,87	−0.46	0.14	−1.52
17	−0.28	1.63	−1.26	−0.95	0.34

(*continued*)

Table 1. *(continued)*

Statement number	Factor 1	Factor 2	Factor 3	Factor 4	Factor 5
	Z-score	Z-score	Z-score	Z-score	Z-score
18	−0.3	−0.17	−1.29	−1.35	0.21
19	−0.5	−0.47	−0.13	0.78	1.14
20	−1	−0.55	−0.95	0.15	−0.34
21	−1,96	−1.74	−1,36	−1,9	−1.48
22	−0.96	−1.7	−0.13	−0.77	−0.55
23	−0.36	−0.43	−0.21	0.52	−0.13
24	0.72	−0.17	0.77	−1.42	−1,99
25	−1.09	−0.63	1.05	0.38	−0.81

Table 2. Factor interpretations.

Factor (a systematic belief system)	Has preference for/likes/considers very important/relevant	Rejects / does not like /does not consider very important/relevant
1	(4) Drone technologies can help identify and fine owners of illegal constructions	(21) Digital integration of land information leads to loss of historically relevant land information
2	(16) Land management agencies can increase their performance when all their activities rely on digital work processes	(3) Land titles should always be made available in both a paper and a digital format
3	(11) Designing rural development plans should be done by making geospatial technologies and data available at local levels	(21) Digital integration of land information leads to loss of historically relevant land information
4	(1) City planning can work better using digital twins	(21) Digital integration of land information leads to loss of historically relevant land information
5	(1) City planning can work better using digital twins	(24) Artificial intelligence is the future of generating new land use plans

5 Conclusions

The major conclusion on the influence of external factors is that technological changes or advancements, such as the use of drones instead of aerial photographs or satellites, are seen as having a positive influence on the transition to make land administration more digitalised. This is remarkable because the data acquisition method or technology itself—using pixel-based and image-based data acquisition is not fundamentally different from conventional satellite or airborne images, but it is most likely the sense of having more direct control of the data collection process and the resulting data. Drones can be handled and controlled by the agency itself, whereas for aerial photos and satellite images one is more dependent on external parties. Additionally, it is perhaps the sense of having access to the latest technological gadget. Reversely, one could argue that in particular drones foster new types of applications and services and are therefore the main driver of digital transformation.

Table 3. Ranking values per statement.

Number	Statement	Ranking value
21	Digital integration of land information leads to loss of historically relevant land information	0,054
9	Tax officers need to have up-to-date digital tax payment information in the field	0,063
7	Geospatial statistics should be accessible via interactive dashboards	0,091
23	Fake information can result in land related protests and advocacy	0,114
6	Smart spatial planning leads to more effective spatial designs	0,143
20	Land agencies are particularly vulnerable to hackers who can change and manipulate land information	0,18
14	Integration of national spatial databases leads to better compliance to plans	0,201
22	Land data can be manipulated to influence who receives sanctions or fines of illegal land use	0,268
5	Citizens can are encouraged to participate in spatial planning via interactive geoportals	0,359
2	Drone technology is crucial to monitor changes in the landscape	0,376
18	Geoportals of land agencies should make information available where actors do not adhere to land use plans or do not pay land tax	0,393
11	Designing rural development plans should be done by making geospatial technologies and data available at local levels	0,424
19	Voluntary geographic information creates openness in and accessibility to land management processes	0,453
13	Digital land administration should be designed to also cater for spatial planning needs	0,493
8	Digital land administration makes inequality in land ownership more transparent	0,506
15	The degree of participation in land management processes strongly depends on the degree of ability to work with geospatial data	0,523
1	City planning can work better using digital twins	0,581
25	Blockchain technology can increase the trust in land administration systems	0,648
12	In digital information ecosystems the interlinkages between land management and disaster risk reduction can be improved	0,717
17	Work flow efficiency through digital technology positively influences the quality of land administration services	1,056

(continued)

Another remarkable result is the low value given to paper-based information and the maintenance of historical records in conventional ways alongside setting up the

Table 3. (*continued*)

Number	Statement	Ranking value
10	Women are more involved in land management when empowered by digital technologies	1,086
4	Drone technologies can help identify and fine owners of illegal constructions	1,131
16	Land management agencies can increase their performance when all their activities rely on digital work processes	1,222
24	Artificial intelligence is the future of generating new land use plans	1,249
3	Land titles should always be made available in both a paper and a digital format	1,941

Table 4. Q scores for statement 21.

Factor 1	Factor 2	Factor 3	Factor 4	Factor 5
−3	−2	−3	−3	−2

Table 5. Factor loadings for statement 3.

Factor 1	Factor 2	Factor 3	Factor 4	Factor 5
2	−3	−1	1	2

fully digital procedures. There is the general belief in both belief system 1 and 2 that digitalisation will make the information better, regardless of how the information is acquired.

References

1. FAO, UNECE, and FIG: Digital transformation and land administration: sustainable practices from the UNECE region and beyond. FIG. 88 (2022)
2. de Vries, W.T.: GeoICT uniformity in flexibility. Analysis of the influence of geoICT coordination on the cooperation between public organisations with geoICT, p. 383. Erasmus University Rotterdam, Enschede (2013)
3. Kurniawan, M., de Vries, W.: Impact of geoICT applications on speed and uptake of local government services. Local Gov. Stud. **41**(1), 119–136 (2015)
4. Pelizza, A.: Developing the vectorial glance: infrastructural inversion for the new agenda on government information systems. Sci. Technol. Human Values **41**(2), 298–321 (2016)
5. Ameyaw, P.D., Vries, W.T.: Blockchain technology adaptation for land administration services: the importance of socio-cultural elements. Land Use Policy **125**, 106485 (2023)
6. de Vries, W.T.: Trends in the adoption of new geospatial technologies for spatial planning and land management in 2021. Geoplanning: J. Geomatics Plan. **8**(2), 85–98 (2021)

7. de Vries, W.T., Piyasena, N.M.: Trends in geospatial technologies in Asia. In: Geospatial Science for Smart Land Management, pp. 13–25. CRC Press (2023)
8. Brown, S.R.: Political subjectivity: applications of Q methodology in political science. Yale University Press (1980)
9. Watts, S., Stenner, P.: Doing Q methodology: theory, method and interpretation. Qual. Res. Psychol. **2**(1), 67–91 (2005)
10. van Exel, J., de Graaf, G.: Q methodology: A sneak preview (2005)
11. Chandran, R., et al.: Conflicting policy beliefs and informational complexities in designing a transboundary enforcement monitoring system. J. Clean. Prod. **105**, 447–460 (2015)
12. de Vries, W.T.: Towards a theory of metagovernance of land: fundamentals and prospects. J. Land Policy and Governance (JLPG) **1**(1), 26–36 (2018)
13. Atkins, M.T.: Creating age-friendly cities: prioritizing interventions with Q-methodology. International planning studies (2020)
14. Watts, S., Stenner, P.: Doing Q Methodological Research : Theory, Method and Interpretation, p. 238. Sage publications, London (2012)

Revolutionizing Oil Palm Replanting: Geospatial Terrace Design for High-Precision Ground Implementation Compared to Conventional Methods

Nursuhaili Najwa Masrol(✉), Nur Hafizah Mohammed, Nur Nadhirah Rusyda Rosnan, Vijaya Subramaniam, and Sim Choon Cheak

SD Guthrie Berhad, Telok Panglima Garang, Pulau Carey, Selangor, Malaysia
nursuhaili.masrol@simedarbyplantation.com

Abstract. Replanting in oil palm cultivation is essential for introducing new planting materials and improving infrastructure such as roads, drainage, terrace design, and planting density. This process is required every 25 years. Adopting a digital replanting blueprint is crucial for the Malaysian Oil Palm industry to tackle challenges like labor shortages and limited expertise. Planning for replanting should begin at least six months in advance. This study focuses on creating a precise replanting blueprint using advanced geospatial technology to maximize yield potential. The blueprint includes detailed designs for roads, drainage, terraces, and planting points, adhering to ARM standards, and undergoes statistical analysis for accuracy. Geospatial and remote sensing technologies, such as LiDAR and ortho imagery, are pivotal in this process. Tools like GIS and high-precision GPS instruments (EOS Arrow Gold and HIPER VR GNSS) ensure design reliability, with a variance of less than 0.5 m in most cases. The study finds that 65% of terraces are constructed at a 12-degree slope, exceeding the minimum requirement. The digital blueprint approach promises optimized land use, increased efficiency, sustainability, and cost reduction. Future efforts will focus on more sophisticated techniques and precision GPS devices to ensure accurate blueprint design and implementation, aiming for automation to save time, workforce, and costs in commercial production.

Keywords: Precision agriculture · Replanting blueprint · Oil palm · Geographic Information System (GIS) · Global Positioning System (GPS)

1 Introduction

1.1 Background

Oil palm cultivation is a vital component of the agricultural economy in tropical countries, particularly in Malaysia. The industry's sustainability and productivity depend on periodic replanting, typically required every 25 years. This allows for introducing superior planting materials and enhancing plantation infrastructure, including roads,

A. Abdul-Rahman et al. (Eds.): GeoWeek 2024, LNGC, pp. 283–299, 2025.
https://doi.org/10.1007/978-3-031-86654-8_22

drainage, and terraces. Replanting, however, presents significant challenges, such as labor shortages and the need for specialized expertise. Previous studies have highlighted the critical role of replanting in maintaining plantation productivity and sustainability (Smith et al., 2018; Lee & Chan, 2019).

The emergence of geospatial and remote sensing technologies offers innovative solutions to these challenges. Digital replanting blueprints, which integrate Geographic Information System (GIS) tools and remote sensing data, can facilitate precise and efficient replanting processes (Jones et al., 2020). By employing high-accuracy technologies like LiDAR for creating Digital Elevation Models (DEM) and ortho imagery for validation, it is possible to achieve detailed planning and implementation, maximizing the potential yield and optimizing land use (Rahman et al., 2021).

This study aims to utilize these advanced technologies to develop a high-precision replanting blueprint for the Malaysian Oil Palm industry. The blueprint incorporates ARM standards to ensure consistency and accuracy in design, covering essential aspects such as road, drainage, terrace design, and planting density. Planning for effective replanting should begin at least six months before the actual process to ensure all variables are carefully considered.

A study was conducted on one of SD Guthrie Berhad's estates (formerly known as Sime Darby Plantation). The Bukit Asahan Estate (Latitude: 2.393517, Longitude: 102.5460292), situated at the border of Melaka and Johor, was selected due to its terrain, which predominantly features high elevations. These conditions align well with the study's objective of achieving high-precision ground implementation of a digital replanting blueprint at the terrace area. Initially, a 48-hectare area of P00M was intended for the study. However, the focus was narrowed to a smaller area of 2.5 hectares within P00M, specifically for terrace and planting point design translation. This 2.5-hectare section was further divided into two areas (POC A and POC B), each covering approximately 1 + hectares, to facilitate easier monitoring during ground implementation (Fig. 1).

This study will discuss the methodology used to create and validate the replanting blueprint, the results of implementing this approach, and its implications for the industry. GIS and high-precision GPS instruments, such as the EOS Arrow Gold and HIPER VR GNSS, will be evaluated for their effectiveness in ensuring the blueprint's reliability compared to traditional methods. This research builds on previous work by demonstrating the practical benefits of integrating geospatial technologies into agricultural planning (Tan et al., 2017; Ali & Yusoff, 2018).

By embracing a digital blueprint approach, the oil palm industry can enhance efficiency, sustainability, and cost-effectiveness in replanting operations. Future initiatives will focus on refining these techniques and incorporating more sophisticated tools to ensure greater precision in the design and implementation phases.

1.2 Problem Statement

Oil palm replanting faces numerous challenges that hinder its efficiency and sustainability. A significant issue is the shortage of skilled labor, which limits the ability to perform replanting tasks effectively. Additionally, many plantations lack the expertise to implement advanced replanting techniques and technologies. Traditional replanting methods often lead to inefficient planning, resulting in suboptimal planting density,

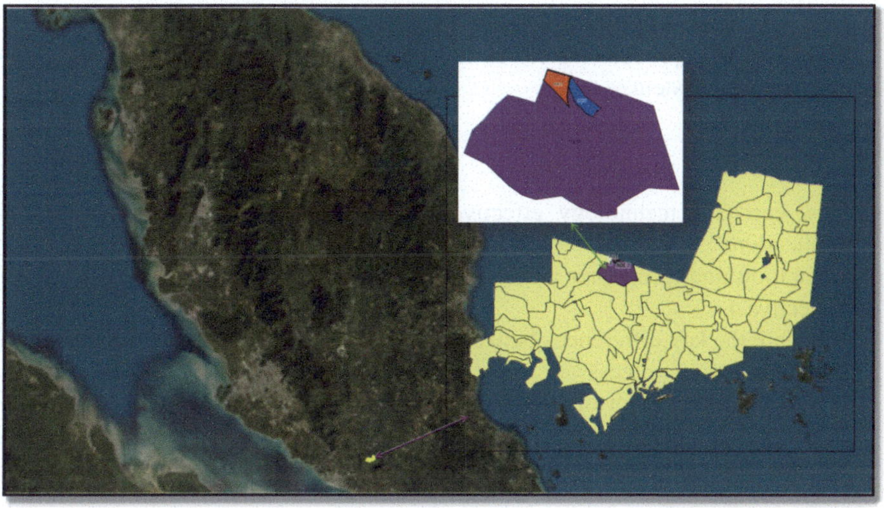

Fig. 1. Study area (Field P00M, Bukit Asahan Estate, Melaka)

poorly designed drainage, and inadequate road systems. This inefficiency can prevent the maximization of yield potential from replanted oil palms. Furthermore, conventional replanting methods are time-consuming and costly, negatively impacting productivity and profitability. Terrain challenges, such as designing terraces and planting points in hilly or uneven areas, add to the complexity without the aid of advanced geospatial tools. Environmental impacts, including soil erosion and other sustainability issues, can arise from poorly planned replanting efforts. Addressing these challenges requires innovative solutions like high-precision geospatial technologies and automated digital blueprints to improve the accuracy, efficiency, and sustainability of oil palm replanting.

1.3 Objective and Research Questions

This study aims to implement GIS technologies as an instrument in producing digital replanting maps for terrace areas by achieving a few key objectives: 1. Producing a precise blueprint based on ARM requirements 2. Improving the planting, road, and drainage system 3. Utilizing GIS technology in translating blueprint design into precise ground implementations 4. Comparing the feasibility of using GIS technology versus conventional methods for replanting.

The key objectives of this study will be achievable based on the research questions posed throughout this study. The research questions for each objective included:

- Objective 1
  - What are the specific requirements of ARM that need to be incorporated into the blueprint?
- Objective 2

  - How can the blueprint be designed to optimize the planting areas?
  - What design improvements can be made to enhance the functionality of the road and drainage system?

- Objective 3

  - How can GIS technology effectively translate blueprint design into ground applications?
  - What tools and techniques are suitable to be implemented?
  - Objective 4
  - What are the key differences between GIS technology and conventional methods in the replanting process?
  - How does the accuracy of GIS technology compare to the conventional methods?
  - What are the potential challenges and limitations of using GIS technology compared to conventional methods?

2 Materials and Methods

This study is segmented into three phases: Data Acquisition, Data Processing and Ground Translation, and Data Analysis and Verification. Advanced Remote Sensing and Geographic Information System (GIS) technologies have been implemented throughout all phases. In the Data Acquisition phase, Remote Sensing technology, specifically Light Detection and Ranging (LiDAR), is utilized for data collection. The Data Processing phase leverages the GIS technology to prepare, analyze, and process the acquired data and produce a proper design to translate on the ground using high-precision Global Positioning System (GPS) instruments. The third phase involves validating actual data translated on the ground with the pre-designed replanting blueprint. The overall process for this study is illustrated in Fig. 2:

2.1 Data Acquisition

The preparation of two types of data, vector (boundary, road, and drain layer) and raster (comprising LiDAR data and RGB drone imagery), is essential for this study. Both types of data acquisition are required. The vector data are obtained through ground survey methods, typically updated each time replanting is conducted within the study area. Vector data is critical for delineating the study area and ensuring all designs are constructed within the specific boundary. Vector data such as roads and drains are used as references in designing the new road and drain for the replanting area. The total hectarage of the study area serves as a reference for calculating the final stand per hectare based on the total number of planting points generated throughout this process. Drone-based LiDAR captures high-resolution images of the planting area before felling. This is accomplished using the MRT-300 drone equipped with an L1 LiDAR sensor to generate the Digital Elevation Model (DEM). The utilization of LiDAR is due to its capability to provide high-accuracy data, particularly for DEM. At the same time, RGB Drone surveys are used to verify the ground conditions after the felling process to ensure data accuracy and completeness (Figs. 3 and 4).

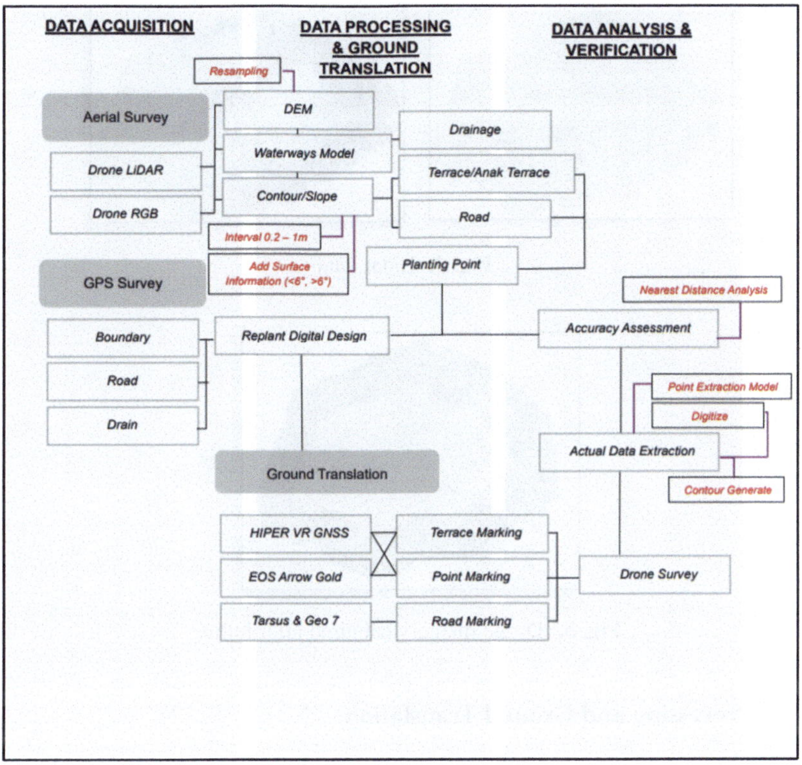

Fig. 2. Overall process for replanting blueprint

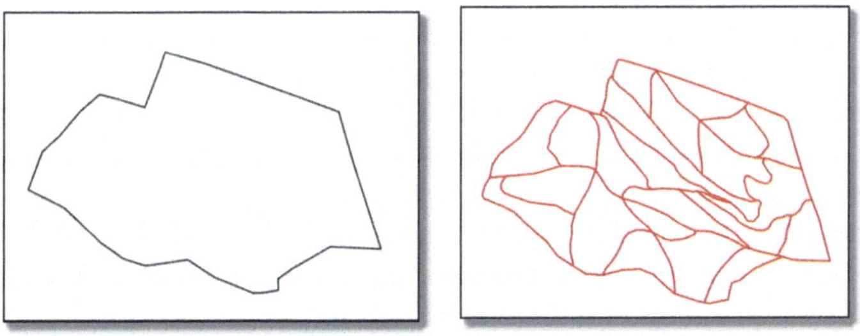

Fig. 3. Vector data (Boundary, Load)

For verification after felling, drone imagery has been captured to assist in precisely verifying the terrain and ground conditions. DJI Phantom 4 was used to capture these images, flying at an altitude of approximately 100 m with a Ground Sampling Distance (GSD) of about 0.5 cm. The images are then corrected with Ground Control Points (GCP) and undergo several processing steps to produce an orthophoto image, which will be used for the design process (Fig. 5).

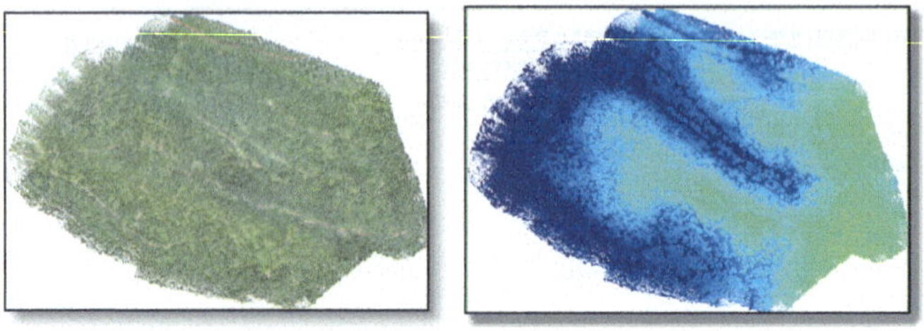

Fig. 4. Lidar data

Fig. 5. Drone imagery captured after felling

2.2 Data Processing and Ground Translation

This phase comprises two key elements: data processing and ground translation or applications.

Data Processing. The initial data processing phase involves generating an orthophoto and Digital Elevation Model (DEM). An orthophoto is a geo-referenced and coordinate-corrected aerial image that will be processed using Agisoft and DJI Terra software tools. The processed imagery and DEM are subsequently utilized in design creation using GIS Software, specifically ArcGIS Pro. Contour lines are derived from DEM data with interval values ranging from 0.2 to 1 m. These contours represent the available terraces within the area. Based on ARM requirements, the distance between two terraces and the distance between planting points should have a total value of 54ft. To meet the requirements, contours at every 24ft interval will be designated as terrace lines. In this study, contours and slope data serve as references for road construction to ensure the roads are not constructed on slopes exceeding 6 degrees and are spaced approximately 200 m apart. Upon finalizing both terrace and road, planting points will be generated based on the distance between each terrace. The stand per hectare (SPH) value for terrace area planting should meet ARM requirements of 148 SPH per hectare. The summary of data processing is illustrated in Fig. 6.

Digital Elevation Model. A Digital Elevation Model (DEM) represents the terrain's surface and is widely utilized in Geographic Information System (GIS) applications. In this study, the DEM was derived from LiDAR and drone imagery. This DEM was employed to generate slope and contour data for terrace areas. The contour information

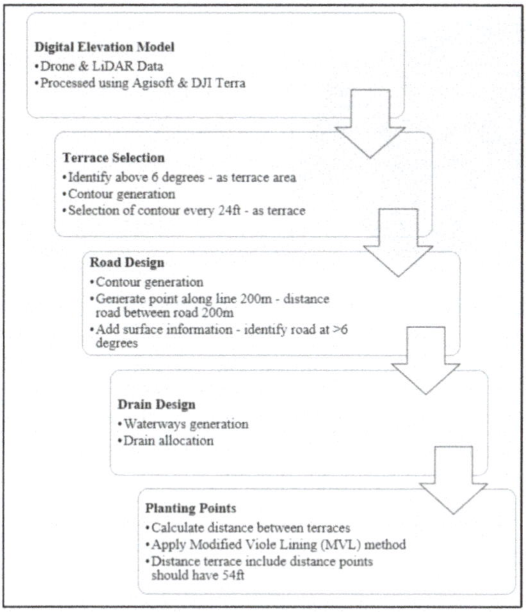

Fig. 6. Data processing steps

is crucial for selecting terrace lines identified from the highest to the lowest elevations (Fig. 7).

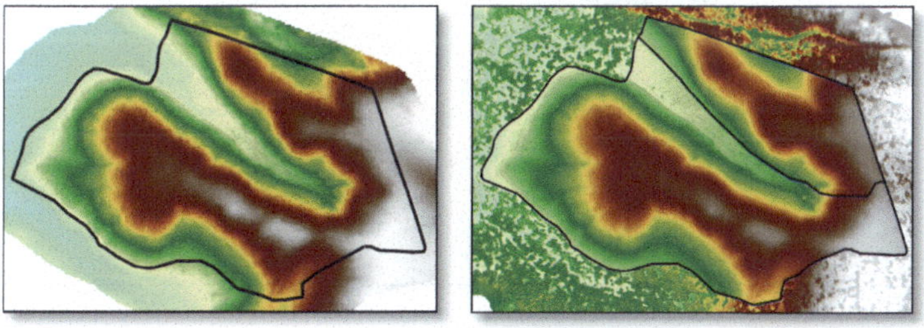

Fig. 7. Digital Elevation Model (DEM)

Terrace Selection. After deriving the contour from the DEM, terraces were delineated using a ruler and generated with a grid spacing of 24 feet. This 24-foot interval was specified based on ARM's requirement for terrace-to-terrace distance. The distance can be adjusted to 26 feet, contingent upon estate consent, to accommodate machinery usage requirements. Several rulers were generated and assigned to each high-elevation point, where the selection of terraces commenced (Fig. 8).

Road Designing. Road design will be constructed by considering two elements: slope degree and distance to the road. The distance between roads should be within 200 to 230 m, while the slope should not exceed 6 degrees. Generation of points along lines with a

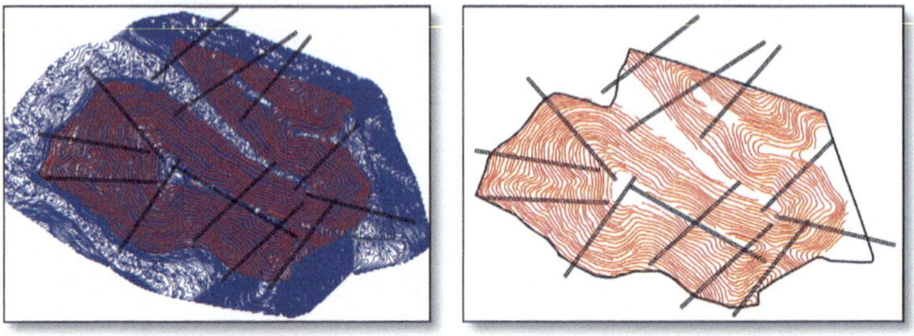

Fig. 8. Implementation of rulers for terrace selection.

200-m distance was implemented using selected terrace lines. Points at 200-m intervals were further analysed by adding raster information to determine the slope degrees. Points located at 200 m with slopes less than 6 degrees were selected to create the road, with reference to the existing road in the area (Fig. 9).

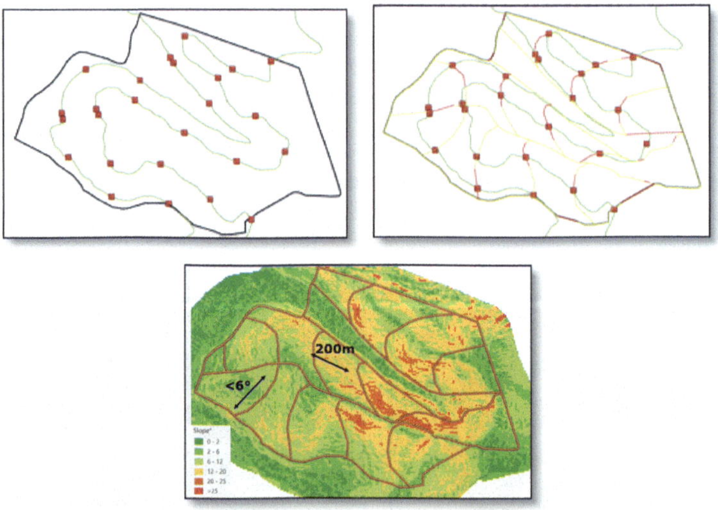

Fig. 9. Road designation using 200 m distance and slope of more than 6 degrees.

Drain Designing. The design of the drainage system begins with generating the waterways model from DEM data. The waterways are produced through spatial analysis using a raster calculator to create the stream network. Waterways illustrated by the waterways model and terrain analysis were used as a reference for creating a suitable drainage system within the area (Fig. 10).

Planting Points. Planting points were generated for two distinct terrain areas: terraces and flat areas. For flat areas, points were generated using a grid system with an inter-point distance of approximately 7.32 m. In terrace areas, the Modified Viole Lining (MVL) method was employed to create points, ensuring that the distance between points and terraces totals 54 feet. The distance between points varies according to the distance

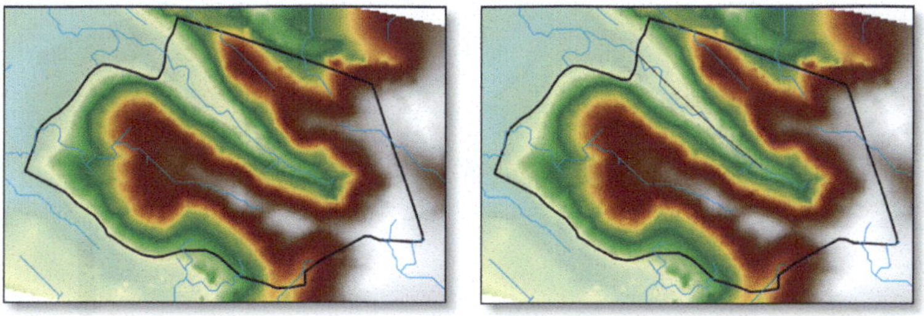

Fig. 10. Waterways model for drainage system design

between terraces. For example, if the distance between terraces is approximately 24 feet, the distance between points should be around 20 feet. Additionally, the location of each point should be 3 feet from the terrace wall (Fig. 11).

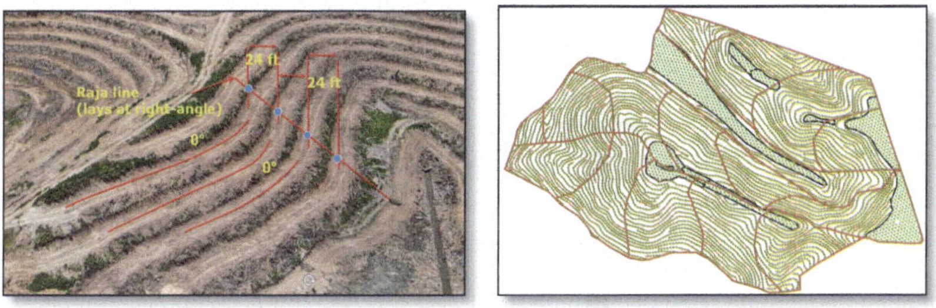

Fig. 11. Modified Viole Lining (MVL) methods for planting points allocation.

Data processing is mainly on the designation of a digital replanting blueprint that begins with DEM generation and continues with slope generation, contour generation, terrace selection, road designing, and planting point generation. The whole process can be summarized in Fig. 12.

Ground Translation. Upon completing the data processing phase, the design will be utilized in the subsequent step, which is ground translation. Ground translation is conducted to assess the reliability and accuracy of the design on the ground, which will then be compared with conventional methods used in the replanting process. This phase also aims to evaluate the compatibility of GPS in providing enhanced accuracy on the ground and identify environmental factors affecting the translation process. This phase comprises several methods: road, terrace, and point marking. During this phase, high-accuracy GPS instruments were utilized, including the HIPER VR GNSS, Geo 7, and EOS Arrow Gold, each with an accuracy range from 0.3 meters to 0.7 meters (Fig. 13).

Road Marking. Data translation on the ground began with road marking. Line data from the design was converted into points at 100-m intervals to expedite the marking process by placing pegs at each generated point. This marking process took approximately two days to complete 48 hectares using Geo 7 instruments. Road marking is the

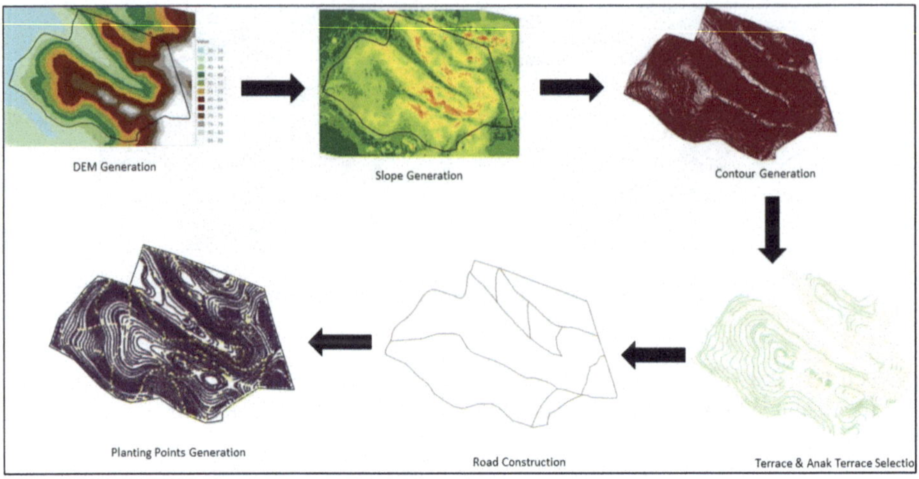

Fig. 12. Replanting blueprint design summary

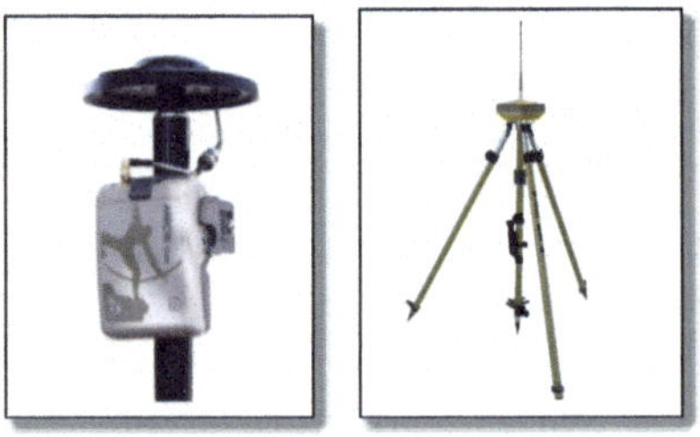

Fig. 13. GPS Instrument for ground marking (EOS Arrow Gold and HIPER VR GNSS)

most critical step in initiating road construction. Completing road construction facilitates subsequent steps, including terrace marking, terrace construction, and planting point marking.

Terrace Marking. After the road was constructed, ground translation proceeded with terrace and planting point marking. These markings, covering about 2.5 hectares, were completed within one week. Two GPS instruments, HIPER VR GNSS and EOS Arrow Gold, were utilized during this marking.

Planting Points Marking. Planting point marking was applied right after the terrace construction was completed. Approximately 245 points were marked over 2.5 hectares using the same instruments: HIPER VR GNSS and EOS Arrow Gold. Pegs were installed at each planting point established during the data processing–designing process.

The data application and ground translation process are important in verifying the suitability of replanting digital design on the ground. Depending on the ground environment, some minor adjustments might happen during this process. The completion of planting point marking will lead to final verification by the estate before the data is captured for further analysis and result generation. The ground translation took around two weeks to complete an area of about 2.5 hectares, while it needed about two weeks for the machine to complete the construction. The completion time for marking and construction might vary depending on the total area of study.

2.3 Data Analysis and Verification

This phase is the final stage of this study, where the design implemented on the ground will be captured through drone imagery for comparison. The nearest distance analysis tool has been employed to assess the accuracy between blueprint design and actual ground data, which will be detailed in Sect. 3.1.

3 Result and Discussion

Results and a comprehensive analysis were conducted for this study. Finalized results were achieved through extensive discussions with team members, estate management, and other relevant departments to gather additional insights and expertise.

3.1 Result and Discussion

The replanting design should be prepared before the completion of land preparation. Initially, two designs were created based on different data sources: Lidar and drone data Fig. 14. Each data had the same process and produced roads, drains, terraces, and planting points. Subsequently, several discussions were held with estate management to determine which blueprint design would be employed for the on-ground implementation. Design from drone imagery was chosen as the final design output for the ground implementation. The design has also been separated into two small portion areas known as POC A and POC B, as mentioned in Sect. 1.1 (Fig. 15).

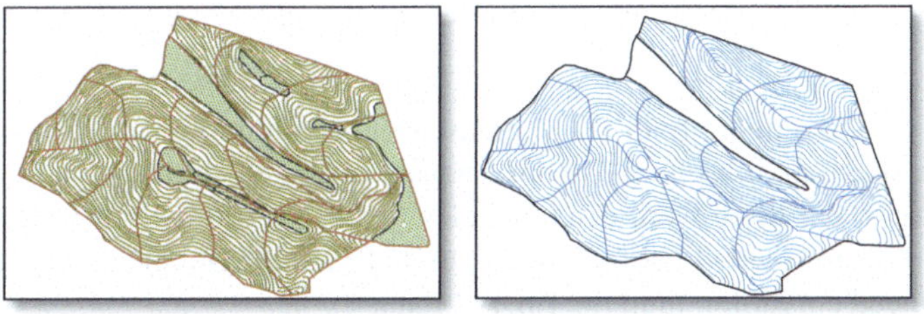

Fig. 14. Blueprint—LiDAR data (left) and blueprint—drone (right)

Once the construction is completed, drone images are acquired to verify the study. The acquisition images cover 48 hectares, and data roads, terraces, and planting points are extracted for verification and comparison (Fig. 16).

This study utilized Digital Terrain Model (DTM) data types: LiDAR and drone data. LiDAR data was initially employed to acquire precise terrain information, leveraging its ability to penetrate beneath palm canopies. Subsequently, during the replanting phase, another terrain dataset from drones was obtained after the post-clearing of palm

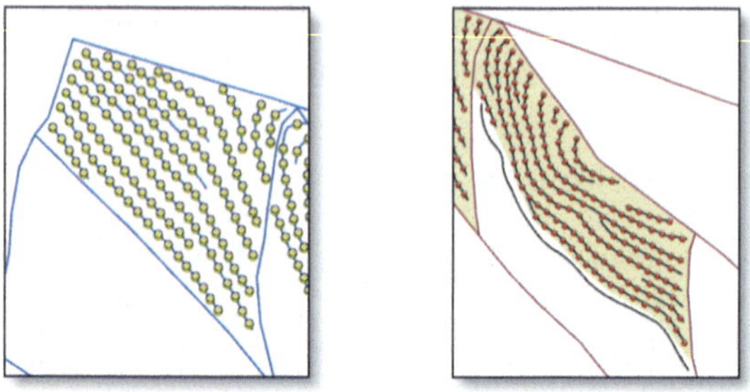

Fig. 15. Blueprint—POC area A & POC area B

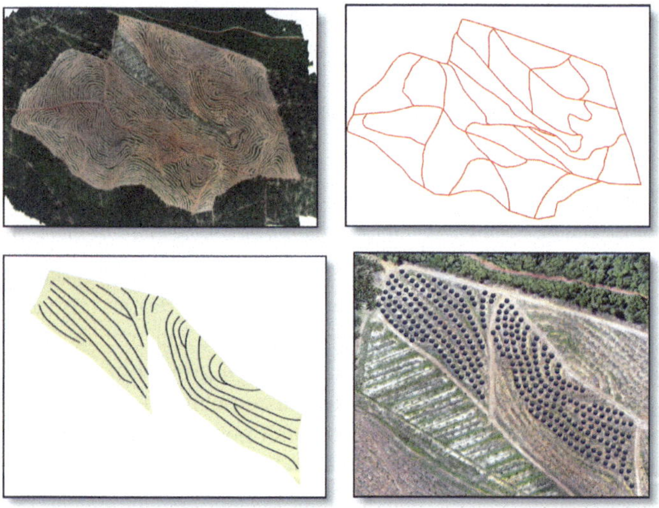

Fig. 16. Drone images with generated roads and terraces.

trees to verify and compare both DTM datasets. The contours from each dataset were compared, as illustrated in Fig. 17. Both datasets exhibited similar contour patterns in undulating areas, but there were discrepancies in steep terrain. Upon comparison, the drone-derived data provided better data due to its finer topographical details, including more pronounced curvature, than the LiDAR data.

Figure 18 shows the results of land preparation for field replanting in P00M (new field 2023C). Based on the result, the actual roads for the 48 hectares constructed were about 14,129 m, while the road designed in the blueprint was only about 8,266 m. This obvious difference is due to the decision of estate management to maintain the existing road due to cost and time constraints. The 2.5 hectares of POC areas were constructed with roads as designed in the blueprint. Therefore, the differences between the design and actual implementation seem minimal. The differences between the design and data can be further observed in Tables 1 and 2.

Tables 1 and 2 provide details of terraces, roads, and planting points for the former field, blueprint, and actual field data. No analysis is required for roads, as the actual

Fig. 17. Undulating area (left) and Steep area (right)

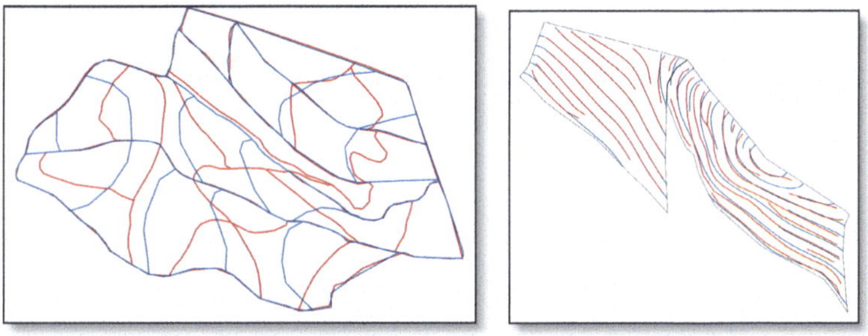

Fig. 18. Overall blueprint (blue) compared to actual (red)

Table 1. Comparison table between ex-field, blueprint, and actual planting for 48 ha area.

2023C – 48 Ha	Ex Field (00m)	Blueprint	Actual
Terrace length (m)	50,098	44,950	47,266
Road length (m)	10,409	8,266	14,129
Planting points	7,052	6,388	6,240

Table 2. Comparison table between ex-field, blueprint, and actual planting for POC 2.5 ha area.

POC—2.5 Ha	Ex Field (00m)	Blueprint	Actual
Terrace length (m)	3,058	2,211.3	2323.7
Road length (m)	6,992.2	1,259.5	2763.9
Planting points	370	249	325

roads do not conform to the blueprint, and the difference between results exceeds 50%. In the Proof of Concept (POC) area, the actual road distances increased by 1504.4 m, while for the whole field of 48 hectares P00M, the actual road distance increased by

5863 m. For terraces, it shows low variances of terrace distances, which are only 112 m for POC areas and 2316 m for the whole 48 hectares when compared to the blueprint.

Using the Near Distance geoprocessing tool, the distances between the terraces in the blueprint and the actual field were calculated, as shown in Table 3. The analysis indicates that in POC Area A, the terraces closely align with the GPS markings, with distance variances mostly under 0.5 m. However, the distances in Area B range from 0 to 6 m. The average variance between the blueprint and actual distances is approximately 1.16 m.

Table 3. Statistical table for near distance analysis between blueprint design and actual.

Near distance (All)	%	Near distance (POC A)	%	Near distance (POC B)	%
0.0–0.5	52.67	0.0–0.5	93.85	0.0–0.5	21.18
0.5–1.0	8.33	0.5–1.0	0.77	0.5–1.0	14.12
1.0–1.5	6.33	1.0–1.5	1.54	1.0–1.5	11.18
1.5–2.0	6.67	1.5–2.0	1.54	1.5–2.0	10.59
2.0–2.5	5.00	2.0–2.5	1.54	2.0–2.5	7.65
2.5–3.0	6.00	2.5–3.0	0.77	2.5–3.0	9.41
3.0–3.5	5.67	3.0–3.5	0.00	3.0–3.5	9.41
3.5–4.0	4.67	3.5–4.0	0.00	3.5–4.0	8.24
4.0–4.5	3.00	4.0–4.5	0.00	4.0–4.5	5.29
5.0–5.5	0.67	5.0–5.5	0.00	5.0–5.5	1.18
5.5–6.0	0.67	5.5–6.0	0.00	5.5–6.0	1.18
6.0–6.5	0.33	6.0–6.5	0.00	6.0–6.5	0.59

Both the POC A and POC B slope and terrace areas are compared in this study. The slopes in both areas vary from 0 degrees to over 26 degrees. According to ARM standards, areas with slopes greater than 6 degrees are necessitated for terrace planting. However, 100% of terrace planting was carried out in both POC areas. This can be attributed to the fact that both areas have ranges of slopes up to 6 degrees slope and about 50% of the area has more than 6 degrees slope. In POC A, 94% of the distance differentials fall within the 0 to 0.5 m range, and approximately 65% of the slopes are at 12 degrees, with none exceeding 20 degrees. Conversely, in POC B, 21% of the distance differentials are within the 0-to-0.5-m range, and 14% fall within the 0.5 to 1-m range. Approximately 57% of the slopes are 12 degrees, but the range extends to 25 degrees.

This underscores the strong connection between steeper slopes and the heightened challenges associated with constructing new terraces, underscoring the critical importance of safety and land conditions in such scenarios. Figure 19 shows the slope comparison for both areas. In the future, terrain slope will be considered for the new blueprint, where more than 20 degrees of slope will follow existing terraces mainly for safety issues.

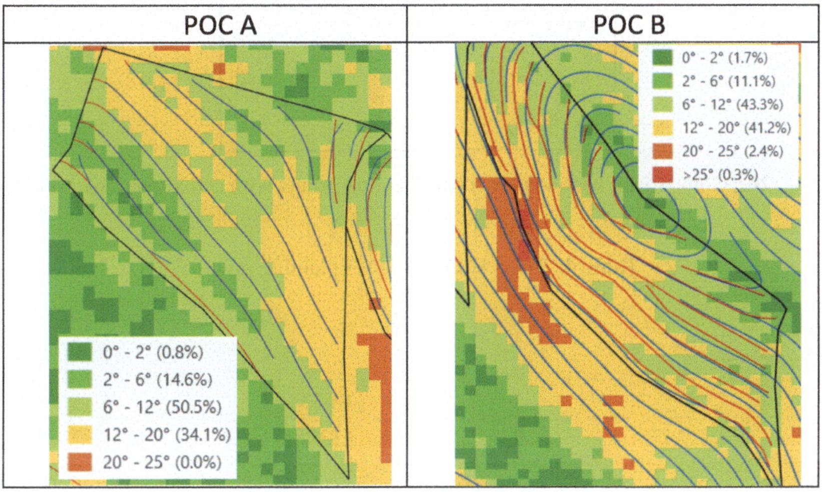

Fig. 19. Slope comparison between POC area

Figure 20 and Table 4 present the actual planting points analyzed compared to the blueprint. The generated blueprint points have been marked with accurate GPS devices on the ground. However, since the generated points were located at the center of the terrace line, adjustments had to be made by the estate personnel, as the back slope of the terraces was not considered; as in the ARM, the planting points are located 3 feet from the back slope. The average distance of points adjusted is about 1.6m, with 17.54% within 1 m and about 50% within 2 m radius. An additional 14 points were detected and compared to the blueprint, as was done by the estate personnel after finalizing the ground points.

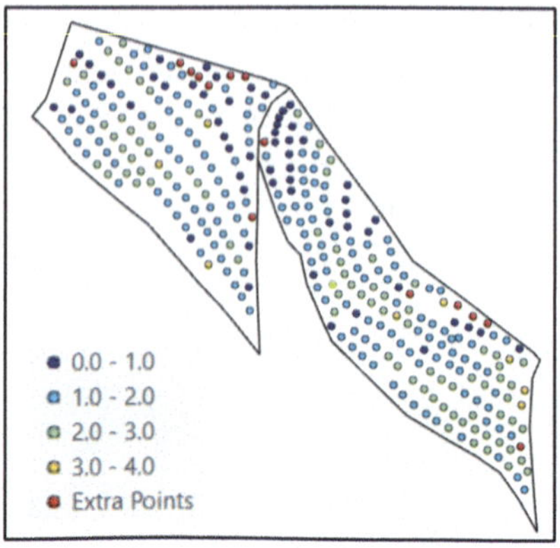

Fig. 20. Comparison between blueprint planting points with actual planting points

Table 4. Nearest Distance for planting points

Near distance (planting points)	%
0.0–1.0	18.33
1.0–2.0	51.77
2.0–3.0	27.33
3.0–4.0	2.57

4 Conclusion

Blueprint replanting represents a promising approach for optimizing land use in agriculture and related fields. Leveraging advanced technologies and precision planning offers numerous benefits, including increased efficiency, improved suitability, and cost savings. Field P00M, or the new replanting field known as 2023C, has been chosen as the study area, and this field case study demonstrated that the practical application of this technique can lead to substantial improvements in agricultural outcomes. To further enhance its effectiveness, future efforts will be required to refine the methodology and integrate it with high-precision GPS devices in an upcoming blueprint replanting study. This strategic evolution aims to ensure accuracy in both the blueprint design phase and the implementation on the ground.

In conclusion, this study successfully fulfilled the objectives outlined at the beginning and provided reliable answers to the research questions. It has demonstrated the application of GIS in designing an effective blueprint based on the standards introduced by ARM. These requirements served as a guideline throughout the replanting blueprint design process. The spatial and non-spatial data obtained during the study benefited estate personnel, enabling them to maximize the design of roads, drainage systems, and planting points for future replanting efforts.

The replanting blueprint acts as a guide for estate personnel to help them understand and identify the best plan for replanting, especially to optimize the functionality of each element in the area. Various spatial analyses were utilized in this study, including the nearest distance analysis, to compare the blueprint design with actual ground implementation. The comparison results inevitably showed some variance due to adjustments made on the ground due to terrain conditions and communication issues among different parties.

GIS technology has significantly streamlined the replanting process by providing various analyses and data without limitations. In contrast, conventional methods required manual ground measurements to determine distances and locations and to mark planting points. This often led to errors due to insufficient measuring instruments or reliance on worker experience. GIS offers high-accuracy information, enabling precise replanting and minimizing errors.

Future advancements in GIS for replanting blueprint design could include automated methods, particularly for implementing the Modified Viole Lining (MVL) technique, which is currently done manually. Additionally, GIS could be further enhanced by integrating automated machinery to navigate points on the ground and auger for planting points. This study has demonstrated the reliability of GIS in providing precise ground implementation of replanting blueprints. However, further improvements are needed to advance digital replanting methods, ensuring they are widely adopted across estates. This could reduce dependence on foreign labor and address future labor issues.

References

1. Smith, J., Brown, R., Johnson, T.: The role of replanting in sustainable oil palm cultivation. J. Agric. Sci. (2018)
2. Lee, H., Chan, K.: Challenges in oil palm replanting: a review. Agric. Forestry (2019)
3. Jones, M., Patel, S., Kim, J.: Integration of GIS and remote sensing in agricultural planning. Geospatial Res. J. (2020)
4. Rahman, A., Abdullah, M., Singh, P.: LiDAR and ortho imagery for precision agriculture. Remote Sensing Appl. (2021)
5. Tan, S., Lim, A., Wong, Y.: Advances in geospatial technologies for oil palm plantations. GIS and Remote Sens. (2017)
6. Ali, N., Yusoff, Z.: High-precision GPS in agricultural replanting. Precision Agric. (2018)

Creative Mapping of a Running Route in Urban Park Settings: A Pilot Study at Marina ParkCity Miri Sarawak Malaysia

Erica Benya Billy[1,5], Abdul Rauf Abdul Rasam[1,2](✉), Cassendra Gilbert[3], and Rafeah Legino[4]

[1] School of Geomatics Science and Natural Resources, College of Built Environment, Seattle, US
rauf@uitm.edu.my
[2] Malaysia Institute of Transport (MITRANS), Shah Alam, Malaysia
[3] Faculty of Sport Science and Recreation, Shah Alam, Malaysia
[4] College of Creative Arts, Universiti Teknologi MARA, 40450 Shah Alam Selangor, Malaysia
[5] Hurricane Geo Inspection Survey Sdn Bhd (HGIS), Kuala Belait, Darussalam, Brunei

Abstract. This study aims to design a thematic map for Marina ParkCity, Miri, which needs a creative running route map using the GIS cartographic method. The proposed map was created using geospatial methods and tested for the map's effectiveness and practicality. The Miri City Council (MCC) recommended using the map from the Berlin Marathon 2010 to guide the map's creation. However, the Marathon map still needs improvement in cartographic elements and design principles such as legend, scale, and visual contrast. The respondents also agreed that the proposed map should be developed in the park, as it meets their criteria for a decent map design and practical needs, particularly for enhancing the runner experience and safety in the park. The study also recommends displaying distance and elevation information on the map for engaging and safe exploration.

Keywords: Cartography · GIS · Mapping

1 Introduction

Sport is a physical activity that involves skills and competition. It is crucial for the National Sports Day in 2015 to promote athletics globally. The Sarawak state government announced 13 new sports projects, including a high-performance sports centre at Petra Jaya [1]. These projects aim to provide integrated sports science components for local athletes, including psychology, nutrition, strength and conditioning, performance analysis, recovery, rehabilitation, and planning. Sports benefit the body by providing cardiovascular fitness, bone health, lower obesity risk, better sleep, and enhanced coordination and balance [2, 3]. However, the absence of a designated jogging path increases the risk of injury for users, necessitating the creation of a safe jogging route map in several parks in Malaysia, including Sarawak. A jogging or running route refers to a predetermined path or course individuals follow when they engage in jogging or running

A. Abdul-Rahman et al. (Eds.): GeoWeek 2024, LNGC, pp. 300–307, 2025.
https://doi.org/10.1007/978-3-031-86654-8_23

activities for exercise, training, or recreational purposes [4, 5]. Planning a route is essential to running because it affects the distance, terrain, and overall running experience [6]. Some critical concepts associated with jogging or running routes include distance, terrain, elevation, safety, scenery, accessibility, loops vs. out-and-back, tracking and GPS, variety, and social considerations.

Many runners use GPS devices or smartphone apps to track their routes, distance, pace, and other metrics [7]. These tools can help users plan and optimize running routes and monitor progress over time. Variety is essential to prevent boredom and plateauing in fitness since having several running routes in your repertoire is a good idea. Social considerations must also be considered while enjoying running with others, such as routes that are popular among local running groups or have designated running paths [8]. Social running can be more motivating and enjoyable. Therefore, creating and selecting suitable geospatial-based jogging or running route guidelines is a personal and practical decision that depends on fitness goals, preferences, and local resources. Whether a user, a novice, or an experienced runner, carefully planning your routes can enhance your running experience and help you achieve your fitness objectives. GIS mapping technology can precisely measure distance and elevation changes along the route, besides identifying safe routes by highlighting well-lit areas, well-maintained paths, and low-traffic zones. The main objectives of this study are to review the effectiveness of the existing jogging route map and create a tested jogging route map using cartography and GIS approach in Marina ParkCity Miri. GIS provides data and analytical tools for generating running routes or tourism maps, while cartography ensures visually appealing, easy-to-understand, and user-friendly results, as conducted by previous studies [10–13]. This technology combination will establish the foundation for mapping and analysis used in research and almost every industry. Users may use GIS mapping to better understand patterns, connections, and their location's context and enhance communication, productivity, and decision-making [14]. An ideal jogging or running route map will also improve the running experience, promote safety, and contribute to a healthier and more active community. It also offers valuable tools for event organizers to plan and execute successful running events.

2 Methodology

2.1 Study Area

Marina ParkCity, Miri, was chosen because of its benefits as an investment and sporting destination (Fig. 1). It is a coastal city in Sarawak. Miri is known for its natural beauty and rich cultural heritage. It offers a wide range of recreational and tourism activities for visitors, including beaches, national parks and nature reserves, cultural attractions, adventure and outdoor activities, island hopping, shopping and night markets, festivals and events, and water activities. Park tourism in Miri also immerses visitors in the region's lush rainforests, unique wildlife, and stunning landscapes. This activity requires an attractive running route map to be produced effectively for promotion and documentation purposes.

Fig. 1. Marina ParkCity, Miri, Sarawak as the study area (*Source* Google Map, 2024)

2.2 Research Framework

Creating an interactive map in ParkCity Miri involved three phases (Fig. 2). The preliminary research is the first stage, focusing on the study's context and data accessibility. It involved reviewing the existing map and potential tools, data, and techniques. The study created a jogging route map using GIS cartography, base maps, jogging information datasets, and multimedia data collected. The primary data used in the study are base maps, jogging info datasets, and multimedia data, including graphical maps, location, building information, and route details (Table 1). In ArcGIS, the acquired data was used as the base map. Then, the base map was added to the ArcGIS online for map production. Next, data processing and map-making involved editing, management, classification, generalisation, and symbolising the cartographic features using ArcGIS and Adobe Illustrator. The map was then published through the ArcGIS service. Each phase is crucial in creating the final map of the suggested running route. Symbolising, for instance, visually marked the jogging map by assigning attributes like shape, orientation, colour, and size to represent geospatial data. While for generalization or simplification, the data should be simplified to make the map more visually appealing. As a result, the existing line path in the map can be digitized to reduce complexity while keeping the path's originality. The online platform was used through the Google form for map testing by ten random respondents. 70% of the respondents are 18 to 30, while 20% and 10% are below 18 and 31 to 40, respectively. 60% of respondents were students, while four respondents, 40%, worked in the government sector.

3 Result and Discussion

3.1 Identifying the Effectiveness of the Existing Jogging Route Map

Some existing maps were reviewed to evaluate the effectiveness of the current jogging route map (Fig. 3). It was found that there is no specific concept for a jogging route map in Malaysia. However, a few marathon maps are available on platforms such as MyPacer and MapMyRun. The study emphasizes the importance of cartography and good map design in creating a practical jogging route map [15–18] (Fig. 4). The ideal cartographic content and principles include visual contrast, legibility, figure-ground, hierarchical organization, and balance [19, 20]. For example, visual contrast refers to

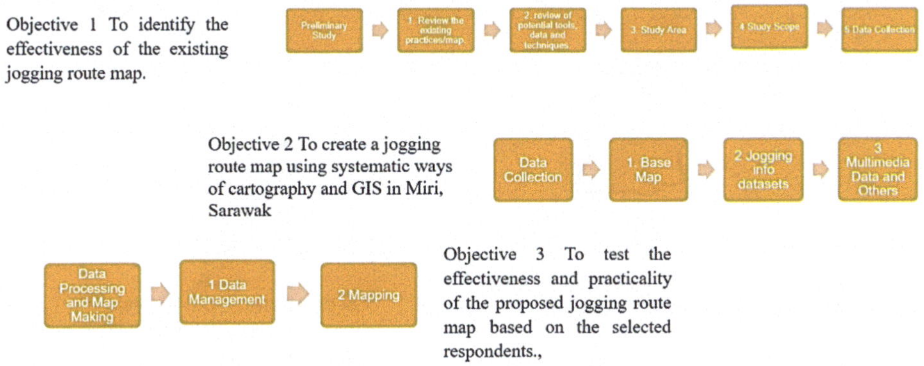

Fig. 2. Research framework

Table 1. The primary data and sources used in the study

No	Data	Sources
1	Graphical map of Miri, Sarawak	Miri City Council official websites
2	Location of Miri, Sarawak	Google maps Open street maps Google Earth
3	Building information Name	Google maps Open street maps ArcGIS online
4	Route Name Length The junctions	Google maps Open street maps ArcGIS onlizne Google earth

how page items and map features contrast with one another and their backgrounds. In cartography, this idea of visual contrast is also applicable.

3.2 Designing and Testing the Proposed Jogging Route Map in the ParkCity

Selected experts evaluated a final running route map before finalizing the proposed map (Fig. 5). The initial map (A) was based on the Berlin Marathon route. Still, it required enhancements in cartographic elements such as the north arrow, legends, and other aspects, as indicated in the revised Map (B). The finalised proposed map (C) was developed after considering criteria for good map design and expert opinions, as the initial map appeared overly generalised and lacked essential features in Map (A). As a pilot study, the final map (C) meets design criteria, including visual hierarchy, element layout, symbolization, and typography lettering. This proposed map consists of essential elements of map layout (title, legend, north arrow, scale, and map body). It also fulfils the criteria of a quality or practical map.

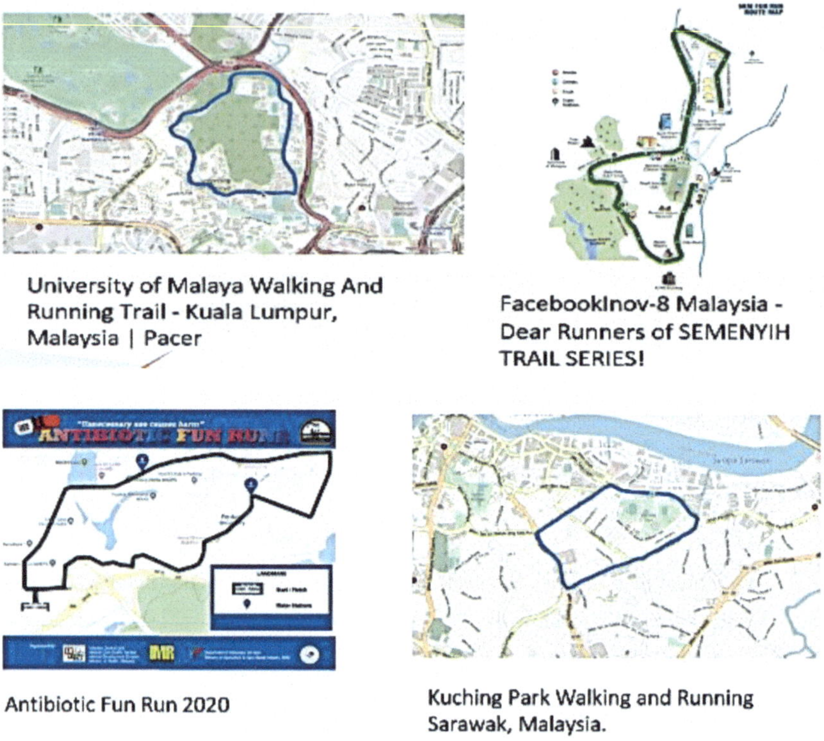

Fig. 3. Existing jogging maps in Malaysia

Map Elements based on ESRI	Qualities of a good map	An example of a good map:
Converting Map Elements to Graphics	Clear idea of what the map is trying to demonstrate	Clear and easy to understand
North Arrow	Easy to interpret legend or key	Colour contrast
Scale Bars	Good use of symbols to demonstrate a clear point	Symbols are simple and easy to understand/interpr et
Scale Text	Good sources/metadata	Simple style
Legends	Clear features, not overcrowded	Not overcrowded
Frames		

Fig. 4. Criteria of a good map design for running route map of Marina ParcCity, Miri

The survey revealed that 90% of respondents agreed with the title concept and the map's beauty, structure, and design. However, 60% preferred improving the web system

Fig. 5. Processes for final map production of the running route in the ParkCity, Miri.

to increase tourism park spot popularity, and 40% preferred enhancing map information such as legend and color selection (Fig. 6). Designing a jogging map with thoughtful cartographic principles and smart GIS techniques makes routes more straightforward, safer, more accessible, and visually appealing, making it easier and more enjoyable for users to choose and follow the best paths [20].

The proposed map benefits fitness enthusiasts, allowing users to plan and explore different routes safely. It also provides variety and motivation, allowing users to plan and explore different routes. This variety can keep your workouts interesting and prevent boredom because users are more likely to stay motivated and committed to their jogging routine when they have different routes. Users can also easily track their progress,

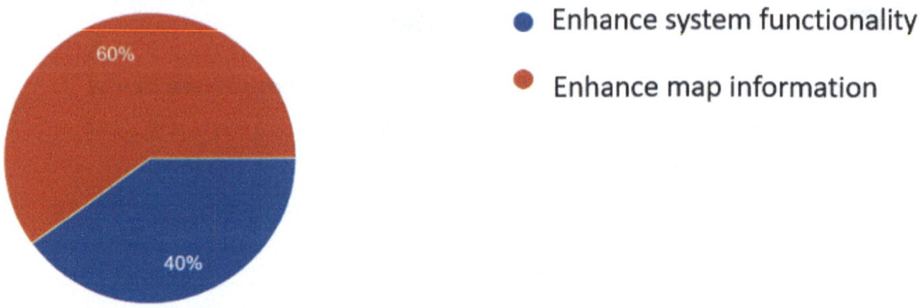

Fig. 6. Understanding the respondents' perspectives on enhancing the park's public popularity or usage.

including monitoring route distance, pace, and time. Tools like GPS watches or fitness apps can help users gather data and assess their improvements [7]. For safety and aesthetics, where a jogging route is planned, users can also select safe and well-lit paths, minimising the risk of accidents or encounters with potentially dangerous situations [9].

4 Conclusion

The absence of a designated jogging route may increase the likelihood of users utilising the road for their daily jogs. As a result, a creative and safe jogging route with relevant information to attract more people was created in Marina Park City in Miri Sarawak, allowing users to plan and use the map independently. The map was also created based on the user's requirements, especially in cartographic elements and design principles. This creative map aids enthusiasts by allowing them to plan and explore various itineraries safely. Incorporating distances and elevation with advanced GIS planning tools could also be applied in future studies for a better map user experience, safety, and user-friendly engagement for the public.

Acknowledgement. The authors acknowledge UiTM Shah Alam for financially sponsoring this research (KEPU—600-RMC/KEPU 5/3 (015/2021)). We also appreciate the Miri City Council (MCC) for valuable suggestions on the proposed map.

References

1. The Borneo Post: The quest to make Sarawak a sports powerhouse by 2024. https://www.theborneopost.com (2019)
2. Eubank, M., Nesti, Mark., Cruickshank, A.: Understanding high-performance sport environments: impact for the professional training and supervision of sport psychologists. Sport & Exercise Psychol. Rev. (2014)
3. Sarawak State Legislative Assembly. Sarawak State Legislative Assembly Official Reports and Debates. https://duns.sarawak.gov.my (2024)
4. Kostrubala, T.: The joy of running: an unconventional approach to distance running. Track & Field Q. Rev. **71**(4), 267–271 (1971)

5. Miller, M.J., Booth, F.W.: Running: the role of distance, terrain, and safety. J. Sports Sci. Med. **8**(3), 567–573 (2009)
6. Higdon, H.: Run fast: how to beat your best time every time. Rodale Books (2000)
7. Fokkema, T., Knobbe, A., De Koning, J.J.: The effectiveness of a smartphone app for injury prevention in recreational runners: a randomized controlled trial. J. Sci. Medicine in Sport **20**(9), 897–902 (2017)
8. Shipway, R., Jones, I.: Running together: the impact of running groups on social identity and community development. Int. Rev. Sociol. Sport **47**(3), 349–364 (2012)
9. Perry, M.J., Jackson, R.E.: Applications of GIS in urban safety and running path planning: analyzing distance, elevation, and environmental factors. J. Urban Plan. Dev. **140**(4), 1–9 (2014)
10. Mustapha, N.I., Rasam, A.R., Saraf, N.M., Idris, R., Wakijan, A.: Cycling route mapping via cartography and GIS techniques. IOP Conf. Ser.: Earth Environ. Sci. **1240**(1), 012008 (2023)
11. Sha'aban, S.N., Abdul Rasam, A.R., Asmat, A.: A 3D GIS mapping for Landscape Geo-visualisation and park management: a case study in Kuala Keniam Station, Taman Negara Malaysia. IOP Conf. Ser.: Earth Environ. Sci. **685**(1), 012010 (2021). https://doi.org/10.1088/1755-1315/685/1/012010
12. Mohd Hasmizi, N.S., Abdul Rasam, A.R., Mohamed Saraf, N.: E-mosque tourism: discovering mosques via geodigital mapping in Kuala Lumpur, Malaysia. Charting the Sustainable Future of ASEAN in Science and Technology, pp. 507–517 (2020). https://doi.org/10.1007/978-981-15-3434-8_44
13. Othman, H., Abdul Rasam, A.R., Jaini, N.: Geopark management and GIS: geospatial tree information inventory system. Charting the Sustainable Future of ASEAN in Science and Technology, pp. 553–567 (2020)
14. Goodchild, M.F.: GIS and public policy: improving decision-making and management through geographic information systems. Int. J. Geogr. Inf. Sci. **22**(6), 877–890 (2008)
15. Christophe, S., Hoarau, C.: Expressive map design based on pop art: revisit of semiology of graphics? Cartographic Perspect. **73**, 61–74 (2012)
16. Gisgeography. 33 Map Elements to Include in Cartographic Design: A 'How to' Guide to Map Making (2015)
17. Field, K.: Cartography. Esri Press, The Definitive Guide to Making Maps (2018)
18. Aileen, B.: Design principles for cartography. https://www.esri.com (2011)
19. MacEachren, A.M.: Visualizing uncertainty in geographic information. Cartographic Perspect. **13**, 10–19 (1992)
20. Slocum, T.A., McMaster, R.B.: A cartographic framework for the evaluation of map design: visual contrast, legibility, and figure-ground. Cartogr. Geogr. Inf. Sci. **31**(2), 99–113 (2004)

Dilemmas and Contradictions in Managing Integrated Land and Water Rights, Restrictions and Responsibilities

Walter Timo de Vries[1](✉), Rizqi Abdulharis[2], and Alfita Puspa Handayani[2]

[1] Technical University of Munich, Munich, Germany
wt.de-vries@tum.de
[2] Bandung Institute of Technology, Bandung, Indonesia
{rabdulharis,alfita}@itb.ac.id

Abstract. In the context of developing a framework for marine governance based on transdisciplinary research supporting the Global Ocean Decade, and to contribute to the development of a Marine Administration Domain model (MADM), there is a need to understand the differences between managing land and water and constructing relevant information models to support this. Given this need, this study aims at generating insights in how which conceptual and practical challenges, opportunities, dilemmas and contradictions exist and/or emerge when converting conventional monodisciplinary management and information management practices to integrated land and water (information) management. We formulate a set of key principles which guide the evaluation of transdisciplinary challenges. We approached the description of the problems and challenges in both land and water management from 5 perspectives: spatial /administrative domain modelling, participation and relevant actors, ontological and conceptual definitions, legislative formulations, practices, contradictions and redundancies, and practical case-based challenges, problems and inconsistencies. We conclude that a conversion to integrated land and water (and forest) management with the associated administrative domain models requires a complete rethinking of transdisciplinary integration and transformation. Nonetheless, the urgency of the global environmental problems requires a rethinking process. The principles of transdisciplinary research can help in this transition.

Keywords: Marine cadastre · Water rights · Marine governance · Land governance · Integrated land and water management

1 Introduction

The difference between land and water boundaries is not always evident. At first sight, seas, rivers, oceans and lakes would clearly identify as water bodies and their boundaries are shared boundaries with adjacent land parcels. But when describing land rights, restrictions and responsibilities, one often experiences that these boundaries are not so evident, neither in shape nor in time. The first problem concerns their physical shape.

A. Abdul-Rahman et al. (Eds.): GeoWeek 2024, LNGC, pp. 308–318, 2025.
https://doi.org/10.1007/978-3-031-86654-8_24

Coastal areas are typically affected by tides, which makes the land and water boundary vary with the tidal and wind forces, and irregular water flow patterns. This causes coastal lines to be more fractal than Euclidian [1]. Secondly, coastal areas are affected by different levels and heights of land and water. Besides the tides one has to refer to the height of the land and/or the height of the water at certain times [2]. Thirdly, rivers and water bodies can either turn into floods in case of surpluses of water, or dry (river, lake, coastal) beds in times of severe droughts. Although these phenomena are mostly temporary, there may be situations where the flood persists so long that the original land owner is no longer considered the de jure land owner, because when land turns into water, a private land ownership may turn into a state or public ownership [3]. The question in such cases is whether the obliterated land is still land, and whether the same rights, restrictions and responsibilities still exist. Fourthly, the shape of land in relation to water may be so specific that it makes more sense to manage and plan the land activities in close connection with those of the water. Islands, peninsulas, straits are for example defined by the interconnection of land with water, and each of these particular geographies have their own type of physical and legal dilemmas, including those of land and water rights, restrictions and responsibilities [4]. A fifth aspect of land and water connections in relation to coastal and marine areas is that construction on the coastal shoreline often directly affects the activities and dependencies of marine activities. Land reclamation, sand mining, mangrove conversion, protection or restoration are activities whereby the boundaries between land and water bodies become ambivalent, and so do the pertaining rights, restrictions and responsibilities. Given these aspects, there is a need to investigate the dilemmas and restrictions of land and water issues in relation to each other in more detail.

The rationale for this study follows from the context of the development of a framework for marine governance based on transdisciplinary research supporting the Global Ocean Decade. In this context this article aims to contribute to the development of a Marine Administration Domain model (MADM). There is a need to understand the differences between managing land and water and constructing relevant information models to support this. Given this need, this study aims at generating an insight in how which conceptual and practical challenges, opportunities, dilemmas and contradictions exist and/or emerge when converting conventional monodisciplinary management and information management practices to integrated land and water (information) management. The land-water information comparisons demonstrate a variety in ontologies and epistemologies in defining and managing land or water. Whereas land units tend to be fixed by measurable boundaries, water units are typically more flexible and variable. Parties and stakeholders in land are better identifiable and recordable than their equivalents in water. RRRs in land tend to be pluriform whereas RRRs in water tend to be more uniform. Based on these differences the spatial and institutional boundaries between land and water (and forest) informational units and land and water (and forest) management are often not evident. In fact, in case of flooding, land reclamation, ecological preservation, sandmining of mangroves, coastal protection, resettlement of fishery communities, amongst others, it is often not clear whether to use the land or water (or forest) units and mandates. This results in conceptual challenges for how and when to define a resource, a spatial unit and a legal entity as land, water, or forest. Furthermore, stakeholder identification and participation in land and water management occurs in different manners, even though there may be strong overlaps and mutual interests in their respective management operations. This is most evident in cases of coastal zone management and land and water management in smaller island states. The emerging boundary and bridge concepts arising from this overview include a range of new possibilities which require further exploration: use of variable / fuzzy land and water units, and integrated land and water parties to which legal entities of RRRs can be allocated dynamically. This requires a timestamp for which the legal rights, restrictions and responsibilities are valid as well as a legal timestamp for when and how which stakeholders should determine a new period of validity. Novel ways of participatory processes whereby land and water implications are evaluated/assessed in an integrated and consultative manner. Inclusion of indigenous and community-based planning and monitoring practices, as legitimate alternatives of spatial planning and integrated land and water management. And, to redesign and revise the pluralist legislation into a single integrated land, water and forest act.

2 Theoretical/Conceptual/Analytical Framework

From the principles of the global ocean decade, we formulate a set of key principles which guide the evaluation of transdisciplinary challenges. As the marine environment is significantly affected by human activities on land and in the ocean, including amongst others polluting industries and resource extraction in and close to the coastal areas, coastal tourism development leading to reduction of biodiversity, land reclamation and urban coastal expansion at the expense of coastal mangrove areas. The key challenge is how to design instruments, strategies, and policies to support marine and land governance—hand-in-hand. This requires, however, transdisciplinary research. Following the recommendations of [5] on transdisciplinary research, we argue that there is a need to reduce the complexity of the inter-relations between the problems and the actors in land and water, to contextualise the problems of both land and water management in a every-day life setting, to achieve integration through open encounters (such as boundary objects and bridge concepts), and to use recursiveness and reflexivity to verify and validate progress in research development.

3 Methodology

Following this theoretical logic of transdisciplinary, we approached the description of the problems and challenges in both land and water management from five perspectives: spatial/administrative domain modelling, participation and relevant actors, ontological and conceptual definitions, legislative formulations, practices, contradictions and redundancies, and practical case-based challenges, problems and inconsistencies. The underlying empirical data derive from different case studies in Indonesia. For the spatial/administrative domain modelling come from the development of a marine extension for LADM for the cases of Maluku Island and Pekalongan (which is constantly flooded by coastal flood); in relation to the elements of participation and the input and claims from relevant actors we use data from participation projects in Lembang and Maluku; for the development and testing of ontological and conceptual definitions we rely on the research on cultural space methodology; for the context of the legal elements we contextualize the research perspective of Indonesia use other practical cases of Indonesia.

4 Results

In Maluku Islands, water units can be classified under two classes, namely territorial and individual unit. Territorial unit exists in the form of indigenous community territory and local government territory [6]. While the characteristics of indigenous community sea territory is mostly a shallow water area that is designated as marine protected area due to its function as fish spawning ground, the boundary of indigenous community territory is defined based on the coastline, the sea boundary between indigenous communities' territories and the line that separates shallow and deep water [7]. The sea boundary between indigenous communities' territories is defined by an object on land [6]. Due to the characteristics of Maluku Islands, such boundaries are visible, although, because of the dynamics of the sea environment, they are flexible [6]. Figure 1 displays the sea boundary of territory of indigenous community of Tulehu, Fig. 2 shows the territory of indigenous community of Nolloth and Fig. 3 portrays the marine protected area or *labuhan* in local language in the territory of indigenous community of Haruku. Figure 4 depicts the mark signing boundary between the territories of indigenous community of Paperu and Booi.

Fig. 1. The sea of territory of indigenous community of Tulehu with a brighter color.

Before 2014, each municipality adjacent to a territorial sea had an authority to manage the area from the baseline up to a third of the breadth of territorial sea, while every province has a right to administer the rest of territorial sea. However, in 2014, the right of municipality to manage the territorial sea was given to provincial governments. In

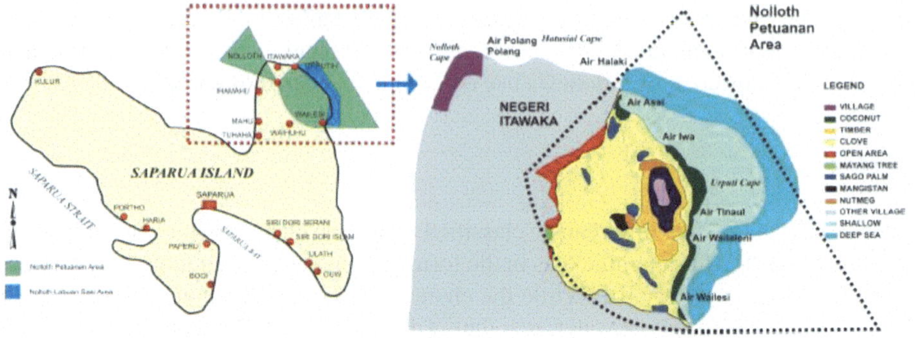

Fig. 2. The territory of indigenous community of Nolloth.

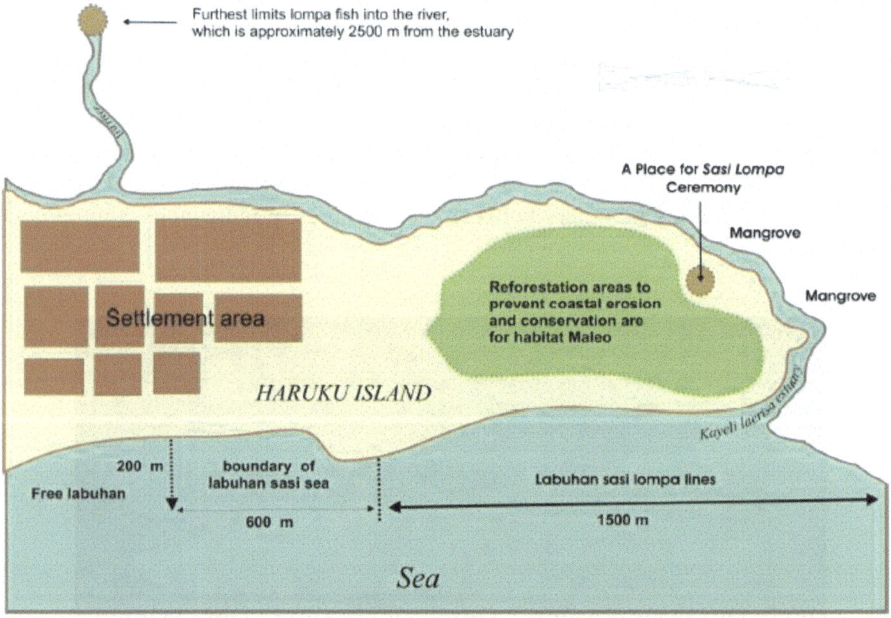

Fig. 3. Marine protected area or *labuhan* in local language in the territory of indigenous community of Haruku.

most cases, the boundaries of local government sea territory cannot be seen and is only mathematically defined.

Individual sea unit in Maluku Islands is dominated by inshore fish plantation [6]. Such a unit is delineated by mathematical planes based on the construction of the plantation. Figure 5 shows an inshore fish plantation in the territory of indigenous community of Tulehu.

Pekalongan. In Pekalongan, in which coastal floods have occurred constantly—resulting even in the facts that several parts of it have been inundated almost constantly by sea water, the sea dikes currently delineate the boundaries between land and water

Fig. 4. The mark signing boundary between the territories of indigenous community of Paperu and Booi.

units. Before the 2000s, the land use of coast of Pekalongan was dominated by coastal fishponds and paddy fields [3]. At that time, the boundary of spatial unit in the coast of Pekalongan was fishpond or paddy field bund. Since the early 2000s, many paddy fields have been inundated by sea water, which had forced the conversion of paddy fields to fishponds. Due to land subsidence and sea level rise, most of coastal fishponds and paddy fields are no longer operational because such areas have been permanently inundated (Ramadhani et al., in press).

Moreover, to cope with flood and coastal flood, there has been development of disaster risk reduction (DRR) infrastructure such as drainage, storage pond and absorption well, as well as DRR facilities for instance dike, pump station, flood gate and revetment. The

Fig. 5. For an inshore fish plantation in the territory of indigenous community of Tulehu.

DRR infrastructures and facilities are also considered as water units, which boundaries are defined based on their dimensions.

Aceh. Coastal areas in Aceh have been customarily administered by an indigenous institution called *Panglima Laot*, which main focus is to manage fishery in Aceh [1]. In the customary administration of inshore areas, the seaward boundary is delineated by the average distance of fishermen boats' half-day trip [1]. The institutional boundary of *Panglima Laot* has been changed over time, particularly due to the advancement of fishermen boat technology. Fig. 6 displays the institutional boundary of *Panglima Laot*, or *Hak Ulayat Laot* in Aceh language, in the Province of Aceh.

Natural and man-made objects. In sea, there are natural and man-made objects that are important to be administered, such as sandbanks, rigs, cables and pipes. Most of the aforementioned objects are delineated not only by their dimensions but also by mathematical planes that represent safety zones. Parties and stakeholders in land are better identifiable and recordable than their equivalents in water.

The case of the Maluku Islands. In Maluku Islands, parties and stakeholders of the territorial and individual sea unit have been clearly defined. The territorial sea unit belongs to indigenous community in question, which boundaries had been defined based on the agreement among the indigenous communities in this area. The boundary of territorial sea unit that belongs to provincial government is defined based on the agreement among provincial governments as well, which is legalised by means of provincial decree. Individual sea unit is mostly located in an indigenous community territory, which belongs to the member of the indigenous community in question.

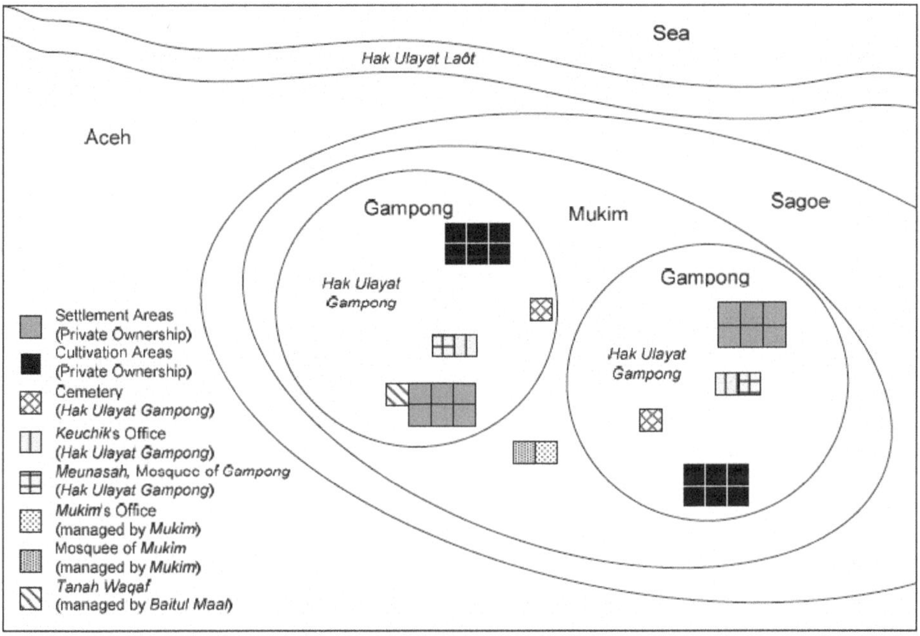

Fig. 6. For the institutional boundary of *Panglima Laot*, or *Hak Ulayat Laot* in Aceh language, in the Province of Aceh.

The case of Pekalongan. In Pekalongan, parties who are responsible on management of DRR facilities and infrastructures may be identified from various regulations at national, provincial and municipal level [7]. On the other hand, several coastal fishponds and inundated paddy fields are already registered, which provides information concerning their parties.

The case of Aceh. After the promulgation of Act number 11 of 2006 concerning Aceh Governance, Panglima Laot has been formally assigned the task to manage fishery in the territory of the Province of Aceh [6]. Panglima Laot has central organisational structure at provincial level and formally represented as well at district level, or lhok in Aceh language. Furthermore, parties and stakeholders are those who perform fishery-related activities in the Province of Aceh.

Natural and man-made objects. In Indonesia, parties and stakeholders of natural and man-made objects is sea may be identified from various regulations.

RRRs in land tend to be pluriform where RRRs in water tend to be more uniform. Maluku Islands. The existing customary marine management system in Maluku Islands has already had a well-defined RRRs for every customary marine spatial unit. As the customary marine management in Maluku Islands has been focusing on maintaining the sustainability of sea environment, every indigenous community institution is responsible on defining customary regulations that provide guidelines on the restrictions and rights of marine resources management at every marine spatial unit [6, 8].

Pekalongan. RRRs concerning flood and coastal flood mitigation in Pekalongan have also been defined by various national, provincial and municipal regulations [9]. Table 1 summarizes the RRRs related to flood and coastal flood mitigation in Pekalongan.

Table 1. RRRs on flood and coastal flood mitigation in Pekalongan.

Subject	RRR			Object
	Right	Restriction	Responsibility	
S1 Individual person S2 Communities S3 Corporation S4 National Government S5 Provincial Government S6 Municipal Government S7 Technical Agencies of National Government S8 Technical Agencies of Provincial Government S9 Technical Agencies of Municipal Government	R1.1 –	R2.1 Littering R2.1 Illegal Building	R3.1 Masterplan Design R3.2 Feasibility Study R3.3 Detail Design R3.5 Construction R3.6 Normalization R3.7 Operation R3.8 Maintenance R3.9 Monitoring R3.10 Evaluation R3.11 Land Acquisition	O1 Parcel O2 Drainage O3 Storage Pond O4 Absorption Well O5 Dike (Sea Dike) O6 Pump Station O7 Inspection Basin O8 Flood Gate O9 Cross Drainage Building O10 Trash Rack O11 Inspection Road O12 Plunge Drainage Building Sand Cacthment Building

Natural and man-made objects. RRRs on natural and man-made objects management in sea have been described by various regulations. In Indonesia, however, an integrated management system is still being developed.

5 Conclusions

A conversion to integrated land and water (and forest) management with the associated administrative domain models requires a complete rethinking of transdisciplinary integration and transformation. Nonetheless, the urgency of the global environmental problems necessitates such a rethinking process. The principles of transdisciplinary research can help in this transition. Assuming that socially relevant problems are by nature complex and can only rely on diversity of stakeholder perspectives to reach a solution, it is necessary to establish relations between different forms of relevant, case-specific and transmissible knowledge which incorporates insights from both land and water epistemic communities.

Spatial units in an existing marine management system may be clearly defined either by using an indicator or mathematical definition. The dynamics of sea environments may require a redefinition of spatial unit at different periods of time. Furthermore, legal framework is considerably important to facilitate the clear definition of spatial unit. Also, a specific legal framework is crucial to define parties and stakeholders and RRRs. This results in conceptual challenges for how and when to define a resource, a spatial unit and a legal entity as land, water or forest. Furthermore, stakeholder identification and participation in land and water management occurs in different manners, even though there may be strong overlaps and mutual interests in their respective management operations. This is most evident in cases of coastal zone management and land and water management in smaller island states. These findings should ultimately contribute to the formulation of a Marine Administration Domain model (MADM).

The emerging boundary and bridge concepts arising from this overview include a range of new possibilities which require further exploration: use of variable/fuzzy land and water units, and integrated land and water parties to which legal entities of RRRs can be allocated dynamically. This requires a timestamp for which the legal rights, restriction s and responsibilities are valid as well as a legal timestamp when and how which stakeholders should determine a new period of validity. Novel ways of participatory processes whereby land and water implications are evaluated/assessed in an integrated and consultative manner. Inclusion of indigenous and community-based planning and monitoring practices, as legitimate alternatives of spatial planning and integrated land and water management. And, to redesign and revision of the pluralist legislation into a single integrated land, water and forest act.

References

1. Piña-García, F., et al.: Determination of geometry and measurement of maritime–terrestrial lines by means of fractals: application to the Coast of Cantabria (Spain). J. Coastal Res. **32**(5), 1174–1183 (2016)
2. Parker, B.B., The difficulties in measuring a consistently defined shoreline—the problem of vertical referencing. J. Coastal Res. 44–56 (2003)
3. Pinuji, S., et al.: Is obliterated land still land? Tenure security and climate change in Indonesia. Land **12**(2), 478 (2023)
4. Van Dyke, J.M.: Legal and practical problems governing international straits 1. Ocean Yearbook Online **12**(1), 109–125 (1996)
5. Pohl, C., Hadorn, G.H.: Principles for designing transdisciplinary research. oekom Munich (2007)
6. Abdulharis, R.: The potential of customary spatial unit administration for sustainable development: the case of Ambon Lease Region, Province of Maluku, Indonesia. 2014, Dissertation, Dortmund, Technische Universität (2012)
7. Hernandi, A., MeiLing, M., Abdulharis, R.: Qualitative Survey for Identifying The Customary Marine Tenure In Maluku Islands, Indonesia, in Coast GIS 2011. In: 10th International Symposium on GIS and Computer Mapping for Coastal Zone Management. Corila, Oostende (2011)

8. Novaczek, I.: An institutional analysis of sasi laut in Maluku, Indonesia, vol. 1648. WorldFish (2001)
9. Ramadhani, D.N.: Pengembangan Land Administration Domain Model untuk Penyelenggaraan Mitigasi Bencana Banjir dan Rob di Wilayah Pesisir. Institut Teknologi Bandung (2024)

Land Administration in Malaysia: An Insight of the 3D-Based Future Development Framework

Ahmad Hamidi Mohamed(✉), Mohamad Hezri Razali, Saiful Aman Sulaiman, and Amir Sharifuddin Ab Latip

College of Built Environment, Universiti Teknologi MARA, 40450 Shah Alam, Malaysia
midi.trg@gmail.com, saifulaman@uitm.edu.my

Abstract. The current development approach in Malaysia is moving towards vertical development with the need for 3D elements in its land administration system. Rapid development with a vertical development approach, which consists of stratified development and multi-layered development, requires a solid foundation on the land administration legal framework and surveying practices. This research is an initial study on the legal framework for 3D development, attempting to meet the needs of such development while providing legal framework insight into the 3D aspect. This research is conducted through fully qualitative methods focusing on the legal aspect with critical reference to the land law in Malaysia involving the National Land Code [Act 828] and assessing the government initiative towards a multi-layered development approach. This research proposed that the current legal framework should include the initiative of 3D development in its foundation for the rights and ownership of the land.

Keywords: Land administration · Stratified development · Multi-layered development · Legal framework · 3D element

1 Introduction

Malaysia's development is currently going rapidly, especially within the central conurbation area. As Malaysia's capital city, Kuala Lumpur faces significant areas of development, so its development is very tightly concentrated. From being horizontally developed, the approach is now going vertically. Similarly, in other conurbations areas, such as the capital city of a state, namely Georgetown in Penang and Johor Bharu in Johor, the design of the vertical development approach has been replicated. Having this kind of development approach, Malaysia has embraced vertical development and recognised stratified development. The legal aspect of the development evolved from a subsidiary title into a strata title to acknowledge the rights and ownership of the stratified development. This ownership forms part of the land ownership in Malaysia, especially for the stratified development of housing and commercial development. In addition, besides focusing on housing and commercial development, Malaysia experienced stratified utility development as well. Such development is equally important for the country as it forms the

A. Abdul-Rahman et al. (Eds.): GeoWeek 2024, LNGC, pp. 319–326, 2025.
https://doi.org/10.1007/978-3-031-86654-8_25

foundation of the development for commercial needs. Utilities, such as roads and rail tracks, are very much needed for urban mobility and access in the conurbation area. In the need for rapid urban development, especially the ownership system in Malaysia, the rights of use and enjoyment should be incorporated in the title or permission to use the land. These rights of use and enjoyment are all important in ownership in Malaysia through its titling system—alienation of land, as well as non-alienation of land such as permit and reservation of specific purpose. Both alienation and non-alienation rights of use and enjoyment are determined by the statutory form, which includes the plan of the land, informing the area involved. Thus, this study intends to explore the use of 3D plans in the titling system and the uses of rights in Malaysia, specifically in the gazette of reservation of land or permit for use of land.

2 Methodology

This research deploys the review of the current literature on 3D development and multi-layered development. The literature reviewed is narrowed down to the legal aspects of the 3D element together with the land administration framework in Malaysia. The statutes and the guidelines published by the government office are examined and reviewed. In addition, the current scenario is screening from the articles available under the topics. Current initiative from the government also forms part of the review as the research will critically assess the guidelines given by the government. The inside first-hand observation and information are gained from the experience of the researcher himself. The experience and research conducted in the government office of *Jabatan Ketua Pengarah Tanah dan Galian Persekutuan (JKPTG)* or the Department of Director General of Land and Mines are included in the findings of this study. Interviews were also conducted to gain input from several personnel from government officials and industry practitioners, such as personnel from the Department of Director General of Lands and Mines (JKPTG), the State's Department of Lands and Mines (PTG), and the Department of Survey and Mapping Malaysia (JUPEM).

3 Findings and Discussions

3.1 Land Ownership and Usufructuary

Land ownership in Malaysia is governed solely by the National Land Code [Act 828] (NLC). The land title confers the ownership and use of the land to the person named in the title. Ownership may be held solely or by shares from the alienation, transfer, and transmission of ownership, such as from the probate administration. This ownership and usufructuary are conferred particularly by Sections 62, 75A, and 76 of the NLC [1]. The utmost right conferred to the land proprietor is through Section 76 of the NLC, which confers not only the usufructuary but also includes the right of ownership and enjoyment of the land. The land proprietor is entitled to all the rights to the land. However, it is subject to the category of the land, its express condition, and restrictions in interest, which come together with titles.

In addition, the rights and enjoyments are also conferred by the NLC through Part Five of the NLC, which refers to the disposal of the underground land. Under Section 92A, the NLC defines the scope of the disposal of underground land by referring to several terms, such as underground land and stratum. These terminologies are important to be interpreted as the ownership slightly different from the usual disposal by alienation of the NLC. Moreover, the same authority for land disposal comes from the Strata Titles Act 1985 [Act 318]. Historically, Act 318 is part of the NLC, which is called the subsidiary title. As the ownership evolves and the management of the subsidiary title needs to be emphasised, this particular element is enacted as a new law in 1985 [2]. According to Act 318, instead of referring to the ownership of land parcels, the ownership of the particular lot is known as the strata parcel. The proprietor of the parcel will have rights and enjoyment similar to the title under the NLC but with additional conditions to the ownership and enjoyment.

3.2 The Plan Requirement for the Land or Parcel

The legal framework discussed above shows that the ownership and the usufructuary are not freely available; rather, they will be subjected to several conditions. The most important element of the ownership is the area—the border of the land or strata parcel. Under the current framework, the area of ownership is shown by the plan of the land or parcel.

Alienation of land. There are particular sections of the NLC, as well as Act 318, that require the plan of the land to be attached to the title, permits, or reservations. As land is disposed of under the NLC, the alienation of the land by the State Authority is subjected to the area approved by them. The application for alienation will be attached together with the plan of the land showing the area approved for the alienation. This requirement is mentioned in the NLC under Section 77, which is cross-referred with Section 396. The land needs to be surveyed or have a certified plan of the land for a final title issued to the proprietor. Similarly, the application for strata titles also needs to be read together with the NLC. Ownership requires the plan of the parcel, and the plan of the parcel is ready at the beginning of the construction of the scheme. The strata title confers the rights of ownership and usufructuary to the parcel proprietor confined to the plan approved for the respective strata scheme.

Disposal of land other than alienation. Apart from the ownership under the NLC and Act 318, the land can also be disposed of in ways other than alienation, such as reservations, temporary occupations, and permits. Those ways also require the plan to be attached to the application. For example, under the usufructuary of a permit to use the airspace, it is mentioned under Section 75B of the NLC as follows:

> *"(b) details of the structure to be erected and maintained, and a plan thereof as approved by the appropriate planning authority; and"*(emphasis added).

This shows that the plan is required to be attached to the application. The plan needs to show the specific details of the intended structures to be constructed and maintained, as well as construction plans authorised by the appropriate planning authorities, such as the District Council or Municipal Council. Thus, this permit to use the airspace is limited to the confined area, which is built within such scale, perhaps the 3D scale, although it is not clearly stated. It is worth noting that this plan, at least in Selangor, clearly mentions that the application, together with the plan, is solely meant for the purpose of establishing, maintaining, and occupying structures across the State land or reserved land as a connection to any building located on adjacent land [3]. The "structure" element, together with the "State land," is the implied element of the application demonstrating that the use of air space is limited to a confined space above the State land or reserved land only.

3.3 The 3D Implementation and Case Study

The implementation of the 3D element in the land administration is rather vague and ambiguous. The 3D element concerned in the land administration is its land plan, which needs to be surveyed and attached to the respective purpose of land records and databases. In brief, the plan in land administration can be classified into two (2) categories, which are certified plan and provisional plan. These categories are with respect to the accuracy of the survey, which is done by a Survey Officer or Licensed Land Surveyor as prescribed under the NLC.

Area and volume . The current development seems to be moving to multi-layered development, which not only caters to the area of the land but also involves the volume of the land. However, the land administration is still using the 2D approach, which can be traced from the plan requirement for several ways of disposing of the land. The "area" connotations still stand widely in the current framework. The 2D approach is more connected to the principle of land administration in Malaysia. It is in line with the principle of ownership in Malaysia, where the landowner will have the right to enjoy the land under his title, below and above the surface of the land, relative to the area of the land involved. This aspect shows that the previous ownership under the NLC from alienation or disposal of land is not strictly mentioned on the "volume" of the land. Until the amendment to the NLC in 1990, which introduced the underground component, the rapid development that inclined to vertical development, either upwards or downwards, took place in a construction environment. Nevertheless, underground alienation is not yet taking place. As information is available, only the state of Kelantan has guidelines related to the alienation of the underground land. On the contrary, the issue of 3D elements arises when the volume of land is required underground for development, such as Light Rail Transit (LRT) and Mass Rapid Transit (MRT). As such, the underground land needs to be surveyed accordingly, using the 3D plan to secure the volume of the underground land.

Current implementation. As such, the study finds that there is a lack of 3D elements in land administration. Some parts of the development are well aware and include the element of 3D in the development, but others are still in the pipeline. 3D implementation awareness refers to the physical developments that use 3D elements in their plans and models. However, the basic foundation of the development, which is the land or development site, is not. Thus, this can emerge as a partial implementation of 3D elements. The only 3D element in land administration is currently attached to the underground development in Malaysia. Underground development requires underground land as its foundation. Nonetheless, the underground development in Malaysia is new and rapidly emerging these years. Some of the underground development in Malaysia can be summarised as follows:

Table 1. Land use for underground land [4].

Purpose	Development projects	Completed
Commercial and parking area	Plaza Dataran Merdeka (Merdeka Square)	1990
Commercial and parking area	Petronas Twin Tower	1997
Rail track tunnel/station	Light Rail Transit (LRT) (Kelana Jaya Line)	1998
Dam	*Pergau* Dam	2000
Road/flood tunnel	SMART Tunnel	2007
Rail track tunnel/station	Mass Rapid Transit (MRT) 1	2016

Referring to Table 1 above, the development of the underground is not only the physical development, but the foundation of the development also involves the 3D element, which is the land matter. Although not all the listed developments are under ownership in their nature, which is not alienated and without title, the land matters involved the 3D element in giving the usufructuary right to the responsible parties. The survey and plan need to be prepared in a 3D scheme. By implementing the 3D scheme survey plan, the rights of use of the respective column of the land or the volume of the land is protected by the law. In addition, the current 3D survey plan is used for the land involved in the compulsory land acquisition and mutual agreement of co-existence. The land that is involved in the compulsory land acquisition under the Land Acquisition Act 1960 [Act 486] requires the said underground land acquired to be surveyed for the purpose of alienation and reservation. Additionally, the remaining surface land, which is left unacquired, needs to be resurveyed for the title in continuation as spelled out in the NLC and Act 486. Thus, this land plan embraced the 3D survey plan, which have been inserted through amendment of the legislation of the land law and administration in 2016 [5, 6].

Case study. *Stratified development.* Stratified development refers to the development which is governed by the Strata Titles Act 1985 [Act 318] and the Strata Management Act 2013 [Act 757]. This can be in contrast to unstratified development, which is horizontal development, or conventional development, which is usually associated with residential properties [7]. In addition, stratified development has evolved from residential development into commercial development, which carries several names, such as service apartments, Small Office Home Offices (SOHO), and Small Office Virtual Offices (SOVO). Hence, the stratified development is basically, at least their ownership and management, in good hands because it is governed by specific acts. Nonetheless, the stratified development seems to be upwards and downwards as long as it is still under one roof—the strata title and management. The important element of the management of strata titles and their ownership is the strata scheme itself, which should be under one scheme. To be precise, the scheme was approved earlier by the initial strata plan. There is no issue with the implementation of 3D as it is governed by Act 318, which is quite similar to the NLC, requiring only a 2D survey plan. However, the issue occurs when the current stratified development tries to connect the scheme through amalgamation with another stratified development. This scenario has been consulted with JKPTG on the case in Penang. As the land administration is not ready with the 3D concepts of amalgamated ownership, the issue has not been resolved yet.

Multi-layered development. Nevertheless, vertical physical development is not only limited to going downwards, which, in a way, has been governed by the underground alienation of the NLC. The vertical development may also go upwards, which is now rapidly initiated by the developer. Similarly, it has surely been governed by Act 318, as discussed earlier. In between, physical development may also be horizontal-vertical in the air that needs the disposal of air space. This disposal of air space, which is named "spatium" by the respective government agency for the time being, is another 3D element that is indeed needed for the 3D implementation as well [8]. Spatium, by its proposed name, is suggested to be a mirror of the underground land concepts and legislation. It is an introduction of a new alienation of rights, which initially forms part of the land as defined by the NLC. Supposedly, the alienation consists of a column of the airspace above the surface of the land – taking the definition of land under paragraph (a) Section 5 of the NLC—forming the surface of the land. As such, it is a totally new concept and worth exploring as another layer of alienation and ownership.

Meanwhile, the other development associated with the 3D elements in land administration is the multi-layered development [9], especially on the public infrastructure. This development can be illustrated as the elevated road, over the rail tracks, crossing a bridge above a river. This development becomes complex as multiple layers of infrastructure exist on a plot of land. This particular case and scenario happens in the alignment of Mass Rapid Transit (MRT). According to one respondent, the multiple layers of infrastructure, particularly public infrastructure, need to be reserved for its right of use and enjoyment. Therefore, the 3D element in the gazette plan is very vital in determining the border of each and every infrastructure for its management and responsibility.

3.4 Constrain and Challenges

The implementation of the 3D element in the land administration seems to have its constraints and challenges.

Legal framework. As observed by the researcher, the biggest challenge is the legal framework. Before discussing the challenges deeper, this research is an extension of a similar study that was presented at a similar platform and forum, FIG Congress, in 2018. As such, the proposed model has been scrutinised, and it has been found that the implementation mechanism is a big hindrance in the current legal framework background [8].

As from the example in Fig. 1, the proposed 3D elements of the development are on the new construction of building labeled as "1", which is the horizontal structure

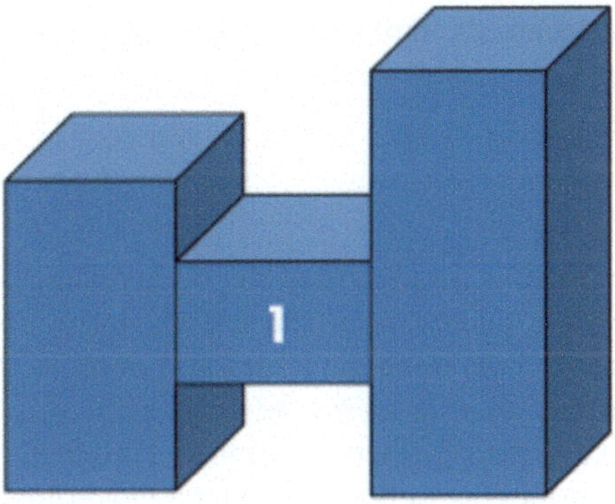

Fig. 1. Example illustration for a proposed 3D element of future development [10].

addition to both already constructed buildings. This illustration comes from an analysis of an actual building found in several states, such as Selangor, Melaka, and Penang. Most of those buildings are constructed above the road, which is considered as reserved land. In discussing the above model, the researcher agrees with the issue raised by one of the respondents. According to the respondent, in the current building, the vertical structure, connecting the horizontal structure, is based on the current legislation of 2D plan requirement [11, 12]. According to another respondent, the 2D plan requirement seems to be above the 2D plan but is not exactly a 3D plan, which is further agreed to be called a 2D plus. This is because vertical development is currently governed by Act 318, which is a strata scheme comprising a plan surveyed for the strata guided by a circular from JUPEM [13]. In the end, although the development can occur, the legislation and legal framework did not support the alienation of airspace area or volume. On the one hand, there is no disposal of air space, and on the other hand, the 3D plan has not been extended to airspace yet.

Technical scrutiny. As such, the constraint to the model is to merge the 2D element and the 3D elements of the plan, to be specific, as the structures are from a different scheme altogether. This differs from the land parcel, in which the amalgamation application for the said lands occupied can solve the issue of the development of adjoining land. Furthermore, it is worth noting that the government's initiative in addressing this issue is already on the table, such as the multi-layered development for transit, also known as Transit-Oriented Development (TOD) [14]. This initiative can be differentiated from the above-mentioned issues. The earlier discussion and issue are on the ownership of the land from the perspective of alienation of the land. Hence, the current initiative is a reservation, non-alienation approach, which is more inclined toward public utilities and infrastructures. This can be illustrated by rail tracks, bus stations, and roads, as well as a combination of those amenities and utilities. Likewise, it is also a challenge to incorporate public amenities, utilities, and infrastructure into the private commercial sector. The difficulties arise in the essence of obligations on maintenance and the ownership of the facilities. It was found that the challenge is related to the 3D implementation of the building structure plan. Again, the building structure plan issue is related to the legal framework of the current land administration.

4 Conclusion

This study indicates that there is awareness of the issue arising on the legal framework instituting the 3D elements in land management and administration. There are also some steps that have been taken to overcome the issues and improve the future development of 3D elements. The current solution seems only enough to provide the guidelines and short to medium-term solutions to the incorporation and implementation of 3D elements. Moreover, the State Authorities, as well as the Federal Government are looking forward to tackling the future development framework through their current jurisdiction and trying to work hand in hand in determining and providing the general solution to the issue. However, the scope and the implementation of 3D-based future development, especially in land administration, still have a lot of room for improvement.

References

1. The Commission of Law Revision M: National Land Code [Act 828]. Attorney General's Chamber, Malaysia (2020)
2. Attorney General Chambers: Strata Title Act 1985 [ACT 318] (2021)
3. Jabatan Ketua Pengarah Tanah dan Galian Prsekutuan: Pengeluaran Permit Untuk Menggunakan Ruang Udara Atas Tanah Kerajaan Atau Tanah Rizab (Disemak 2007) (1985)
4. Abu MZZ: Penilaian Harta Tanah Strata Dan Tanah Bawah Tanah. Kelana Jaya (2023)
5. Attorney General Chamber: National Land Code (Amendment) Act 2016 (2016)
6. Attorney General Chamber: Land Acquisition (Amendment) Act 2016. Laws of Malaysia, Malaysia (2016)
7. Khoo, N.: Living in Non-Stratified Residential Properties. EdgeProp.my (2020)
8. Bahagian Dasar dan Konsultasi, J.K.P.T.G.: Laporan Soal Selidik Pengenalan 4 Konsep Pindaan Kanun Tanah Negara. Putrajaya (2021)
9. Choon, T.L., Hussin, K., Ernest, K.H.O.: Making 3D Property Legislation Feasible in Malaysia. In: International Symposium and Exhibition on Geoinformation 2009 (2009)
10. Lim, C.K., Chee Hua, T., Loo, D.K.B.: Proposed Legislation For 3d Cadastre In Malaysia 1. In: Persidangan Pentadbir Tanah Malaysia Ke-24 (2018)
11. Choon, T.L., Hussin, K., Ernest Hock, K.: Towards 3D Property Legislation in Malaysia. In: Universiti Teknologi Malaysia Postgraduate Seminar (2009)
12. Choon, T.A.N.L., Hussin, K.B., Ernest, S., Hock, K.: 3D Property Situation in Malaysia-Initiatives towards 3D Cadastre. Building (2010)
13. Jabatan Ukur dan Pemetaan Malaysia: Garis Panduan Pecah Bahagi Bangunan atau Tanah Untuk Pengeluaran Hakmilik Strata (2015)
14. PTG Selangor: Dasar Pembangunan Berorientasikan Infrastruktur Awam Di Negeri Selangor (2017)

UAV based Mapping

Assessing the Performance of Continuous Operating Reference Station (CORS) in UAV Mapping

Bakeri Samsudin[1](✉), Sulaiman Budin[1], Mickcy Mandang Anak Morgan[1], Siti Hayatunissa Saminan[1], Dzuren Hamzah[1], and Wan Anom Wan Aris[2]

[1] Lands and Surveys Department Sarawak, Sarawak, Malaysia
bakeris@sarawak.gov.my
[2] Geomatic Innovation Research Group, Faculty of Built Environment and Surveying, University Technology Malaysia, Johor Bahru, Malaysia

Abstract. This paper investigates the performance of CORS in UAV mapping in the State of Sarawak, Malaysia. The location of the study area is near equatorial, which is heavily affected by the troposphere conditions. As Sarawak moves into using a geocentric datum from a topocentric datum, no study has been conducted on the performance of CORS in UAV mapping. The dependency on coordinates obtained from the topocentric datum is still widely used for UAV missions. This study analyses the comparative differences contributing to the distance-dependent factor and attempts to tabulate the mapping applications that can be used based on the outcomes.

Keywords: CORS · UAV · Mapping

1 Introduction

1.1 Overview

With the advent of GPS technology becoming available in the 90s to the public, it has become an essential tool for determination of position with ease of use. Nowadays is incorporated in most of the commonly found devices. Even though geocentric mathematical models existed before, space technology made determining the geocentric positioning easier. During the initial times, the realization was that to fully utilize the space technology, the datum being used for land surveying purposes had to migrate from the traditional topocentric base to a geocentric-based datum.

Sarawak uses the Borneo Rectified Skewed Orthomorphic datum for its cadastral surveys, which is topocentric, and the realization to move to a geocentric datum was necessary to fully utilize the available space technologies such as Global Navigation Satellite System (GNSS) through multitude of navigation satellites available for public use.

A. Abdul-Rahman et al. (Eds.): GeoWeek 2024, LNGC, pp. 329–342, 2025.
https://doi.org/10.1007/978-3-031-86654-8_26

The move by Sarawak to a geocentric datum is still relatively recent and is still in its infancy. Existing studies on coordinate comparisons and biases between the topocentric and geocentric coordinates are very limited for Sarawak.

This study explores the dynamic integration of Unmanned Aerial Vehicles (UAVs) and Continuous Operating Reference Stations (CORS) within the evolving field of geospatial technology. By analyzing each technology's individual capabilities and then examining their synergistic effects, the research aims to demonstrate how this combination can significantly enhance the accuracy and efficiency of mapping operations. The focus will also include an evaluation of the performance of CORS in the context of UAV mapping.

This study focuses on the Continuous Operating Reference Station (CORS) network, which consists of strategically positioned Global Navigation Satellite System (GNSS) receivers across diverse geographic areas. CORS stations continuously collect satellite signals and serve as reference points to enhance the accuracy of GNSS-based positioning. CORS mitigates errors arising from atmospheric conditions and satellite clock discrepancies by meticulously tracking satellite signals and providing real-time corrections. The core idea behind CORS is its capacity to enhance the unprocessed GNSS data that user devices get by sending correction signals that raise the location accuracy to centimeter-level accuracy. This networked infrastructure is a priceless tool for geospatial applications like mapping, surveying, and navigation.

In parallel, the development of UAVs has transformed mapping by providing a flexible platform for gathering data in a variety of environments. Unmanned aerial vehicles, or UAVs for short, are planes without human occupants outfitted with advanced GPS receivers, cameras, and sensors. By overcoming the constraints of conventional mapping, these tools make it possible to collect high-resolution geographical data with previously unheard-of efficiency and flexibility. UAVs with sophisticated GNSS receivers use satellite signals to pinpoint their location and orientation. Positional awareness enables UAVs to collect precise three-dimensional spatial data when combined with onboard sensors such as LiDAR (Light Detection and Ranging) or high-resolution cameras. Because of their portability and accessibility, unmanned aerial vehicles (UAVs) are useful for a variety of tasks, including surveying, mapping, urban planning, infrastructure inspection, environmental evaluation, and agricultural monitoring.

The issues inherent in UAV-based mapping are addressed by integrating CORS and UAV technology. Even though unmanned aerial vehicles (UAVs) are excellent at recording fine spatial information, atmospheric distortions and satellite signal obstacles can cause positioning mistakes. The total accuracy of spatial information produced by UAVs is greatly improved when CORS, with its network of reference stations, steps in to correct these mistakes. The combination of CORS and UAVs works especially well when standard GNSS corrections are insufficient. The dependability of raw GNSS data is frequently compromised in urban contexts with tall buildings, canopies of dense foliage, or difficult topography. Under challenging circumstances, CORS serves as a corrective mechanism, improving the accuracy of UAV-derived datasets by offering real-time fixes customized to particular locals.

Beyond just increasing accuracy, this integration is significant. The efficiency of data-collecting procedures is improved by CORS-enabled UAV mapping. The integrated technologies simplify mapping workflows and increase efficiency and cost-effectiveness, which is important for time-sensitive applications like environmental monitoring and catastrophe response. This study aims to evaluate the effects of CORS integration in UAV mapping in a range of environmental and geographic scenarios. The study seeks to provide a thorough knowledge of the benefits and drawbacks of this combination approach by methodically examining positioning accuracy, data quality indicators, and mapping efficiency. The research findings have potential applications in various fields, such as environmental monitoring, infrastructure development, urban planning, and geomatics. The CORS-enabled UAV mapping paradigm has the potential to significantly impact the direction of mapping technology in the future, as the need for accurate and trustworthy geospatial information keeps increasing.

In summary, the combination of Unmanned Aerial Vehicles (UAVs) and Continuous Operating Reference Station (CORS) technology is a ground-breaking development in the fields of mapping and geospatial engineering. The precision-enhancing qualities of CORS complement the intrinsic capabilities of UAVs to record high-resolution spatial data, creating a synergistic impact that surpasses the limitations of individual technology. Through a methodical assessment of CORS's effectiveness in UAV mapping settings, this study seeks to provide important insights for maximizing the application of this integrated technique. A thorough grasp of the revolutionary potential of CORS-enabled UAV mapping will become evident as we work through the specifics of the methodology, findings, and discussions in the following sections. This will open the door to improved accuracy, dependability, and efficiency in the acquisition of geospatial information. The objective of the research is to evaluate positioning accuracy—assesses the accuracy of CORS-derived positioning data in UAV mapping applications by comparing it to ground truth measurements or high-precision reference data sources.

2 Evaluation of Methodology

(a) Selection of sites for study purposes
(b) Determination of set criteria to be evaluated for the purpose of research
(c) GCP and UAV data acquisition
(d) Data processing and Analysis
(e) Data Interpretation
(f) Comparative analysis of results against study objectives.

3 Literature Review

CORS (Continuously Operating Reference Station) stations serve as vital infrastructure for precise positioning and navigation applications by laying out a GNSS reference of certain distribution and density through continuous satellite signal tracking. This space technology has been widely applied in point positioning, navigation and a wide spectrum of studies. CORS networks have been implemented worldwide, aiding in precise positioning and navigation. These networks, consisting of strategically placed stations, support a range of applications and geophysical research. CORS networks have become integral to global geospatial infrastructure, facilitating technological advancements and improving societal outcomes.

3.1 CORS Around the Globe

The NOAA CORS Network, managed by National Oceanic and Atmospheric Administration's National Geodetic Survey, provides essential GNSS data across the US. It supports various applications like surveying, GIS, and engineering by offering precise positioning and meteorological data. Fiducial control stations within the network enhance position accuracy and align with the National Spatial Reference System. CORS operates as a multi-purpose initiative through collaboration with government, academic, and private entities. Stations contribute GNSS/GPS data to NOAA for analysis and dissemination, enabling accurate positioning data for users at no cost. This supports applications with post-processed coordinate accuracies of a few centimeters' horizontally and vertically nationwide [3].

The core of Canadian Spatial Reference System (CSRS) includes Canadian Active Control System (CACS) and the continuously operating GNSS receiver network, which was established and managed by Natural Resources Canada (NRC). In Germany, a system known as *Satellitenpositionerungsdienst* (SAPOS) unifies differential GPS in various agencies. It includes 200 or more permanent GPS reference stations, providing centimetre-level real-time positioning and millimetre-level high-accuracy positioning services. In Asia, the GPS observational network was established in 1993 by The Geographical Survey Institute of Japan (GSI). More than 1200 continuous GPS stations with a mean separation distance of 25 km covering the whole country were processed by GSI.

3.2 Development of CORS in Sarawak, Malaysia

The Lands and Surveys Department of the Sarawak government initiated a strategic effort in 2016 (Kueh 2016). The goal was to identify areas where datum infrastructures are to be modernized. The development of CORS in Sarawak consists of two phases. In Phase 1, seven roof-based stations were established. Phase 2, twenty-one ground-based stations have been distributed throughout the states. The location between stations shows the interstation distance between 70 and 120 km, covering most of the Sarawak region. When the stations are fully operational, a Sarawak Zero Order Network formation is established, which will fit with the Global International Terrestrial Reference Frame geodetic framework.

3.3 Studies on Performance and Benefit of CORS

The advantages and contributions of CORS development in several fields, such as science, urban planning, and geodesy are undeniable. Numerous research projects have been undertaken globally. The study's findings can be used to support industrial research and development projects that make greater use of GNSS and CORS technology.

(a) Efficiency of Earth Gravitational Model 2008 in Northeast China

The Earth Gravitational Model 2008 demonstrates good adaptability, where cleared areas of Northeast China offer higher global accuracy. The model achieves a maximum error of 0.030 m and a mean square error of 0.022 m in the Chinese Geodetic Coordinate System

2000 (CGCS2000) after accounting for system deviation and constant deviation. The model has proven to be an efficient method in enhancing the accuracy and reliability of GPS height measurements by the Jiangsu Land-based CORS (JLCORS) system [9].

(b) Transformation Algorithm for Coordinate Systems

The application of a transformation algorithm between China Datum (CGCS2000) and Beijing54 coordinates highlights the effectiveness of using the surface approximation method to calculate the seven two-dimensional parameters. It offers a simple yet effective way of obtaining coordinates, particularly in the event of unknown or less accurate geodetic elevation. For achieving high accuracy over large areas, it is recommended to calculate transformation parameters by partition [9].

(c) Role of CORS in Geodetic Applications

The compatibility of WGS84 highlights the continuous operation and reliability of CORS coordinates with CGCS2000 coordinate system used by JLCORS. Under optimal observation situation and signal coverage, rover stations achieved high positional accuracy with a single GPS receiver. Additionally, the swift acquisition of coordinates in different systems and the provision of orthometric or normal height using local geoids illustrate CORS's efficiency, accuracy, and versatility. These systems are set to substantially impact various fields such as control surveys, cadastral surveys, land management, urban planning, and topographic mapping, offering high reliability and low costs [9].

(d) Performance Evaluation of GNSS Stations and Recommendations for Inclusion in the IGS Continuous Network

Based on a study in Mexico, the IGS station INEG was found to exhibit better performance according to both IGS and regional Mexican recommendations. However, it showed high cycle slip values. Conversely, the MPR1 station, mounted on an aerial control antenna, demonstrated the worst behavior. Four other stations, MEXI, MTY2, OAX2, and TOL2, have met similar quality indicators as INEG, with the same values of multi-path, strong signal-to-noise ratio (SNR), and integrity exceeding 95%. These stations could be considered for inclusion in the IGS continuous network based on their observed behaviors. The study observed a high multi-path effect, particularly in systems with outdated receiver firmware, especially when the L2 signal is degraded [10].

(e) Differences in GNSS Station Compliance: IGS Versus RVM Standards

Values recommended for multipath effect and cycle slips tend to be more lenient in the RVM standards than the IGS standard. Also, the RVM standards are notably stricter for signal-to-noise ratio and integrity values. Interestingly, the global average of fulfilled variables based on IGS recommendations was 3.30 out of 6 possible variables across 65 stations. However, adherence to RVM standards resulted in agreement with 3.54 variables on average. Despite a seemingly minor numerical difference between IGS and RVM standards, more stations met the RVM recommendations, fulfilling 4–5 variables [10].

(f) GNSS Station Behavior and Implications for Positioning Precision: Insights from Mexico

The findings indicate an inverse relationship between signal-to-noise ratio (SNR) and multipath (MP) for stations such as INEG, MEXI, MTY2, OAX2, and TOL2, while a direct correlation between MP and SNR is observed in 60 stations, particularly pronounced in L1 signal. Despite efforts to mitigate multipath errors through careful site selection and antenna usage, complete mitigation remains unachievable. Hence, selecting optimal locations for GNSS continuous observations is crucial. The study underscores the favorable behavior of CGPS/CORS in Mexico, with over 85% of stations meeting at least 3 quality indicators. Furthermore, a significant correlation is identified in integrity, cycle slips, and daily observation time. Shorter observation periods lead to reduced integrity and increased cycle slips per thousand observations, impacting positioning precision. Nonetheless, velocity estimations have remained consistent with the interaction of Mexico's principal tectonic plates.

3.4 Unmanned Aerial Vehicle (UAV) and CORS

Continuous Operating Reference Station (CORS) technology, consisting of a network of GNSS receivers that are permanently deployed, has become essential for improving positioning accuracy. Simultaneously, Unmanned Aerial Vehicles (UAVs) have revolutionized mapping through their agility and capacity to acquire high-resolution data. The incorporation of CORS with unmanned aerial vehicle mapping has the potential to address innate difficulties and improve mapping accuracy. Navigation, geodesy, and surveying have all benefited from the technological underpinnings and historical development of CORS. Its role in improving GPS accuracy is essential to overcome obstacles in dynamic contexts. Geospatial location accuracy is ensured by CORS, a network that offers real-time adjustments.

Theoretically, tackling issues unique to UAV mapping forms the foundation for merging CORS and UAV technologies. CORS adjustments significantly increase accuracy in urban, topographically complicated, and vegetated areas. Case studies demonstrate this synergy's effectiveness in a variety of mapping contexts. Przybilla and Bäumker proposed to use RTK/PPK, depending on the application. For topography mapping, where the accuracy range is between 0.010 and 0.020 m, they proposed the use of RTK/PPK; for cadastre mapping, accuracy range between 0.010 and 0.030 m, they proposed the use of RTK/PPK with the GCPs; and, for engineering surveying (accuracy range 0.010 m), the use of GCPs was recommended. The study by Eker, Alkan and Aydın highlights the importance of considering the distance between base station and rover (UAV platform) as well as elevation difference between them. It suggests that accuracy can be compromised with distances over 10 km, with vertical accuracy being more affected than horizontal accuracy. Elevation differences exceeding 500 m can also degrade accuracy significantly (White Paper, 2018). The research also compares the accuracy of photogrammetric analysis using Real-Time Kinematic (RTK) and Post-Processed Kinematic (PPK) methods. It indicates that PPK-GNSS (M2) has shown the best accuracy in camera locations, while PPK-CESM (M4) resulted differently due to deviations beyond recommended values in the distance to the rover and elevation difference [12]. The study's outcome also evaluates the impact of using Ground Control Points (GCPs) together with cameras in the photogrammetric analysis process. It notes that while using GCPs can reduce errors, especially in the Z direction, the reduction is not significant in X and Y component. Furthermore, GCPs together with cameras did not reduce total georeferencing errors to less than 10 cm for long-baseline methods (M4 and M5), indicating limitations in their effectiveness for certain scenarios [12].

4 Data Collection

4.1 Case Study Area and Related Data

An area of approximately 52 hectares in size is located at Kampung Santubong, Kuching Sarawak, Malaysia (Latitude: 1°43′06.45″ N, Longitude: 110°19′12.14″ E), with ground elevation between 2 and 15 m. Two reference stations were used for measurement for at least 2 h using a Spectra SP80. The GNSS receiver configuration was set accordingly, position dilution of precision (PDOP) mask of 4.0, an elevation mask of 15° and a logging interval of 1 s. GNSS data from measured reference stations were processed using the Trimble Business Center and Satlab post-processing software. The accuracy level of the reference stations was determined as 0.02 m in horizontal and 0.05 m in vertical. The coordinate system for horizontal is based on SDeG20 Datum. Nine (9) GCPs points were established and selected based on the surveyed boundary marks. The GCPs were observed using the GPS/GNSS static technique via SGeDnet (Fig. 1).

Fig. 1. The study area

4.2 GCPs Measurement and Establishment

GNSS Fieldwork was carried out on 2nd December 2023 with 9 h of continuous observation for 3 base stations (A108, A118 and A121) and followed by parallel observation of 10 GCPs. Base station A118 was a benchmark established by JUPEM(K312) with height value. With 3 operational CORS stations and 5 receivers used during the fieldwork, the baseline network established is depicted in Figs. 2 and 3.

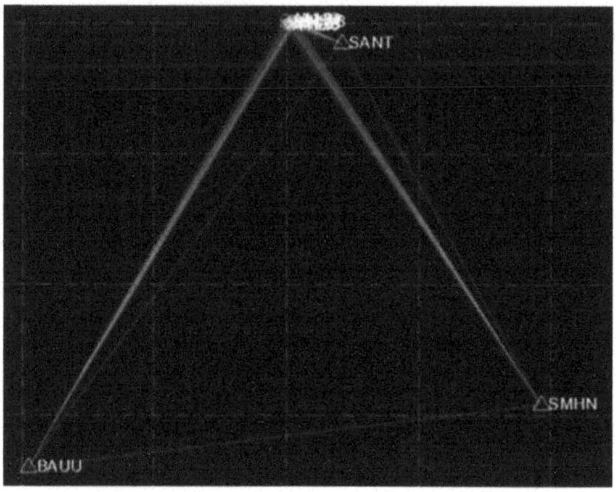

Fig. 2. Baseline SANT-BAUU-SMHN

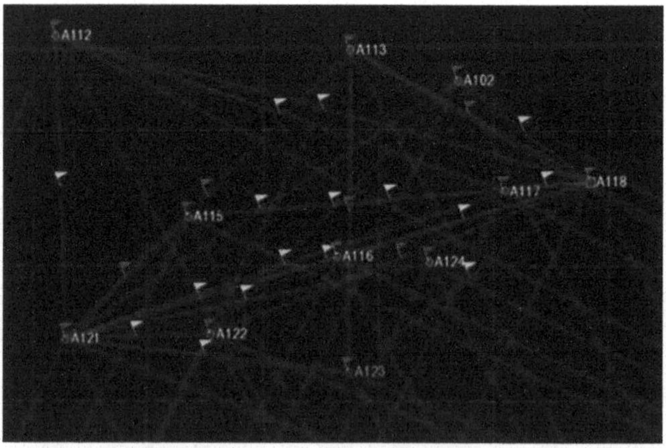

Fig. 3. Baseline network

Observation data for base station A108 could not be retrieved and only 2 base stations were utilized for data processing. Allowable tolerance of geocentric coordinate after fully constraint adjustment applied, must not exceed 3 cm in x and y, and 6 cm in orthometric height (Table 1, Fig. 4).

Table 1. Final coordinate of GCPs in geocentric datum, SGeD

Point	X(m)	Y(m)	Z(m)	Ellipsoidal height (m)
A102	−2214226.61995	5978460.94100	190199.46010	45.687
A112	−2213568.42320	5978692.63280	190264.53000	36.353
A113	−2214048.79170	5978519.79210	190244.28600	40.464
A115	−2213788.34610	5978619.23835	189997.53285	35.872
A116	−2214029.94570	5978531.56530	189938.12845	35.772
A117	−2214303.28140	5978432.84050	190035.80945	41.051
A122	−2213827.81090	5978611.90215	189824.22685	37.499
A123	−2214051.68480	5978530.87095	189768.12615	37.573
A124	−2214180.13250	5978477.45005	189930.44095	36.953
G118	−2214447.15330	5978387.34480	190050.06140	48.783
G121	−2213589.71880	5978698.60900	189816.50020	35.907

Fig. 4. GCP locations

4.3 UAV Image Acquisition

The aerial survey and image acquisition were captured on 4th December 2023, using a eBee X with a senseFly eBee S.O.D.A RGB 20 megapixel camera with GNSS. The UAV was flown approximately 120 m above the ground, collecting images at nadir with a forward overlap of 75% and a side lap 75% with an average of ground sample distance (GSD) of 3 mm. The flight speed was controlled between 13 and15 m/s, the surrounding temperature is 30 °C, and the camera start capturing images at 10:00 a.m. local time. The drone flew for 30 min and manage to capture 262 photos. A software named eMotion was used during image collection, with an ability to construct an autopilot flight path based on input parameters, including flying height, forward lap, and side lap. CORS SMHN was used as a base station for GNSS-PPK survey. The observations duration took about 1 h, and the base station located approximately 34 km from the case study area, known as a non-obstructed line of sight for UAV operation to connect with base station (Fig. 5).

Fig. 5. UAV planned flight path

4.4 Post-process Kinematic (PPK) Direct Geopositioning

The UAV was equipped with a third-generation Loki system from GeoCue, which estimated a priori X, Y, Z positions for each image after postprocessing in eMotion. An accuracy of 0.02 m was achieved for all images in both horizontal and vertical components. The imageries were processed in SGeD20 coordinate systems.

4.5 Continuously Operating Reference Stations (CORS)

The CORS used in our evaluation is SMHN, BAUU, TBDU, and TEMU (Table 2).

Table 2. CORS used in project

Station	Sample rate	Approx. Distance from study area (km)	Location
SMHN	1 s	35.17	Samarahan: N1° 27′ 15.1024″, E110° 29′ 28.7912"
BAUU	1 s	39.24	Bau: N1°24′ 37.2348″, E110° 08′ 34.5108″
TBDU	1 s	77.88	Tebedu: N1° 0′ 57.2099″, E 110° 21′ 17.6738″
TEMU	1 s	142.08	Temudok: N1° 9′ 38.6046″, E 111° 27′ 59.0921″

4.6 UAV Image Processing

Pix4DMapper software is used to process image datasets captured by UAV. Generally, a framework of Pix4DMapper includes automatic aerial triangulation (AAT), bundle block adjustment (BBA), point cloud, DSM, and orthophoto creation. The UAV manufacturer delivers an approximation of adjusted interior orientation parameters in the data processing stage.

5 Result and Analysis

5.1 Indirect Georeferencing (Arial Triangulation) and Direct Georeferencing

Image processing using Pix4D for Indirect Georeferencing using CORS(Samarahan) is shown as in Tables 3 and 4.

Localization accuracy per GCP and mean errors in the three coordinate directions. The last column counts the number of calibrated images where the GCP has been automatically verified versus manually marked.

Table 3. Indirect georeferencing (Aerial Triangulation)

Min error [m]	Max error [m]	Geolocation error X [%]	Geolocation error Y [%]	Geolocation error Z [%]
–	−3.46	0.00	0.00	0.00
−3.46	−2.77	0.00	0.00	0.00
−2.77	−2.07	0.00	0.00	0.00
−2.07	−1.38	0.00	0.00	1.91
−1.38	−0.69	0.00	0.00	12.98
−0.69	0.00	48.47	44.27	43.13
0.00	0.69	51.53	55.73	25.19
0.69	1.38	0.00	0.00	11.83
1.38	2.07	0.00	0.00	4.96
2.07	2.77	0.00	0.00	0.00
2.77	3.46	0.00	0.00	0.00
3.46	–	0.00	0.00	0.00
Mean [m]		0.006591	0.006472	−0.028043
Sigma [m]		0.214733	0.202587	0.723131
RMS error [m]		0.214834	0.202691	0.723674

Table 4. GCPs accuracy

GCP name	Accuracy XY/Z [m]	Error X [m]	Error Y [m]	Error Z [m]	Projection error [pixel]	Verified/marked
A102 (3D)	0.020/0.020	−0.008	0.010	−0.023	1.223	17/17
A112 (3D)	0.020/0.020	0.000	0.011	0.058	1.375	12/12
A113 (3D)	0.020/0.020	−0.042	−0.020	0.066	0.951	11/11
A115 (3D)	0.020/0.020	0.013	0.021	−0.039	1.111	16/16
A116 (3D)	0.020/0.020	0.009	0.017	−0.065	1.358	17/17
A117 (3D)	0.020/0.020	0.030	−0.012	−0.082	0.846	15/16
A118 (3D)	0.020/0.020	0.015	−0.003	−0.008	0.995	7/7
A122 (3D)	0.020/0.020	0.025	−0.025	−0.004	1.049	13/13
A123 (3D)	0.020/0.020	−0.013	−0.005	0.009	0.999	11/11
A124 (3D)	0.020/0.020	−0.023	−0.002	0.106	1.125	16/16
Mean [m]		0.000726	−0.000797	0.001820		
Sigma [m]		0.021288	0.014684	0.056545		
RMS Error [m]		0.021300	0.014705	0.056574		

6 Conclusion

The study has provided valuable insights into the performance of Continuous Operating Reference Stations (CORS) in UAV mapping applications. The key objective of the study was achieved by understanding the accuracy, reliability, consistency, and limitations of CORS-derived data in supporting UAV-based surveying and mapping tasks. An evaluation of positioning accuracy indicates that CORS-derived data generally aligns well with ground truth measurements or high-precision reference data sources with differences of less than 1.8 m.

References

1. Richard, A.S., Soler, T.: Continuously operating reference station (CORS): history, application, and future enhancements. J. Surv. Eng. **134**, 95–104 (2008)
2. Jawan, J., Musa, T.A., Aris, W.A.W., Putit, R., Abdillah, R., Usop, M.I.M.: Borneo triangulation 1968: a review and the way forward. Surv. Rev. (2023). https://doi.org/10.1080/00396265.2023.2207327
3. National Oceanic and Atmospheric Administration. http://www.ngs.noaa.gov/CORS/. Accessed 09 Feb 2024
4. McMahon, C., Mora, O.E., Starek, M.J.: Evaluating the performance of sUAS photogrammetry with PPK positioning for infrastructure mapping, Drones (2021)
5. Przybilla, H.-J., Bäumker, M.: RTK and PPK: GNSS-Technologies for direct georeferencing of UAV image flights. In: Proceedings of the FIG Working Week 2020, Amsterdam, The Netherlands, 10–14 May 2020, p. 52
6. The Canadian Spatial Reference System, CSRS. http://www.nrcan.gc.ca/earth-sciences/geomatics/geodetic-reference-systems/9052#vlbi. Accessed 14 Sept 2015
7. Li, B., Teunissen, P.J.G.: GNSS antenna array-aided CORS ambiguity resolution. J. Geodesy **88**(4), 363–376 (2013). https://doi.org/10.1007/s00190-013-0688-2
8. Tetsuro, I., Yoshiro, N.: GPS Earth Observation Network (GEONET) of Japan. 2011. http://fig.net/resources/proceedings/fig_proceedings/korea/full-papers/pdf/session11/imakiire-nakahori.pdf. Accessed 14 Sept 2015
9. Wu, Q., Kang, J., Li, S., Zhen, J., Li, H.: GNSS positioning by CORS and EGM2008 in Jilin Province, China. Published 4 December 2015
10. Hernandez-Andrade, D., Romero-Andrade, R., Sharma, G., Trejo-Soto, M.E., Cabanillas-Zavala, J.L.: Quality assessment of Continuous Operating Reference Stations (CORS)-GPS stations in Mexico. Geodesy Geodyn. **13**, 275e287 (2022)
11. Žabota, B., Kobal, M.: Accuracy Assessment of UAV-photogrammetric-derived products using PPK and GCPs in challenging terrains: in search of optimized rockfall mapping. Remote Sens. **13**, 3812 (2021)
12. Eker, R., Alkan, E., Aydın, A.: A comparative analysis of UAV-RTK and UAV-PPK methods in mapping different surface types. Eur. J. Forest Eng. **7**(1), 12–25 (2021)

Optimal Post-Processing Kinematic UAV Photogrammetry Mapping for Volumetric Sandy Beach Mapping

Nursyafiqah Jessica Georgeson[1](✉), Ahmad Razali Yusoff[2], Radzuan Sa'ari[1], Ilya Khairanis Othman[1], and Mohd Radhie Mohd Salleh[1]

[1] Department of Water and Environmental Engineering, Faculty of Civil Engineering, Universiti Teknologi Malaysia, UTM Skudai, 81310 Johor, Malaysia
nursyafiqahjessica@gmail.com

[2] Department of Geotechnics and Transportation, Faculty of Civil Engineering, Universiti Teknologi Malaysia, UTM Skudai, 81310 Johor, Malaysia

Abstract. Sandy beaches are well-known for their dynamic and complex environments. Traditional survey methods in these areas have limitations, such as requiring significant manpower, being time-consuming, and potentially expensive. These limitations have introduced the use of a more innovative and cost-effective, called Unmanned Aerial Vehicle (UAV). Generally, UAV Photogrammetry mapping requires Ground Control Points (GCPs). However, placing GCPs on beaches can be challenging due to accessibility issues and environmental factors. One practical method to overcome the issue is Post-Processing Kinematics (PPK) UAV Photogrammetry. The purpose of this study is to determine the optimal altitude for PPK UAV Photogrammetry beach mapping. The study uses CHCNAV AA450 for PPK UAV Photogrammetry and Light Detection and Ranging (LiDAR) mapping since the sensor has a built-in camera for aerial mapping. Three altitudes for PPK UAV Photogrammetry were selected, specifically 60 m, 100 m, and 140 m, over a horizontal distance of approximately 2 km. Data processing uses CoPre for LiDAR and Agisoft for PPK UAV Photogrammetry, while data analysis uses Global Mapper for volume estimation. The volume estimations calculated from Triangulated Irregular Network (TIN) and the results from PPK UAV Photogrammetry were compared to the result from LiDAR, as LiDAR can provide precise elevation data using high-resolution 3D point clouds acquired directly during data collection. The volume analysis demonstrates that the lowest altitude does not necessarily correspond to the optimal altitude for data collection. The accuracy assessment reveals that the 100 m altitude yields the most reliable results for volumetric measurements.

Keywords: Beach mapping · Unmanned Aerial Vehicle (UAV) · LiDAR · Post-Processing Kinematic (PPK)

A. Abdul-Rahman et al. (Eds.): GeoWeek 2024, LNGC, pp. 343–355, 2025.
https://doi.org/10.1007/978-3-031-86654-8_27

1 Introduction

Sandy beaches are often marked by continuous morphological changes due to natural processes and exacerbated by human activities. These characteristics make accurate measurements on the sandy beach challenging, especially for assessing beach changes over time. Beforehand, beach mapping was conducted using traditional methods such as ground survey using Total Stations (TS) and Global Positioning System (GPS) [1, 2] and transect survey using leveling instruments (Auto Level and Dumpy Level) [3]. However, it is well known that these methods are very time-consuming and require much manpower. These limitations have led to the idea for the discovery of new technology, namely the usage of Unmanned Aerial Vehicle (UAV) and Light Detection and Ranging (LiDAR) for ground survey and mapping, including beach mapping [4–12]. Since the introduction of the UAV, this device has been able to provide various data and analysis results, for example, in the form of beach volume [13]. Accordingly, the results of the beach mapping data can be used by many parties, such as local authorities, especially to monitor coastal erosion.

Generally, there are several types of UAV measurement techniques used for beach mapping purposes, including conventional UAV that require Ground Control Points (GCPs), UAV with Real-Time Kinematic (RTK), and UAV with Post-Processing Kinematic (PPK) as presented in Table 1. Conventional UAV captures aerial imagery and georeferenced the images using GCPs with known coordinates. Notably, this method can be tedious and time-consuming, particularly in remote or potentially hazardous areas. Using RTK UAVs can improve positioning accuracy with less or without GCPs by making real-time data corrections to the UAV from a base station or network. Unfortunately, UAV RTK requires stable and constant correction by the Radio Technical Commission for Maritime Services (RTCM). Structures, trees, and weather can impair signal reception, causing multipath errors [14, 15]. Meanwhile, PPK UAV records Global Navigation Satellite System (GNSS) raw data during the flight and applies corrections during post-processing. The corrections are implemented post-flight using data from a base ground station, ensuring exceptional accuracy even under difficult conditions [16–18].

The UAV Photogrammetry technology can benefit many areas, including beach mapping. Apart from capturing aerial photo images, UAVs can also produce data that can also be used for volume estimations. Corrected aerial photo images can be processed to obtain point cloud data using a photogrammetric method called Structure from Motion (SfM) [19]. The SfM method involves producing a Digital Surface Model (DSM), which can then be converted to a Digital Terrain Model (DTM) or Digital Elevation Model (DEM) using Agisoft software.

DEM is a numerical representation of topography comprised of equal-sized grid cells with elevation values. The simple data structure and universal availability of the DEM have made it a popular land characterization tool [20]. DEM can be constructed from regular or irregular point clouds representing the earth's surface. One of the most popular models for transforming point clouds into a continuous surface representation is the Triangulated Irregular Network (TIN) [20, 21]. Generally, TIN creates the triangle network using Delaunay triangulation. This method assures that no vertex is inside any triangulation triangle's circumcircle. It avoids skinny triangles by maximizing the minimum angle of all triangulation triangle angles.

Table 1 summarizes the previous research conducted using UAV sensors for beach mapping purposes.

Table 1. Literature review on UAV beach mapping

Literature	Method of UAV photogrammetry	GCP distribution
[22–27] [28, 29]	Conventional UAV	With GCP
[30, 31]	RTK UAV	With GCP
[15]	PPK UAV	With & without GCP
[4]	PPK UAV	With GCP

Several research projects have conducted a limited number of beach mapping utilizing PPK UAV Photogrammetry, namely in references [4, 15]. This study aims to conduct research using PPK UAV Photogrammetry to obtain accurate measurements of sandy beach volume, with the goal of further investigating the mapping of linear beach volume without GCP distribution. This study also includes the determination of the optimal altitude for PPK UAV Photogrammetry flight to get the volume estimation of sandy beaches at different altitudes. The volume estimation results were compared to the volume estimation obtained from LiDAR. LiDAR can provide precise elevation data of coastal terrain, allowing for accurate mapping of shoreline positions, beach profiles, and volumetric changes [4, 12, 32]. For ease of reference, the term 'PPK UAV' will be used in this study to denote the complete method name 'PPK UAV Photogrammetry'.

In addition, the UAV itself has some technical issues that need to be considered before planning a UAV mapping project. Several issues highlighted are battery usage during each flight, flight duration, flight regulations, data processing issues involving a large number of captured UAV images, storage requirements size, and the need for high-powered computing devices to process extensive data. The issues need to be considered and minimized. Therefore, it is crucial to determine the optimal altitude for UAV flights.

2 Study Area

The study area in Fig. 1 is 2 km along the beach at Damat Beach, Bachok, Kelantan (approximate latitude and longitude of 60°04′45.2″N 102°23′40.0″E). Damat Beach is located near Irama Beach, Bachok, Kelantan. This coastal area is renowned for its ecological, economic, and cultural significance to the local communities. The coastal economic sector encompasses activities such as fishing and producing dried seafood. It serves as a means of livelihood for the local communities [33]. Given the significance of this location to the local economy, it is crucial to conduct a comprehensive study and research on the beach mapping of this area. This will enable the local authorities to implement conservation and mitigation measures effectively.

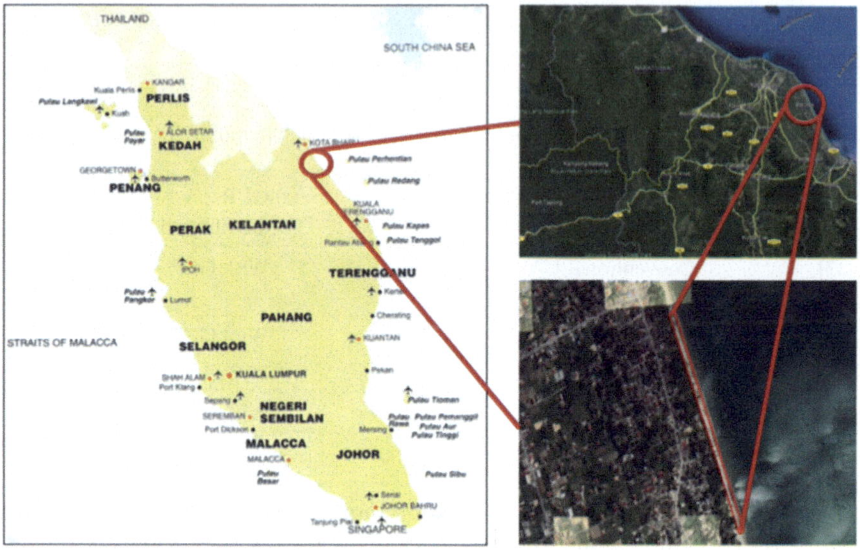

Fig. 1. The study area at Damat Beach, Bachok, Kelantan.

3 Research Tools

In this study, the CHCNAV AA450 was used for both PPK UAV and LiDAR mapping since the tool has a built-in camera for photogrammetry mapping (see Fig. 2). The DJI Matrice 300 in this study is solely as a carrier for the AA450. Table 2 summarizes the specifications for each tool.

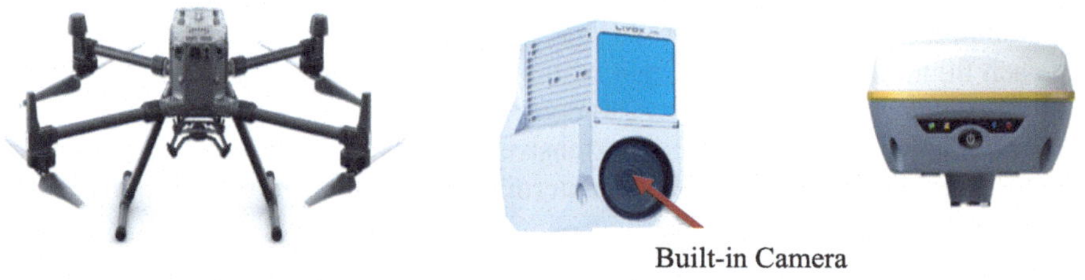

(a) (b) (c)

Fig. 2. UAV, LiDAR, and PPK-GNSS module: (a) DJI Matrice M300; (b) AA450 LiDAR; and (c) South G7 GNSS receiver.

4 Research Methodology

This study consists of several phases, namely, preliminary study, data collection, data processing, and data analysis. The preliminary study needs to be conducted through a thorough reading of UAV Photogrammetry and LiDAR methods. As for data collection,

Table 2. Specification of the tools used in the study.

Type	Specification
DJI Matrice M300	GPS, GLONASS, BeiDou, Galileo Accuracy; Horizontal 1 cm + 1 ppm, Vertical 1.5 cm + 1 ppm
AA450 LiDAR	Dual frequency GNSS GPS, GLONASS, BeiDou, Galileo Accuracy: Horizontal 0.010 m RMS, Vertical 0.020 m RMS GPS, GLONASS, BeiDOu, Galileo, SBAS, IRNSS, QZSS High-Performance Livox Avia laser scanner Built-in calibrated Camera: Sony A5100 Camera pixels: 24.3 MP, 11 fps Resolution camera: 6000 × 4000 Focal length: 16 mm
South G7 GNSS receiver	Accuracy; Horizontal 2.5 mm + 0.5 ppm RMS, Vertical 3.5 mm + 0.5 mm RMS (Static)

Fig. 3. South G7 GNSS receiver with pole and bipod at the base station.

this study was initiated by placing a base or static station in an open and obstacle-free place, as displayed in Fig. 3. Prior to recording the static data, the GNSS receiver is used to establish a connection with the JUPEM RTKNet system. This is a crucial step in determining and observing the coordinates of the base station or GNSS receiver position in their true coordinates and defining the Z value relative to the Mean Sea Level (MSL). This step is necessary since the GNSS receiver position and base station coordinates would be expressed in WGS84 format [34].

For this study, data collection took approximately two hours and thirty minutes, and each flight was estimated to take less than 30 min due to UAV battery constraints. In this study, three UAV flight altitudes were used: 60 m, 100 m, and 140 m (see Fig. 4). The proposed minimum height for this study is 60 m to avoid any obstacles during the data collection for safety reasons. The next two altitudes are selected based on intervals of every 40 m. This phase focuses on collecting and processing data to generate DTM/DEM for volume calculation at these varying flight altitudes. This is to detect the changes in volumetric data calculation obtained from three different flying altitudes.

UAV images involve the SfM photogrammetric method to transform overlapping images into a DTM/DEM. The process involves collecting numerous overlapping photos, identifying unique characteristics, estimating camera positions and orientations, generating a sparse 3D point cloud, increasing its density, transforming it into a seamless surface, and projecting the 3D surface onto a grid [19]. Meanwhile, LiDAR sensors directly collect a high-resolution 3D point cloud by measuring the distance between the sensor and the ground or objects [35]. The data obtained were processed using CoPre for LiDAR and Agisoft for UAV photogrammetry. Consequently, they analyzed the volume estimation within the Area of Interest (AOI) using Global Mapper software to pre-determine the optimal flight altitude of the UAV photogrammetry.

In this study, the AOI is divided into six sections from the total study distance (2 km). For more accurate ground analysis, vegetated and sea areas must be avoided during the selection of the AOI. The research methodology flowchart is depicted below in Figs. 5 and 6, illustrating the selection of the AOI.

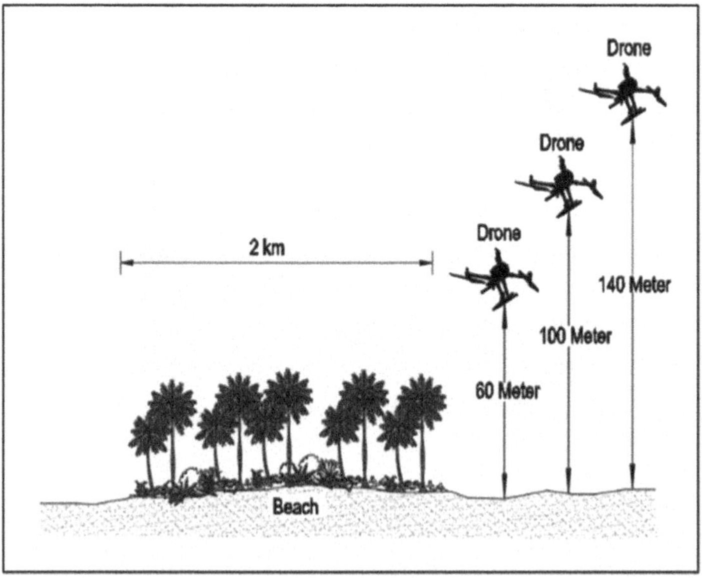

Fig. 4. Illustration of UAV data collection using different flight altitudes.

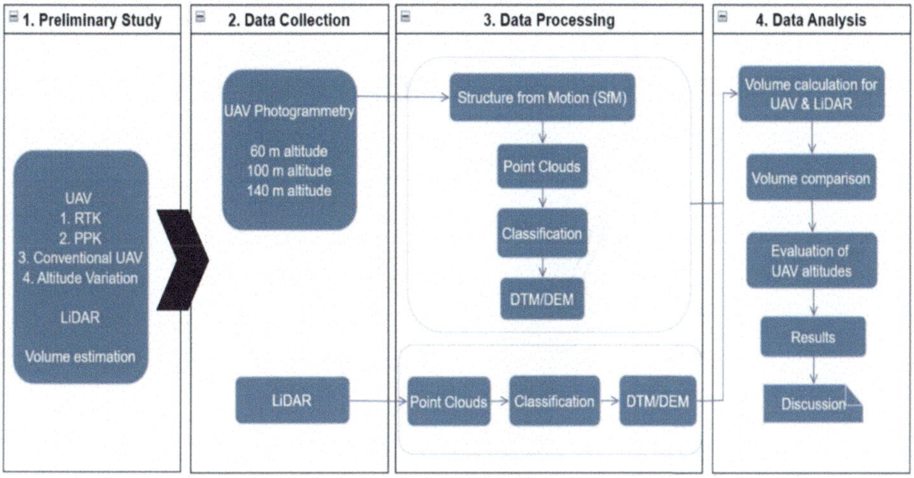

Fig. 5. Flowchart of the research methodology for this study.

5 Results and Analysis

During data collection, which was conducted on 5th March 2024, DSM was successfully produced for further analysis to analyze volume estimation at each PPK UAV's altitude of 60 m, 100 m, and 140 m. The DSM for each altitude was then converted to DEM for volume estimation using the TIN approach. During data collection, 695 images were captured from 60 m altitude, 597 images were captured from 100 m altitude, and lastly, 452 images were captured from 140 m altitude. This finding also proved that the lower the flight altitude, the more images are captured. Figure 7 illustrates the sample of processing results of the DEM from each PPK UAV and LiDAR mapping at various altitudes within the same AOI.

To calculate the volume represented by a TIN, the volume between a TIN surface and a reference plane can be determined by adding up the volume of every single vertical triangle. The reference plane could be based on MSL or a specified reference elevation plane. In this study, a reference plane at an elevation of 0 m above MSL was selected for volume estimation purposes.

Theoretically, according to Blue Marble Geographics [42], the volume estimation process using Global Mapper software involves several steps. The step involves calculating the surface area, finding the average elevation difference, calculating the volume for a single triangle, and finally, summing the total volume of each triangle within each AOI. The formula for surface area uses the Shoelace theorem, also known as Gauss's area formula, which is often applied to cartesian coordinates in a plane to determine the surface area of a polygon [36]. The calculation for volume estimation using TIN [37, 38] within AOI is described below in Eq. (1).

$$V = \sum_{i=1}^{n} (A_i \times \Delta h_i) \tag{1}$$

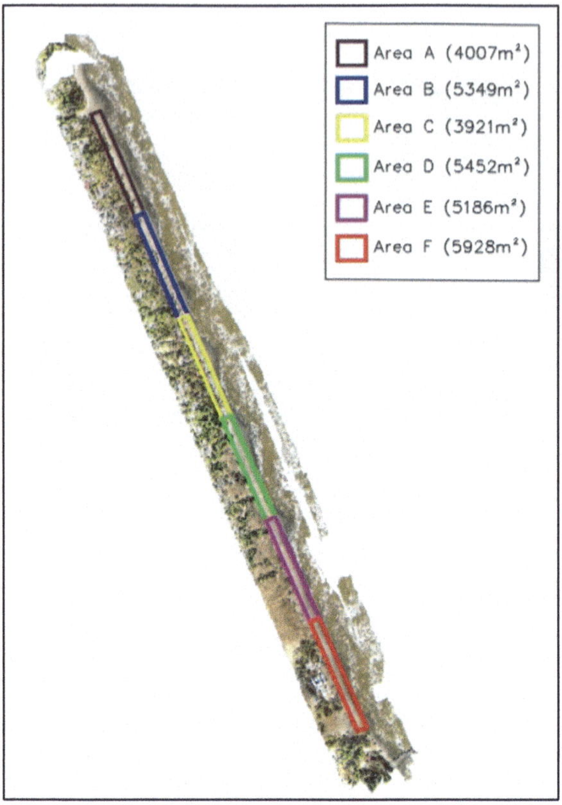

Fig. 6. DSM of the site area and the AOI (6 areas).

where V is the total volume, n is the total number of triangles built by TIN, A_i is the surface area of a single triangle ($i = 1, 2, 3, \ldots, n$), and Δh_i is the average height based on the difference between the elevation values of the triangle's vertices and the reference plane elevation. The results of volume estimation within the AOI for PPK UAV at 60 m, 100 m, 140 m, and for LiDAR at 60 m are presented in Table 3 and Fig. 8.

To evaluate the accuracy for each altitude, further analysis was conducted to determine which PPK UAV's altitude has the nearest volume estimation value to the reference value (LiDAR's volume). This was calculated using the formula of Mean Absolute Error (MAE) and the Mean Absolute Percentage Error (MAPE) analysis. MAE assesses the precision of a model's estimation in regression tasks by indicating how close the volume estimations are to the reference volume value, with lower values representing better accuracy [39, 40]. Meanwhile, MAPE calculates the error by representing it as a percentage of the true value [41]. The formula of MAE and MAPE are given in Eqs. (2) and (3) below, respectively.

$$MAE = \frac{1}{n}\sum_{i=1}^{n}|y_i - \hat{y}_i| \quad (2)$$

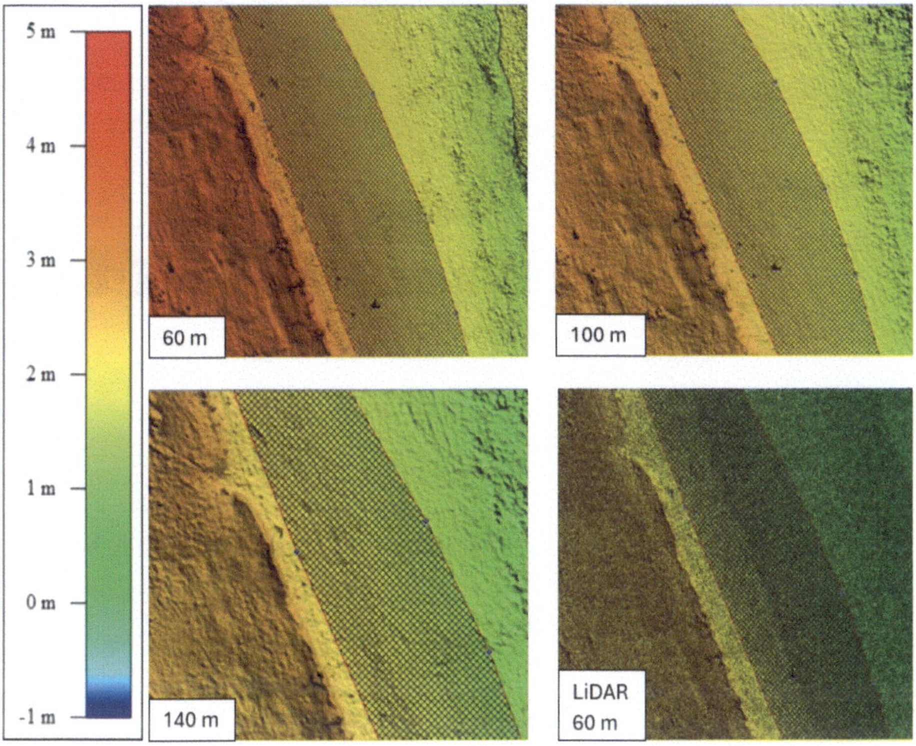

Fig. 7. Sample DEM from PPK UAV and LiDAR mapping.

Table 3. Results of volume estimation for PPK UAV and LiDAR, based on different flight altitudes.

Altitudes (m)	Total Vol. (m^3)	A (m^3)	B (m^3)	C (m^3)	D (m^3)	E (m^3)	F (m^3)
60	58791	7680	10944	7802	10990	10266	11109
100	58699	7771	10924	7769	10905	10271	11059
140	59064	7972	10919	7817	11001	10275	11080
60 (LiDAR)	57544	7899	10752	7691	10859	10661	9682

$$MAPE = \frac{1}{n}\sum_{i=1}^{n}\left|\frac{y_i - \hat{y}_i}{y_i}\right| \times 100 \quad (3)$$

where n is the total number of the AOI, for y_i, $i = 1, 2, \ldots, n$. In this formula, y_{i} is the reference volume estimation from LiDAR. Meanwhile, $\hat{y}_{\text{i}}$ is the volume estimation within each of the AOI for PPK UAVs. The result of MAE and MAPE calculation is represented in Table 4.

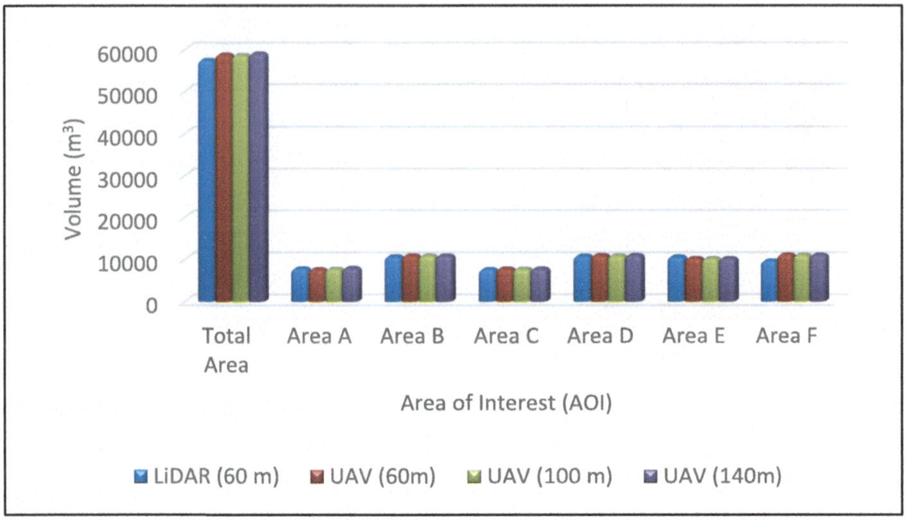

Fig. 8. Volume comparison between PPK UAV at different altitudes with LiDAR.

Table 4. The result of MAE and MAPE calculation.

UAV altitude (m)	MAE (m^3)	MAPE (%)
60	412.9	2.1
100	365.3	2.0
140	382.2	2.7

6 Discussion and Conclusion

In this study, although 60 m is the lowest altitude, it does not indicate that the lowest altitude is the most optimal for PPK UAV flights. Based on the results of the data analysis in Table 4, it reveals that the 100 m altitude of PPK UAV has the lowest MAE (365.3 m^3) and MAPE (2.0%). It concludes that the volume estimation using data from 100 m altitude is the nearest to the volume estimation from LiDAR (reference data) and represents better accuracy compared to 60 m and 140 m altitudes. Furthermore, at an altitude of 100 m, PPK UAV flights for photogrammetry mapping strike an optimal balance between image quality and processing efficiency. This altitude is sufficiently high to cover larger areas with more endurance and fewer images compared to lower altitudes, such as 60 m, requiring more images to achieve the same coverage. This concludes that, in this study, the 100 m altitude of PPK UAV is the most optimal and good enough for beach mapping purposes.

7 Limitations and Recommendations

This research was only conducted at three different altitudes: 60 m, 100 m, and 140 m. For future research, it is recommended to conduct comprehensive studies comparing various altitudes (e.g., 60 m, 80 m, 100 m, and 120 m) in terms of image resolution, processing time, and data accuracy. Such studies should include diverse coastal environments to identify the optimal altitude for different coastal features. Another recommendation for future study is to suggest the research of the application of advanced data processing techniques such as machine learning and Artificial Intelligence (AI) to improve the accuracy and efficiency of image stitching and feature extraction at various altitudes. This could also include automated anomaly detection in coastal environments.

Acknowledgment. The authors would like to express their appreciation to Universiti Teknologi Malaysia for funding this research through the NEXUS Young Researcher scheme (UTMER grant number Q.J130000.3822.42J03).

References

1. Lee, J.-M., Park, J.-Y., Choi, J.-Y.: Evaluation of sub-aerial topographic surveying techniques using total station and RTK-GPS for applications in macrotidal sand beach environment. J. Coast. Res. **65**, 535–540 (2013). https://doi.org/10.2112/si65-091.1
2. Huang, J., Jackson, D.W.T., Cooper, J.A.G.: Morphological monitoring of a high energy beach system using GPS and total station techniques, Runkerry, Co. Antrim, Northern Ireland. J. Coast. Res. **36**, 390–398 (2002). https://doi.org/10.2112/1551-5036-36.sp1.390
3. Ravindran, A., Ramanujam, N., College, V.O.C.: Experimental design and continuous monitoring of costal hydraulic study of beach ground water table at Vellappatti Shoreline, Thoouthukudi **1**, 2–10 (2013)
4. Shaw, L., Helmholz, P., Belton, D., Addy, N.: Comparison of UAV lidar and imagery for beach monitoring. Int. Archiv. Photogramm., Remote Sens. Spatial Inf. Sci. - ISPRS Archives, 589–596. International Society for Photogrammetry and Remote Sensing (2019)
5. Pinton, D., Canestrelli, A., Moon, R., Wilkinson, B.: Estimating ground elevation in coastal dunes from high-resolution UAV-LIDAR point clouds and photogrammetry. Remote Sens. **15** (2023). https://doi.org/10.3390/rs15010226
6. Zanutta, A., Lambertini, A., Vittuari, L.: UAV photogrammetry and ground surveys as a mapping tool for quickly monitoring shoreline and beach changes. J. Mar. Sci. Eng. **8** (2020). https://doi.org/10.3390/JMSE8010052
7. Huang, C., Zhang, H., Zhao, J.: High-efficiency determination of coastline by combination of tidal level and coastal zone DEM from UAV tilt photogrammetry. Remote Sens. **12** (2020). https://doi.org/10.3390/rs12142189
8. Yin, Z.: Analysis of coastal erosion between different flights of UAV (2021)
9. Gonzales, J.J., Pingel, T., Dura, T., Shao, Y.: Comparing UAV and pole photogrammetry for monitoring beach erosion (2021)
10. Taveneau, A., Almar, R., Bergsma, E.W.J., Sy, B.A., Ndour, A., Sadio, M., Garlan, T.: Observing and predicting coastal erosion at the langue de barbarie sand spit around saint louis (Senegal, West Africa) through satellite-derived digital elevation model and shoreline. Remote Sens. **13** (2021). https://doi.org/10.3390/rs13132454

11. Hoffmann, T.K., Pfennings, K., Hitzegrad, J., Brohmann, L., Welzel, M., Paul, M., Goseberg, N., Wehrmann, A., Schlurmann, T.: Low-cost UAV monitoring: insights into seasonal volumetric changes of an oyster reef in the German Wadden Sea. Front. Mar. Sci. **10** (2023). https://doi.org/10.3389/fmars.2023.1245926
12. Wang, J., et al.: An overview of Shoreline mapping by using airborne LiDAR. Remote Sens. **15**, 1–28 (2023). https://doi.org/10.3390/rs15010253
13. Ponte Lira, C., Taborda, R., Silva, A.N., Andrade, C.: Challenges and new strategies in assessing multidecadal shore platform sandy beach evolution from aerial imagery. Mar. Geol. **436** (2021). https://doi.org/10.1016/j.margeo.2021.106472
14. Czyża, S., Szuniewicz, K., Kowalczyk, K., Dumalski, A., Ogrodniczak, M., Zieleniewicz, Ł: Assessment of accuracy in Unmanned Aerial Vehicle (UAV) pose estimation with the REAL-time kinematic (RTK) method on the example of DJI Matrice 300 RTK. Sensors **23**, 2092 (2023). https://doi.org/10.3390/s23042092
15. Taddia, Y., Stecchi, F., Pellegrinelli, A.: Coastal mapping using DJI phantom 4 RTK in post-processing kinematic mode. Drones **4**, 1–19 (2020). https://doi.org/10.3390/drones4020009
16. Dinkov, D., Kitev, A.: Advantages, disadvantages and applicability of GNSS Post-Processing Kinematic (PPK) method for direct georeferencing of UAV image proceedings. In: 8th International Conference on Cartography and GIS, Vol. 1, pp. 747–759 (2020)
17. Tamimi, R., Toth, C.: Assessing the viability of PPK techniques for accurate mapping with UAS. Int. Arch. Photogramm. Remote Sens. Spat. Inf. Sci. - ISPRS Archiv. **48**, 479–488 (2023). https://doi.org/10.5194/isprs-archives-XLVIII-1-W1-2023-479-2023
18. Zhang, H., Aldana-Jague, E., Clapuyt, F., Wilken, F., Vanacker, V., Van Oost, K.: Evaluating the potential of post-processing kinematic (PPK) georeferencing for UAV-based structure-from-motion (SfM) photogrammetry and surface change detection. Earth Surf. Dyn. **7**, 807–827 (2019). https://doi.org/10.5194/esurf-7-807-2019
19. Villanueva, J.K.S., Blanco, A.C.: Optimization of ground control point (GCP) configuration for unmanned aerial vehicle (UAV) survey using structure from motion (SFM). Int. Archiv. Photogramm., Remote Sens. Spat. Inf. Sci. - ISPRS Archiv. 167–174. International Society for Photogrammetry and Remote Sensing (2019)
20. Bandara, K.R.M.U., Samarakoon, L., Shrestha, R.P., Kamiya, Y.: Automated generation of digital terrain model using point clouds of digital surface model in forest area. Remote Sens. **3**, 845–858 (2011). https://doi.org/10.3390/rs3050845
21. Yilmaz, N.: Comparing the volume methods through using digital elevation models created by different interpolation methods. Fresenius Environ. Bull. **26**, 4734–4741 (2017)
22. Yusoff, A.R., Darwin, N., Majid, Z., Razali, A.F., Mohd Ariff, M.F.: Beach volume measurement on variation of UAV altitude mapping. In: 2020 IEEE 10th International Conference System Engineering Technology ICSET 2020—Proceedings M, pp. 145–149 (2020). https://doi.org/10.1109/ICSET51301.2020.9265390
23. Abdurrahman, U., Pratyaksa, I.F., Suprijo, T., Park, H.: Beach slope distribution mapping using UAV in the Cirebon coastal area. In: E3S Web of Conferences. EDP Sciences (2021)
24. Di Paola, G., Minervino Amodio, A., Dilauro, G., Rodriguez, G., Rosskopf, C.M.: Shoreline evolution and erosion vulnerability assessment along the central adriatic coast with the contribution of UAV beach monitoring. Geoscience **12** (2022). https://doi.org/10.3390/geosciences12100353
25. Laporte-Fauret, Q., Marieu, V., Castelle, B., Michalet, R., Bujan, S., Rosebery, D.: Low-Cost UAV for high-resolution and large-scale coastal dune change monitoring using photogrammetry. J. Mar. Sci. Eng. **7** (2019). https://doi.org/10.3390/jmse7030063
26. Yoo, C.I., Oh, T.S.: Beach volume change using UAV photogrammetry Songjung beach, Korea. Int. Archiv. Photogramm. Remote Sens. Spat. Inf. Sci. - ISPRS Archiv. 1201–1205. International Society for Photogrammetry and Remote Sensing (2016)

27. Vecchi, E., Tavasci, L., De Nigris, N., Gandolfi, S.: Gnss and photogrammetric UAV derived data for coastal monitoring: a case of study in Emilia-Romagna, italy. J. Mar. Sci. Eng. **9** (2021). https://doi.org/10.3390/jmse9111194
28. Burdziakowski, P., Specht, C., Dabrowski, P.S., Specht, M., Lewicka, O., Makar, A.: Using UAV photogrammetry to analyze changes in the coastal zone based on the sopot tombolo (Salient) measurement project. Sensors (Switzerland) **20**, 1–21 (2020). https://doi.org/10.3390/s20144000
29. Lu, C.H.: Applying UAV and photogrammetry to monitor the morphological changes along the beach in Penghu islands. In: Int. Archiv. Photogramm. Remote Sens. Spat. Inf. Sci. - ISPRS Archiv. 1153–1156. International Society for Photogrammetry and Remote Sensing (2016)
30. Casella, E., Drechsel, J., Winter, C., Benninghoff, M., Rovere, A.: Accuracy of sand beach topography surveying by drones and photogrammetry. Geo-Marine Lett. **40**, 255–268 (2020). https://doi.org/10.1007/s00367-020-00638-8
31. Bertin, S., Stéphan, P., Ammann, J.: Assessment of RTK quadcopter and structure-from-motion photogrammetry for fine-scale monitoring of coastal topographic complexity. Remote Sens. **14** (2022). https://doi.org/10.3390/rs14071679
32. Robertson, Q., Wozencraft, J., Dong, Z., Pierro, T., Zhang, K.: Demonstrated coastal engineering applications using LIDAR data. J. Coast. Res. **39**, 334–346 (2023). https://doi.org/10.2112/JCOASTRES-D-22-00036.1
33. Wannahari, R., Abdullah, N.A., Nordin, M.F.M., Muhammad, M.: Evaluation of heavy metal in coastal water at Kelantan. Am. J. Environ. Sci. **9**, 505–510 (2022). https://doi.org/10.3844/ajessp.2013.505.510
34. Zainal, M.Z., Talib, K.H., Mohamed, A.: Orthometric height determination through MyRTKnet virtual reference station (2015)
35. Pozarowski, P., Holden, E., Darzynkiewicz, Z.: Laser scanning cytometry: principles and applications-an update. Methods Mol. Biol. **931**, 187–212 (2012). https://doi.org/10.1007/978-1-62703-56-4_11
36. Lee, Y., Lim, W.: Shoelace formula: connecting the area of a polygon and the vector cross product. Math. Teach. **110**, 631–636 (2017). https://doi.org/10.5951/mathteacher.110.8.0631
37. Siriba, D.N., Matara, S.M., Musyoka, S.M.: Improvement of volume estimation of stockpile of earthworks using a concave hull-Footprint. Int. Sci. J. Micro Macro Mezzo Geoinf. **5**, 11–25 (2015)
38. Yongxiao, W., Xitun, Y., Xiongfei, Y., Hang, Z., Qi, H.: Research on earthwork calculation based on TIN model. E3S Web Conf. **248**, 3–6 (2021). https://doi.org/10.1051/e3sconf/202124803001
39. Willmott, C.J., Matsuura, K.: Advantages of the mean absolute error (MAE) over the root mean square error (RMSE) in assessing average model performance. Clim. Res. **30**, 79–82 (2005). https://doi.org/10.3354/cr030079
40. Hodson, T.O.: Root-mean-square error (RMSE) or mean absolute error (MAE): when to use them or not. Geosci. Model Dev. **15**, 5481–5487 (2022). https://doi.org/10.5194/gmd-15-5481-2022
41. Tofallis, C.: A better measure of relative prediction accuracy for model selection and model estimation. J. Oper. Res. Soc. **66**, 1352–1362 (2015). https://doi.org/10.1057/jors.2014.103
42. Calculating Cut-and-Fill Volumes for Area and Line Features. https://www.bluemarblegeo.com/knowledgebase/global-mapper-20/index.htm. Last Accessed 23 July 2024

Cross-Sensor Registration Between Ground-Based Mobile Lidar and Airborne UAV Images

Tee-Ann Teo(✉), Hao Yu, and Pei-Cheng Chen

National Yang Ming Chiao Tung University, Hsinchu 30010, Taiwan
{tateo,yuhao2001.en12,a4623217.en07}@nycu.edu.tw

Abstract. This study investigates a cross-sensor point cloud registration methodology, utilizing roadmarks as robust semantic features due to their consistent visibility and uniform distribution in urban areas. The approach integrates data from ground-based mobile mapping systems (MMS) and unmanned aerial vehicles (UAVs). Key steps include selecting roadmarks, establishing graph nodes at their centroids, forming geometric structures based on node distances and angles, applying the Hungarian algorithm for bipartite graph matching, and minimizing an energy function to ensure geometric consistency. Experimental results, using data from a Velodyne VLP32 laser scanner and a DJI P4P UAV, show that the method achieves a 70% success rate in matching after extensive rotation and scaling tests. For real data, the maximum error between two sensors was reduced from 81 cm to 18 cm post-registration, demonstrating improved accuracy and reliability in multi-sensor integration.

Keywords: Cross sensor · Graph-matching · MMS · UAV

1 Introduction

Cross-sensor integration enables the creation of more comprehensive point cloud data by combining information from various platforms. For example, mobile mapping systems (MMS) provide detailed vertical facade data, while unmanned aerial vehicles (UAVs) offer top-view surface data. Integrating airborne and ground-based sensors has the advantage of constructing complete three-dimensional data [3]. There are some challenges for these ground-based and aerial sensors. For example, there is a lack of GNSS signal for MMS in urban corridors with dense urban buildings; the viewing angles of MMS and UAV are different. Therefore, the dataset taken by these two platforms cannot be integrated directly.

The core of point cloud registration is feature extraction and matching, which typically employs three types of geometrical features: feature-based, line-based, and surface-based. Feature-based methods, the most common, involve detecting and describing feature points and then utilizing linear and surface features to address disparities in point cloud density between datasets. Such as [11] apply voxel-based regularization to the

A. Abdul-Rahman et al. (Eds.): GeoWeek 2024, LNGC, pp. 356–365, 2025.
https://doi.org/10.1007/978-3-031-86654-8_28

point clouds and calculate the plane features within each voxel. Subsequently, the least squares method was applied to minimize the corresponding plane features and achieve cross-sensor integration of airborne and mobile laser scanning (MLS) systems. However, the following challenges make point cloud registration tasks difficult: (1) Point cloud registration requires good initial values; (2) A large amount of 3D point cloud data leads to high computational time costs, requiring efficient processing; (3) Differences in data coverage due to varying viewpoints make it difficult to find strict feature correspondences [6]. Aerial imagery and vehicle-mounted lidar are heterogeneous data sources, with UAV cameras taking 2D images and MLS providing 3D point clouds. Therefore, transforming 3D point clouds into 2D grided data could be a solution for registration between UAV cameras and mobile laser scanners.

Previous studies have shown that building areas often serve as effective geometrical features, with contours becoming graph nodes that successfully address the registration of different datasets from different viewpoints and achieving a registration accuracy of 1 to 2 pixels [8]. Li et al. [6] applied the simplified building boundaries as linear features to find corresponding feature line segments to integrate photogrammetry and MLS point clouds. The matched feature line segments can be used to integrate cross-sensor viewpoints and scale differences.

In particular, there might be a lack of suitable building information in urban areas. Instead, roadmarks are unique features in advanced urban areas. They can be captured by both UAV and MLS sensors, making them robust features for analysis. Javanmardi et al. [5] segmented the roadmarks into patches as features and used normal distribution transformation (NDT) to locally registration the MMS data to the reference aerial imagery. This approach reduced the 2D error from 99.7 cm to 11.6 cm, which is smaller than a single pixel (i.e., 12 cm/pixel) in the aerial imagery. Instead of converting multi-view aerial images into point clouds, [4] rasterized the MLS road points into a 2D intensity map. The idea was to find the 2D correspondences in image space. They detected feature points using Harris corner detection and described features with LATCH (Learned Arrangements of Three Patch Codes). Perform feature matching with the clipped tiles of aerial orthoimages. Although the results achieved high accuracy from matched features, the matched features showed more significant errors than the manually digitized checkpoint.

The lidar interpolated data represent intensity, while airborne orthoimages provide spectral color information, resulting in substantial grayscale value discrepancies. Due to their distinct characteristics, traditional methods struggle to perform feature matching on heterogeneous 2D images. Therefore, additional constraints, such as geometrical constraint, have to be added to the cross-sensor matching. Graph-based matching is an essential image registration method utilizing topological relationships between nodes. This approach ensures overall structural consistency on a global scale, thereby increasing the robustness of the registration process [7].

This study aims to establish cross-sensor data matching by integrating data collected from different platforms, i.e., ground-based mobile lidar mapping and airborne UAV photogrammetry systems. The approach involves finding the optimal geometric relationship by combining semantic roadmark features with a graph structure, enabling feature-oriented graph matching. Finally, data co-registration is achieved by finding the transformation parameters of the corresponding points.

2 Methodology

The proposed method leverages the distinct advantages of roadmarks over building features for cross-sensor matching. Unlike buildings, roadmarks are more consistently and accurately captured by various sensors. Their generally uniform and sample distribution in urban areas makes them robust semantic features for the purposed method. Besides, the effect of relief displacement of roadmark is usually smaller than building for center perspective image. The input data are airborne orthoimages from UAV and ground-based intensity maps from MLS for the road region. The proposed method for this study is shown in Fig. 1.

By focusing on roadmarks and utilizing a combination of geometric structures and advanced matching algorithms, the proposed method provides a robust framework for cross-sensor point cloud registration. This approach addresses the common challenges of feature consistency and geometric alignment, resulting in improved accuracy and reliability in multi-sensor integration. The proposed method comprises several key steps:

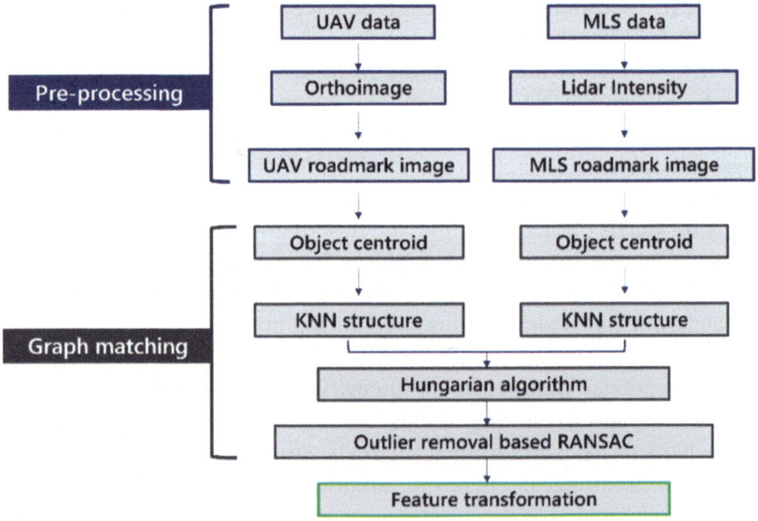

Fig. 1. Workflow of the proposed scheme

(1) Data preprocessing: The original UAV photogrammetry and ground-based laser scanning system datasets are 2D images and 3D point clouds. Therefore, these two data forms need to be unified. Then, UAV multi-view images are processed through aerial triangulation and orthorectification to generate orthoimage. The lidar point clouds are interpolated into grided data with the point cloud intensity value.

(2) Road marks extraction: The goal is to perform cross-sensor matching using roadmark features. The preprocessing stage uses DeepLab v3+ [1] semantic segmentation to train two deep learning models to extract road features from color images and lidar intensity images, respectively. A comprehensive training dataset is generated by manually annotating the color orthoimage and intensity maps with ground truth roadmark labels for accurate roadmark extraction. Types of roadmark include lane line, stop line, direction line, pedestrian crossing line. The annotated dataset spans diverse scenarios, encompassing various road types and marking geometries. During training, the deep learning model learns to correlate specific image patterns with corresponding roadmark classes, producing robust roadmark extractions.
(3) Geometric structure formation: The first step of graph matching [9] is graph node establishment. The centroids of these roadmarks are identified and used as graph nodes. This step simplifies the representation and comparison of features across datasets. These object centroids are used to calculate the adjacent distance between objects. We establish geometric structures based on the distances and angles between these nodes. This step creates a relational framework that can be analyzed to identify correspondences between different datasets.
(4) Bipartite graph matching: We apply the Hungarian algorithm for bipartite graph matching to identify corresponding features between heterogeneous datasets. This algorithm efficiently finds the optimal matching by minimizing the total cost of assignments [10]. Finally, we resolve the graph-matching problem by minimizing an energy function that accounts for the geometric discrepancies between the ground-based and airborne sensors. This function ensures that the matched features are geometrically consistent, leading to accurate registration.
(5) Data co-registration: The centroids of matched roadmarks represent the tie points between color orthoimage and lidar intensity image. This study uses an affine transformation to convert the MLS to UAV coordinates system. Moreover, RANSAC [2] is applied to remove the outliers. The remaining tie points are used to calculate the precise coefficients for data co-registration.

3 Experimental Results

This study used airborne orthoimage as a reference map, and the ground-based point clouds were registered with airborne orthoimage. The ground-based point clouds were rasterized into an intensity map for image-matching purposes. Then, A graph-based matching algorithm was adopted to find corresponding roadmarks across two different sensors. The experiment includes a robustness test, data co-registration, and accuracy analysis.

3.1 Material and Study Area

The test data were acquired by ground-based Velodyne VLP32 and aerial DJI P4P UAV system. The UAV images were oriented by ground control points, and image dense matching was performed to obtain airborne point clouds and orthoimages. The VLP point clouds were initially aligned by POS (Positioning and Orientation System) without ground control points. Table 1 shows details of the parameters with VLP32 and DJI P4P sensors. The experimental area for this study was located in Taichung City, Taiwan. This area was approximately 84 hectares, including the Suinan self-driving bus experimental site, with high-precision map and other related information. In this study, a 330 m × 150 m area was subseted to evaluate the proposed schema. The UAV orthoimage and VLP32 Lidar intensity are shown in Fig. 2a, b. Both datasets have a resolution of 2 cm. The DeepLab v3+ model was applied to perform semantic segmentation for roadmark extraction (Fig. 2c–f), ensuring the consistency of features between VLP and UAV datasets for subsequent processing.

Table 1. Details of the sensor parameters

VLP32		UAV	
Item	Parameters	Item	Parameters
Number of lasers	32	Effective pixels	20 million pixels
Horizontal angle	360°	Field of view	84°
Vertical angle	+10.67° ~ −30.67°	CMOS	1 inch
Scanning point	694444.8 points/s	Aperture	f/2.8 to f/11
Scanning range	70 m	Focal length	8.8 mm/24 mm

3.2 Graph-Based Matching and Robustness Testing

Figure 3 shows the results of graph-based matching using extracted road marks. The blue line indicates correct matches, while the red line indicates incorrect matches. After the initial matching, we used RANSAC to remove outliers, resulting in 115 successful matches out of 162. It is evident that this method relies heavily on the quality of image semantic segmentation. For instance, in UAV orthoimages, shadows or vehicle occlusions on road marks can lead to poor or missing road marks. In the case of lidar intensity, the density of the scans can affect segmentation results, leading to fragmented outcomes. These factors are the primary reasons for the difficulties in cross-sensor matching. In order to evaluate the robustness of the proposed method, this study simulated different rotation and scaling errors between these two data. Table 2 shows the quantitative analysis of rotation and scale effects. After extensive rotation and scaling testing, the success rate of correct matching reached 70%, which confirms that the graph matching method is robust to rotating and scaling. It is important to demonstrate that this method can be applied to matching data across different scales.

3.3 Data Co-registration

After identifying the corresponding road marks through graph-based matching, this study calculated the coefficients of affine transformation between the two coordinate

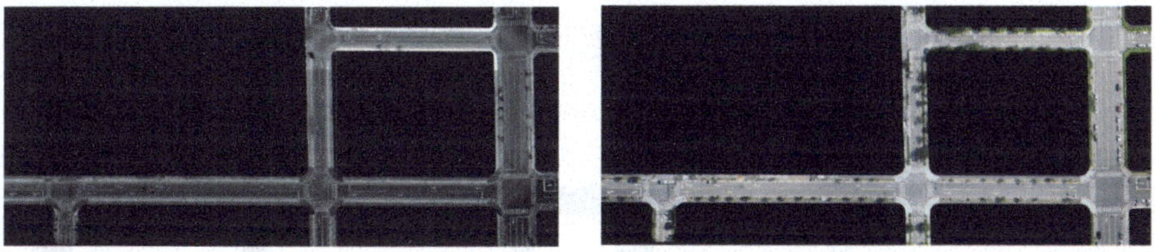

(a) Lidar intensity (b) Aerial orthoimage

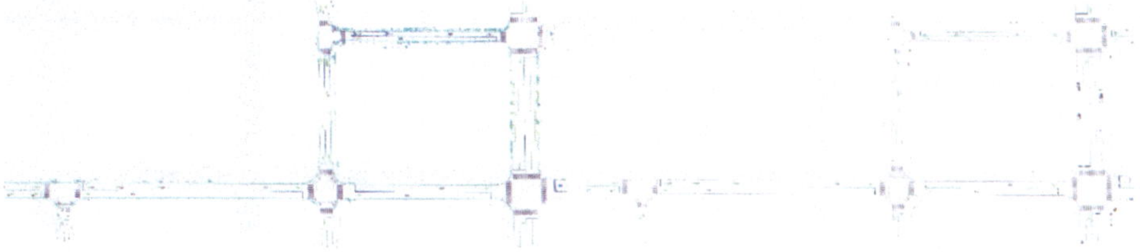

(c) Extracted roadmarks from lidar intensity (d) Extracted roadmarks from aerial orthoimage

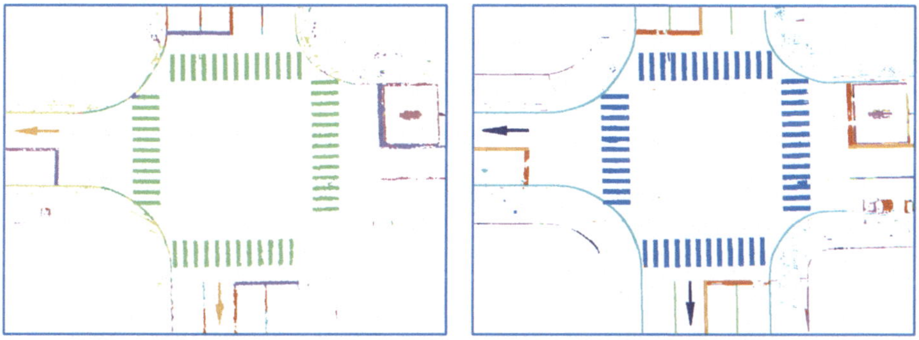

(e) Extracted roadmarks from lidar intensity(Zoom-in) (f) Extracted roadmarks from aerial orthoimage (Zoom-in)

Fig. 2. VLP and UAV test data

systems. Figure 4 displays the overlap of these two datasets. In the figure, red indicates the road marks from the orthoimage, and green indicates the areas where the two datasets overlap. Figure 4a, b compare the results before and after the co-registration. Figure 4c, d provide an enlarged view of the intersection for better visualization. Before the data co-registration, there was a noticeable shift between the two sensors. After the data co-registration, the two datasets overlapped appropriately. The results demonstrate the proposed method's effectiveness in synchronizing these two datasets, ensuring accurate overlay, and integrating the road marks from different sensors.

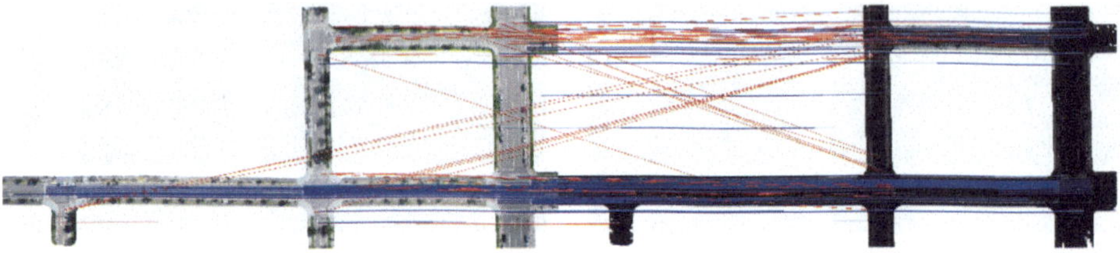

(a) Before outliers removal

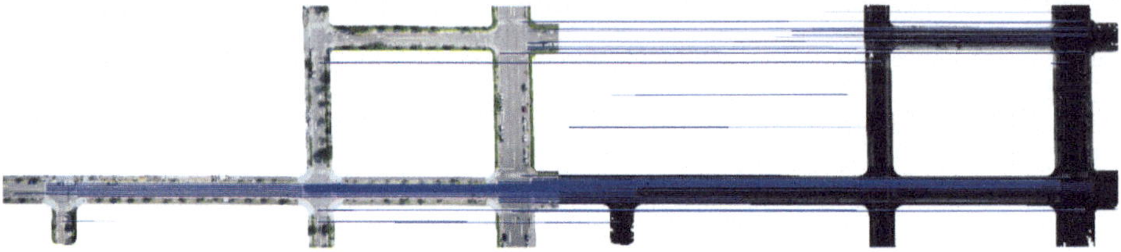

(b) After outliers removal

Fig. 3. Results of graph matching (the left side is color ortho image; the right side is lidar intensity map).

Table 2. Correctness for rotation and scaling simulation test (S = scale R = rotation angle)

	R = 0°	R = 10°	R = 20°	R = 30°	R = 40°	R = 50°
S = 1.0	71%	71%	71%	71%	71%	71%
S = 0.9	71%	70%	70%	69%	70%	70%
S = 0.8	71%	70%	70%	70%	70%	70%
S = 0.7	70%	69%	70%	69%	70%	69%
S = 0.6	70%	70%	69%	69%	70%	69%
S = 0.5	71%	70%	71%	70%	70%	70%

3.4 Accuracy Assessment

The quantitative evaluation was conducted by manually identifying checkpoints from two datasets. In this study, eight checkpoints were selected at each of five road intersections within the test data, resulting in a total of 40 checkpoints. The distance error was defined as the point distance between the two datasets. Table 3 presents the data before and after registration, including the maximum error, average error, and standard deviation. Before registration, the overall average error between the two datasets was approximately 0.288 m. After applying the proposed registration method, the overall average error was reduced to 0.08 m. This improvement demonstrates the effectiveness of the registration method in aligning the datasets. The histogram of the error distribution is shown in Fig. 5, illustrating the changes in error values before and after registration. Before registration, the maximum error was 0.814 m. Following the cross-sensor registration process, the maximum error was significantly reduced to 0.186 m.

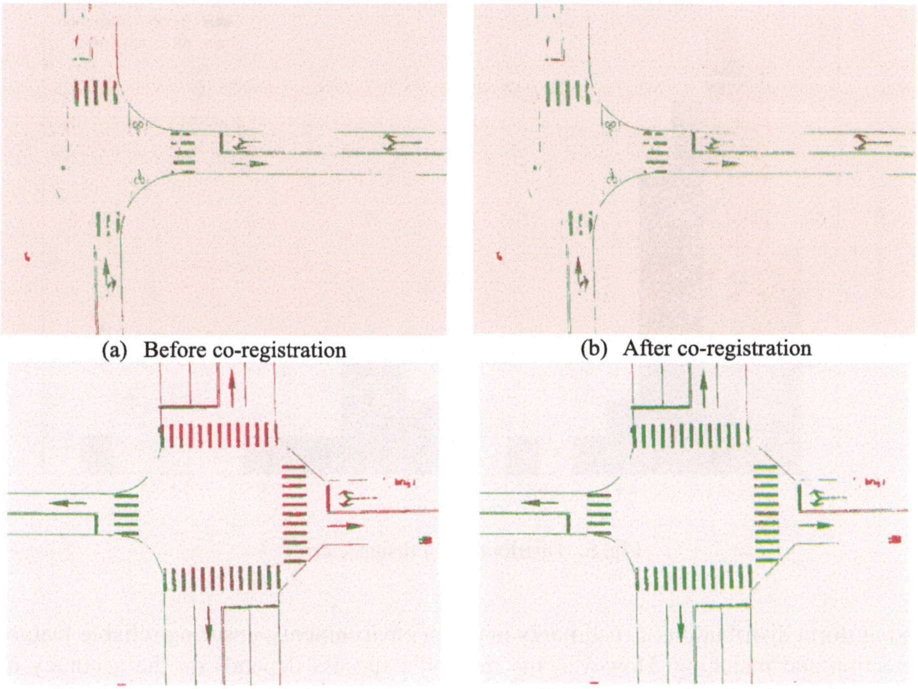

(a) Before co-registration (b) After co-registration

(c) Before co-registration (zoom-in) (d) After co-registration (zoom-in)

Fig. 4. Results of data co-registration (red color is roadmarks from color orthoimage; green color indicates the areas where the two datasets overlap)

This substantial reduction in both average and maximum errors highlights the enhanced accuracy achieved through the registration process, ensuring more precise alignment between the datasets.

Table 3. Evaluation using check points.

Unit: meter	Distance error		
	Max	Avg	Stdev
Before registration	0.814	0.288	0.182
After registration	0.186	0.087	0.046

4 Conclusions and Future Work

This study demonstrates the efficacy of using roadmarks for cross-sensor point matching. The proposed method, which leverages graph-based matching with the Hungarian algorithm, is invariant to rotation and scaling, thereby providing robust initial localization of cross-sensor data. This approach effectively utilizes the consistent visibility

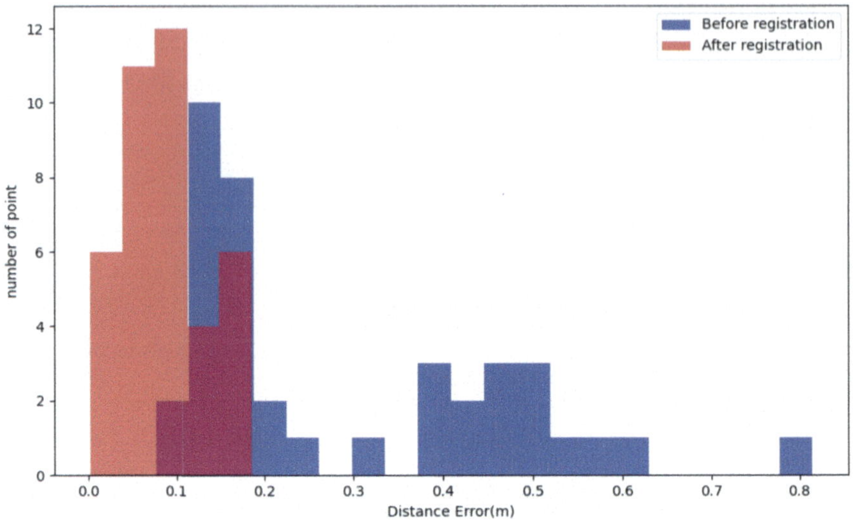

Fig. 5. Distribution of distance error

and uniform distribution of roadmarks in urban environments, ensuring reliable feature detection and matching. However, the method's success depends on the accuracy of image segmentation and the completeness of roadmarks in the datasets. Any inaccuracies in segmentation or incomplete roadmarks can adversely affect the registration process. Despite these challenges, the approach offers a promising solution for integrating data from different sensors, paving the way for more accurate and comprehensive point cloud reconstructions. Future work could focus on enhancing segmentation techniques and exploring additional semantic features to improve registration accuracy and robustness further.

Acknowledgments. This research was partially supported by the National Science and Technology Council, Taiwan. The authors would like to thank CECI and ITRI, Taiwan, for providing the experimental data.

References

1. Chen, L.C., Zhu, Y., Papandreou, G., Schroff, F., Adam, H.: Encoder-decoder with atrous separable convolution for semantic image segmentation. In: Proceedings of the European Conference on Computer Vision (ECCV), pp. 801–818 (2018)
2. Fischler, M.A., Bolles, R.C.: Random sample consensus: a paradigm for model fitting with applications to image analysis and automated cartography. Commun. ACM **24**(6), 381–395 (1981)
3. Gruen, A., et al.: Joint processing of UAV imagery and terrestrial mobile mapping system data for very high resolution city modeling. Int. Arch. Photogramm. Remote. Sens. Spat. Inf. Sci. **40**, 175–182 (2013)

4. Hussnain, Z., Oude Elberink, S., Vosselman, G.: Automatic feature detection, description and matching from mobile laser scanning data and aerial imagery. Int. Arch. Photogramm. Remote. Sens. Spat. Inf. Sci. **41**, 609–616 (2016)
5. Javanmardi, M., Javanmardi, E., Gu, Y., Kamijo, S.: Towards high-definition 3D urban mapping: road feature-based registration of mobile mapping systems and aerial imagery. Remote Sens. **9**(10), 975 (2017)
6. Li, S., Ge, X., Li, S., Xu, B., Wang, Z.: Linear-based incremental co-registration of MLS and photogrammetric point clouds. Remote Sens. **13**(11), 2195 (2021)
7. Lin, X., Niu, D., Zhao, X., Yang, B., Zhang, C.: A novel method for graph matching based on belief propagation. Neurocomputing **325**, 131–141 (2019)
8. Ling, X., Qin, R.: A graph-matching approach for cross-view registration of over-view and street-view based point clouds. ISPRS J. Photogramm. Remote Sens. **185**, 2–15 (2022)
9. Livi, L., Rizzi, A.: The graph matching problem. Pattern Anal. Appl. **16**, 253–283 (2013)
10. Munkres, J.: Algorithms for the assignment and transportation problems. J. Soc. Ind. Appl. Math. **5**(1), 32–38 (1957)
11. Teo, T.A., Huang, S.H.: Surface-based registration of airborne and terrestrial mobile Lidar point clouds. Remote Sens. **6**(12), 12686–12707 (2014)

Author Index

A. Abdul-Rahman et al. (Eds.): GeoWeek 2024, LNGC, pp. 367–368, 2025.
https://doi.org/10.1007/978-3-031-86654-8

The manufacturer's authorised representative in the EU is Springer Nature Customer Service Centre GmbH, Europaplatz 3, 69115 Heidelberg, Germany. If you have any concerns regarding our products, please contact ProductSafety@springernature.com

Printed and bound by CPI Group (UK) Ltd, Croydon, CR0 4YY
01/10/2026
02225850-0017